RESCUE BREATHING

ADULTS AND CHILDREN
(5 years and older)

1 Place victim on back; lift neck and head all the way back (**see sketch** clear mouth of any obstruction.

2 Pinch nose and blow vigorously through mouth to make chest expand (**see sketch (b)**).

3 Inflate 12 times per minute without interruptions.

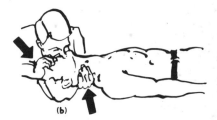

4 If chest does not expand, turn the victim on his side and strike several sharp blows between the shoulder blades using the palm of your hand. Recheck the mouth for obstructions and resume blowing.

(b)

INFANTS AND SMALL CHILDREN
(through age 4)

1 Place child on back; lift neck and tilt head back—**do not force (see sketch (c)).**

2 Tightly seal child's mouth and nose with your mouth.

(c)

3 Puff to make chest expand.

4 Inflate 20 times per minute without interruptions.

5 If chest does not expand, invert child and strike between shoulders (**see sketch (d)**); resume puffing.

(d)

Source: U.S. Department of Health, Education, and Welfare

preventive maintenance of electrical equipment

Charles I. Hubert

professor of electrical engineering
united states merchant marine academy

preventive maintenance of electrical equipment

second edition

McGraw-Hill Book Company

New York   St. Louis   San Francisco   London   Sydney   Toronto   Mexico   Panama

preventive maintenance
of electrical equipment

Library of Congress Catalog Card Number 69-13221

30839

1 2 3 4 5 6 7 8 9 0 M A M M 7 6 5 4 3 2 1 0 6 9

preface

The objective of this book is to set forth up-to-date methods for preventive maintenance, to provide logical methods by which the more common troubles may be identified and localized, to recommend emergency repairs that will keep the equipment in operation until it can be scheduled out of service, to suggest operating procedures, and to outline inspection programs that will help ensure safe, efficient, economical, and dependable operation.

The book has been completely rewritten, enlarged, and reorganized, with a view toward greater adaptability to formal class and laboratory programs in electrical training and technology, as well as to meet the needs of industry-sponsored courses in an age of expanding automation. To accomplish these goals, the principles underlying equipment operation have been added to the well-established how-to-do-it features of the book.

The addition of an introductory chapter on electric and magnetic phenomena provides a minimum of preliminaries necessary to an understanding of instrumentation, testing, and trouble-shooting procedures and further enhances the book's use in self-study programs. Supporting principles pertinent to the operation of specific equipment are woven into the text at appropriate places. The book is extensively cross-referenced to provide the reader with quick access to related or dependent material in other chapters. All topics in the first edition have been retained and updated, where required, to reflect new techniques, improved testing equipment, and the latest IEEE, NEMA, and USAS standards.

New chapters have been added on industrial control, overcurrent protection, distribution systems, and the economics of preventive-maintenance programs.

This second edition provides considerable flexibility in course planning. For example, a course designed specifically for maintenance and trouble shooting in the area of dc machinery and control would include Chaps. 1, 2, 8, 9, 10, 11, 12, 13, and 14. Similarly a course in the area of ac machinery and control would include Chaps. 1, 2, 4, 5, 6, 7, 12, 13, and 14. Students who have an adequate background in electric and magnetic phenomena may omit Chap. 1 without impairing the effectiveness of the remaining chapters. Questions and problems are included at the end of each chapter, and answers to all problems are in the back of the book.

The many additional topics included in this expanded edition should make it ideal for a variety of industrial training programs, for motor and industrial control inspectors, and as a general reference for operating engineers and maintenance electricians.

The author takes this opportunity to acknowledge with gratitude the many contributions made by his colleagues, friends, and students in the preparation of this second edition. He is particularly indebted to Prof. W. H. McDonald of the United States Merchant Marine Academy for his extensive review of the manuscript and his many detailed suggestions and criticisms.

The author is also indebted to Mr. Oliver J. Ruel, P.E., Head, Department, Industrial Electricity, Electronics, and Instrumentation, of the David Ranken School of Mechanical Trades, St. Louis, Missouri, for his very detailed chapter-by-chapter analysis of the manuscript and his many helpful suggestions.

Special mention should be made of Lieutenant J. A. Giaquinto, who taught several courses using draft copies of the second edition and provided many helpful comments.

Mr. W. H. Fifer, formerly Assistant Head Engineer, Electrical Branch, Bureau of Ships, Navy Department, offered many very helpful suggestions that have, to a good measure, been instrumental in expanding the content of this book. Professor R. C. Panuska and Lieutenant N. J. Maroney also provided many suggestions for improvement.

The author is especially indebted to his sweet wife, Josephine, who devoted many tedious hours typing the manuscript, the revisions, and all the related correspondence.

Charles I. Hubert

contents

chapter 11   trouble shooting and emergency repairs of dc machinery   277

chapter 12   mechanical maintenance of electrical machinery   297

introduction

what is preventive maintenance?

Preventive maintenance is the orderly routine of inspecting, testing, cleaning, drying, varnishing, adjusting, and lubricating electrical apparatus.

A good preventive-maintenance program provides for planned shutdowns during periods of inactivity or least usage for the purpose of major overhauls. This ensures continuity of operation and lessens the danger of breakdown at peak loads. Through such a program, troubles can be detected in the early stages, and corrective action can be taken before extensive damage is done.

The relatively high cost of downtime, compared with that of an adequate preventive-maintenance program, has placed greater emphasis on prevention. This is particularly true in processing plants, assembly lines, power plants, etc., where the failure of a relatively minor component can disrupt the entire operation; the total cost of downtime and emergency around-the-clock repairs can be staggering.

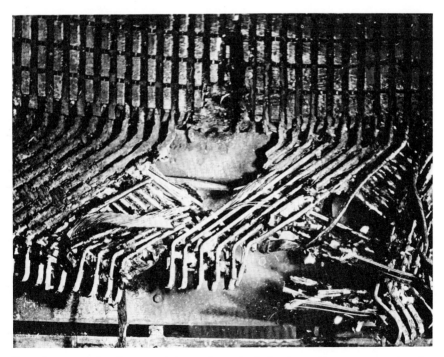

Burned core and coils caused by a short circuit in the stationary armature of a waterwheel generator. The generator had been in service for six years without cleaning. (*General Electric Co.*)

preventive maintenance of electrical equipment

one

electric and magnetic phenomena

A prerequisite to the study of electrical maintenance is an understanding of the effects produced by the movement of extremely small negatively charged particles called *electrons*. By controlling the number and movement of these electrons, directing them in prescribed paths, the designer of electrical equipment has provided mankind with electric lights, motors, generators, heaters, and the great variety of things that depend on an abundance of electric power. It is the function of the electrical-maintenance technician to ensure that these electrons follow the prescribed path, with no shortcuts or unplanned outages.

1-1 Electric Charge, Electric Current, Driving Voltage

The negatively charged electron is the smallest particle of electric charge, and it is the movement of these electric charges that constitutes an electron current (electric current). A current of one ampere is equivalent to 6.28×10^{18} electrons (6.28 million million million) moving unidirectionally past a point in one second.

Any material that allows the essentially free passage of current when connected to a battery or other source of electric energy is called a *conductor*. In metallic conductors "free electrons" wander aimlessly about the crystal structure of the material, resulting in an average current equal to zero. However, if some means is provided to drive these "free electrons" in one direction, the average current will be nonzero. This is illustrated in Fig. 1-1 for a section of metallic conductor with and without

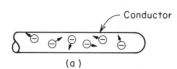

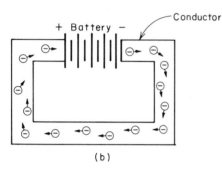

FIG. 1-1 Electron motion in a conductor. (a) Random motion of electrons when no driver is applied; (b) unidirectional motion of electrons caused by the application of a driver.

a driving voltage. The driving voltage in this case is supplied by a battery. Connecting a driver (battery or other electric generator) to the ends of the wire causes a good number of "free electrons" to be driven in the direction of the driving voltage. The direction of this movement is from the negative (−) terminal of the driver, through the conductor, to the positive (+) terminal of the driver, and constitutes an electric current. The closed loop formed by the battery and the conductor is called an *electric circuit*. The driver does not supply the electrons. The electrons are always present in the conductor as "free electrons," and are caused to move by the application of a driving voltage. The negative terminal of the driving source repels the "free electrons," and the positive terminal attracts them. As long as the circuit is closed, the electrons continue to circle round and round the loop. This unidirectional movement of electrons is called *direct current*. The total number of "free electrons" in the conductor remains the same; each time an electron enters the positive terminal of the driver, another electron leaves the negative terminal for its journey around the loop.

The application of an alternating voltage to a conductor causes the electrons to flow first in one direction and then in the opposite direction, oscillating continuously about their middle position. This repetitive back-and-forth movement of electric charges constitutes an *alternating current*. Thus, any apparatus that generates a repetitive alternating driving voltage is logically called an *alternating current (ac) generator*.

1-2 Conductors for Energy Transmission

Conductors selected for the efficient transmission of electric energy must have good electrical conductivity; that is, they must be able to pass the current with little opposition. The relatively high electrical conductivity and good heat-conduction capability of copper compared with other metals of similar cost result in its extensive use in motors, generators, controls, switches, cables, buses, etc. Buses are solid or hollow conductors, generally of large cross-sectional area, from which taps are made to supply three or more circuits. Figure 14-8 illustrates buses that were damaged by magnetic forces caused by excessive current. Aluminum, because of its light weight, good casting properties, and high conductivity (although not so good as copper), is used in squirrel-cage rotors (see Sec. 5-1), cables, bus bars, and long-distance transmission lines.

The ampacity (current-carrying capacity) of a wire or cable depends on the thermal stability of the insulation covering the conductor, the heat-dissipation capability to the surrounding media, and the duty cycle.[1]

Cadmium and silver are used as contacts in control equipment, and gold plating is sometimes used to provide the excellent mating surfaces of bayonet types of contacts in electronic equipment. Although silver has a higher conductivity than copper, the relatively high cost makes its general use prohibitive. Carbon, in very short lengths and large cross-sectional areas, has applications as sliding contacts for rheostats, commutators, and slip rings.

For safety, all water must be treated as having good conductivity. Although distilled water is not a conductor, any water around electrical equipment, tap water, rain water, etc., is contaminated with salts and conducts electricity.

1-3 Conductors for Limiting Current

Conductors of low conductivity are used in those applications where the magnitude of the current must be limited or where it is desired to convert electric energy to heat energy (toaster, electric range, etc.). Examples of low-conductivity materials are iron, tungsten, nickel alloys, compositions of carbon and inert materials, etc. Figure 1-2 illustrates a resistor box containing resistor elements used to limit the starting current of a motor. An adjustable resistor whose resistance can be changed without opening the circuit in which it is connected is called a *rheostat*. The standard color code for small tubular resistors is given in Appendix 11.

1-4 Resistance

When electrons are driven through a conductor, they collide with one another and with other parts of the atoms that make up the material.

---

[1] Ampacity tables for different conditions of insulation and surrounding media are given in Appendixes 1 and 2. More extensive tables for copper and aluminum conductors are available in book form from the Insulated Power Cable Engineers Association.

FIG. 1-2 Resistor box for limiting current. (*Cutler-Hammer Inc.*)

Such collisions interfere with the free movement of the electrons and generate heat. This property of a material that limits the magnitude of a current and converts electric energy to heat energy is called *resistance*.

For a given length and cross-sectional area, materials of low conductivity have a higher resistance to current than materials of high conductivity; thus, to cause the same current in both, the material of low conductivity (higher resistance) requires a higher driving voltage.

Resistance is generally measured with an ohmmeter and is expressed in ohms or megohms. One megohm is 10^6, or one million, ohms. The resistance of all conductors is temperature-dependent to some extent. Hence, valid measurements call for the determination of conductor temperature at the time of resistance measurement. Most conducting materials, such as copper, aluminum, iron, nickel, and tungsten, increase in resistance with increased temperature. Carbon, on the other hand, decreases in resistance with increased temperature. The temperature coefficients of resistance of some of the more commonly used conductors are given in Table 1-1. Multiplying the temperature coefficient of the conductor material by its change in temperature results in the percentage of change in conductor resistance.

Although the temperature coefficients in Table 1-1 are for 20°C, they can be used in calculations over the normal range of operating temperatures associated with electrical machinery without introducing significant errors. For example, the temperature coefficient for annealed copper[1] is 0.00393 at 20°C and 0.0038 at 100°C, a difference of only 0.00013.

---

[1] Annealed copper (heat-treated) is used in all electrical equipment, such as motors, generators, transformers, etc. Hard-drawn copper is used in some telephone lines to reduce the sag between poles.

Table 1-1 Temperature Coefficients of Resistance of Conductor Materials

Material	Temperature coefficient, ohms per centigrade degree per ohm at 20°C
Aluminum	0.0039
Antimony	0.0036
Bismuth	0.004
Brass	0.002
Constantan (60% Cu, 40% Ni)	0.000008
Copper, annealed	0.00393
Copper, hard-drawn	0.00382
German silver	0.0004
Iron	0.005
Lead	0.0041
Magnesium	0.004
Manganin (84% Cu, 12% Mn, 4% Ni)	0.000006
Mercury	0.00089
Molybdenum	0.0034
Monel metal	0.002
Nichrome	0.0004
Nickel	0.006
Platinum	0.003
Silver (99.98% pure)	0.0038
Steel, soft	0.0042
Tin	0.0042
Tungsten	0.0045
Zinc	0.0037

SOURCE: Values taken from Smithsonian Physical Tables.

If the resistance of a conductor at a given temperature is known, its resistance at a higher or lower temperature may be determined by substituting in the following formula:

$$R_H = R_L[1 + \alpha(T_H - T_L)] \tag{1-1}$$

where R_H = resistance at higher temperature, ohms
R_L = resistance at lower temperature, ohms
T_H = higher temperature, °C
T_L = lower temperature, °C
α = temperature coefficient of resistance

If the temperatures are measured in Fahrenheit, see Appendix 12 for centigrade-Fahrenheit conversion formulas.

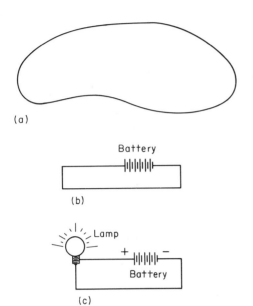

FIG. 1-3 Examples of simple electric circuits. (a) Loop of wire with no driver; (b) loop of wire with a driver; (c) loop of wire with a driver and a useful load.

• *Example 1-1.* The copper field winding of a motor has a resistance of 100 ohms at 20°C. What will be the resistance if the temperature of the machine rises to 70°C? From Table 1-1 the temperature coefficient for annealed copper is 0.00393. Substituting into Eq. (1-1),

$R_{70} = 100[1 + 0.00393(70 - 20)]$
$R_{70} = 119.65$ ohms

As Example 1-1 indicates, a 50°C increase in temperature causes an approximately 20 percent increase in the resistance of annealed copper.

1-5 The Electric Circuit and Ohm's Law

An electric circuit is any arrangement of conducting parts that form a closed loop. A few examples of simple electric circuits are shown in Fig. 1-3. A closed loop constitutes a circuit whether or not it includes a useful load or a driver. In the case of Fig. 1-3a there is no current in the loop unless it is induced by a changing magnetic field, such as that caused by moving magnets, a lightning discharge, or a changing current in a nearby circuit.

The current in the electric circuits in Fig. 1-3b and c depends on the magnitude of the driving voltage and the resistance of the circuit. This relationship, called *Ohm's law,* is expressed by

$$I = \frac{E}{R} \qquad (1\text{-}2)$$

where E = driving voltage, volts
     R = resistance, ohms
     I = current, amp

The resistance of copper and aluminum wire in ohms per 1,000 ft is given in Table 1-2.

• *Example 1-2.* A coil wound with 200 ft of No. 16 tinned copper wire is connected to a 12-volt battery. Determine the current. From Table 1-2, the resistance per 1,000 ft of No. 16 tinned copper wire at 25°C is 4.26 ohms. Thus, the ohms per foot is

$$\frac{4.26}{1,000} = 0.00426$$

The resistance of 200 ft is

$$0.00426 \times 200 = 0.852 \text{ ohm}$$

The current, using Ohm's law, is

$$I = \frac{E}{R} = \frac{12}{0.852} = 14.08 \text{ amp}$$

Circular-mil area. The unit of cross-sectional area commonly used in determining conductor sizes is the circular mil. One mil is 0.001 in., and one circular mil is the area of a circle whose diameter is one mil. The following formulas are useful for determining the circular-mil area of a conductor when its dimensions are given in inches:

$CM = 1.272A\,(10^6)$ for any cross section
$CM = D^2(10^6)$ for circular cross section

where A = cross-sectional area of conductor, sq in.
      D = diameter of conductor, in.

Resistivity. The resistivity of a material is the resistance of a specified length and cross section of that material. The resistance of a one-foot length of annealed copper, one circular mil in cross section, is 10.371 ohms (measured at 20°C). Simply stated, the resistivity of copper is 10.371 ohms per cir-mil ft. Thus, to calculate the resistance of any length of copper cable, bar, or tubing, substitute in the following formula:

$$R = 10.371 \frac{l}{CM} \quad \text{ohms at 20°C}$$

where l = length of copper conductor, ft
    CM = circular-mil area of copper

Table 1-2 Properties of Conductors

Size AWG MCM	Area, cir mils	Concentric lay stranded conductors — No. wires	Concentric lay stranded conductors — Diam. each wire, in.	Bare conductors — Diam., in.	Bare conductors — Area, sq in.*	Dc resistance, ohms/m ft at 25°C (77°F) — Copper† Bare cond.	Copper† Tin'd cond.	Aluminum†
18	1620	Solid	0.0403	0.0403	0.0013	6.51	6.79	10.7
16	2580	Solid	0.0508	0.0508	0.0020	4.10	4.26	6.72
14	4110	Solid	0.0641	0.0641	0.0032	2.57	2.68	4.22
12	6530	Solid	0.0808	0.0808	0.0051	1.62	1.68	2.66
10	10380	Solid	0.1019	0.1019	0.0081	1.018	1.06	1.67
8	16510	Solid	0.1285	0.1285	0.0130	0.6404	0.659	1.05
6	26240	7	0.0612	0.184	0.027	0.410	0.427	0.674
4	41740	7	0.0772	0.232	0.042	0.259	0.269	0.424
3	52620	7	0.0867	0.260	0.053	0.205	0.213	0.336
2	66360	7	0.0974	0.292	0.067	0.162	0.169	0.266
1	83690	19	0.0664	0.332	0.087	0.129	0.134	0.211
0	105600	19	0.0745	0.372	0.109	0.102	0.106	0.168
00	133100	19	0.0837	0.418	0.137	0.0811	0.0843	0.133
000	167800	19	0.0940	0.470	0.173	0.0642	0.0668	0.105
0000	211600	19	0.1055	0.528	0.219	0.0509	0.0525	0.0836
250	250000	37	0.0822	0.575	0.260	0.0431	0.0449	0.0708
300	300000	37	0.0900	0.630	0.312	0.0360	0.0374	0.0590
350	350000	37	0.0973	0.681	0.364	0.0308	0.0320	0.0505
400	400000	37	0.1040	0.728	0.416	0.0270	0.0278	0.0442
500	500000	37	0.1162	0.813	0.519	0.0216	0.0222	0.0354
600	600000	61	0.0992	0.893	0.626	0.0180	0.0187	0.0295
700	700000	61	0.1071	0.964	0.730	0.0154	0.0159	0.0253
750	750000	61	0.1109	0.998	0.782	0.0144	0.0148	0.0236
800	800000	61	0.1145	1.030	0.833	0.0135	0.0139	0.0221
900	900000	61	0.1215	1.090	0.933	0.0120	0.0123	0.0197
1000	1000000	61	0.1280	1.150	1.039	0.0108	0.0111	0.0177
1250	1250000	91	0.1172	1.289	1.305	0.00863	0.00888	0.0142
1500	1500000	91	0.1284	1.410	1.561	0.00719	0.00740	0.0118
1750	1750000	127	0.1174	1.526	1.829	0.00616	0.00634	0.0101
2000	2000000	127	0.1255	1.630	2.087	0.00539	0.00555	0.00885

* Area given is that of a circle having a diameter equal to the overall diameter of a stranded conductor.

† The resistance values given in the last three columns are applicable only to direct current. When conductors larger than No. 4/0 are used with alternating current the multiplying factors in Appendix 13 should be used to compensate for skin effect.

SOURCE: The values given in the table are those given in Handbook 100 of the National Bureau of Standards except that those shown in the eighth column are those given in Specification B33 of the American Society for Testing and Materials, and those shown in the ninth column are those given in Standard No. S-19-81 of the Insulated Power Cable Engineers Association and Standard No. WC3-1964 of the National Electrical Manufacturers Association.

1-6 Measurement of Resistance

Most resistance measurements of conductors may be made with an all-purpose measuring instrument like the volt-ohm-milliammeter shown in Fig. 1-4. The instrument uses internal batteries to supply power for resistance measurements. After selecting the resistance range, for example, $R \times 1$, $R \times 100$, or $R \times 10,000$, and plugging one end of each test lead into the minus and plus terminals respectively, the loose ends of the leads should be clipped together to form a closed loop. The "zero ohms" adjustment knob should then be turned to obtain a zero indication on the "ohms" scale of the instrument. If the instrument pointer cannot be brought to zero, one or more of the batteries must be replaced. The zero adjustment compensates for the resistance of the test leads and the gradual decay of the battery. When the pointer shows zero, the test leads should be unclipped and connected to the unknown resistance for measurement.

If a high degree of accuracy is needed, special-purpose resistance-measuring instruments, called *bridges,* are available.

FIG. 1-4 Volt-ohm-milliamme-ter. (*Simpson Electric* Co.)

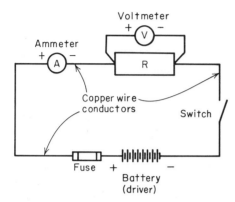

FIG. 1-5 Voltmeter-ammeter method for measuring resistance.

Another way to measure resistance is by the voltmeter-ammeter method shown in Fig. 1-5. This is an indirect method, the resistance being determined from Ohm's law by dividing the voltmeter indication by the ammeter indication. Thus the resistance in ohms is equal to

$$R = \frac{E}{I} = \frac{\text{voltmeter reading, volts}}{\text{ammeter reading, amp}}$$

Whenever an ammeter or voltmeter is connected to a circuit, the ammeter must always be connected in series with the element whose current it is to measure, and the voltmeter must always be connected in parallel with the element whose voltage it is to measure. A fuse is used to prevent damage to the circuit in the event of excessive current (see Sec. 14-1). To prevent a reversed deflection of the instrument pointer, which may bend it, the positive terminal of each instrument should be connected to the wire leading from the positive terminal of the driving source.

Because the resistance is temperature-dependent, the measurements should be made rapidly to avoid errors due to heating.

1-7 Power Loss in Conductors

The forced movement of electrons through a conductor, by the application of a driving voltage, results in the generation of heat. The heating is caused by electron collisions with atomic particles of the material. The rate at which electric energy is converted to heat energy, as a result of these electron collisions, is a function of the resistance of the conductor and the square of the current; that is,

$$P = I^2R \qquad \text{watts} \qquad\qquad (1\text{-}3)$$

where P = rate of conversion of electric energy to heat energy, watts
(joules per sec)
I = current through conductor, amp
R = resistance of conductor, ohms

Equation (1-3) provides the most convenient method for determining the heat-power loss caused by electron collision in electric cables. However, in other applications, such as determining the heat-power loss in resistors and other nonrotating equipment, the following formulas may be more convenient:

$$P = \frac{E^2}{R} \qquad P = EI \qquad\qquad\qquad (1\text{-}4)$$

where E = voltage measured across resistor, volts
I = current through resistor, amp

Formulas (1-4) were derived by substituting Ohm's law equation into the power equation $P = I^2R$.

Although Eqs. (1-3) and (1-4) are very useful for calculating the rate at which electric energy is converted to heat energy, the total energy expended is determined by multiplying the power by the time. Thus, with the power expressed in watts and the time in seconds, the expended energy is

$$W = Pt \qquad \text{watt-sec}$$

also

$$W = I^2Rt \qquad \text{watt-sec}$$
$$W = \frac{E^2}{R}t \qquad \text{watt-sec}$$
$$W = EIt \qquad \text{watt-sec}$$

If the power is in kilowatts (kw) and the time in hours (hr), the energy expended may be expressed in kilowatthours (kwhr). One kilowatt is equal to one thousand watts.

• *Example 1-3.* Given a cable whose resistance is 1 ohm. Determine which of the following combinations of current and time result in greater energy losses: 100 amp for 1 sec, 50 amp for 3 sec, 25 amp for 20 sec.

• *Solution*

$W = I^2Rt$ watt-sec
$W_1 = (100)^2(1)(1) = 10,000$ watt-sec $= 10$ kw-sec
$W_2 = (50)^2(1)(3) = 7,500$ watt-sec $= 7.5$ kw-sec
$W_3 = (25)^2(1)(20) = 13,500$ watt-sec $= 13.5$ kw-sec

One kilowatt (kw) is equal to one thousand watts. The 25-amp current for 20 sec produces the most heat loss, and the 50-amp current for 3 sec the least heat loss.

Depending on the I^2R losses in the conductor, the total operating time, and the environmental conditions, the temperature of the conductor can range from warm to the melting point of the material. A well-known example of heat loss in conductors is the heating of extension cords and cables when overloaded. Almost everyone has observed the heating of a plug where it mates with a wall receptacle for such household appliances as toasters, grills, coffee pots, etc.; if the contact resistance between the two mating surfaces is excessively high, the heat produced can start a fire. Mating surfaces for plug-in devices and contactors depend on spring pressure and clean surfaces for low contact resistance. Corroded or dirty surfaces, worn contacts, or weak spring pressure due to loss of temper results in a higher than normal contact resistance.

1-8 Electrical Insulation

Materials that possess extremely low conductivity are classified as noncon-ductors and are better known as *insulators* or *dielectrics*. A vacuum is the only known perfect dielectric.[1] All other insulating materials, such as paper, cloth, wood, bakelite, rubber, plastic, glass, mica, ceramic, shel-lac, varnish, air, and special-purpose oils, are imperfect dielectrics. The function of electrical insulation is to prevent the passage of current be-tween two or more conductors. To do this, the conductors may be covered with cotton, plastic, or rubber tape, mica tubing, ceramic tubing, etc., depending on the electrical and mechanical stresses involved in the applica-tion. The twin conductors in ordinary 120-volt lamp cord are separated by a rubber or plastic coating, but the conductors on long-distance high-voltage transmission lines are insulated from one another by air. The desired spacing of the transmission-line conductors is obtained with porce-lain insulators.

The electrons in an insulating material are tightly bound to the parent atoms and are not free to wander within the materials, as they do in a conductor. Hence, at normal room temperatures, the application of

[1] In vacuum tubes, such as diodes, triodes, and other multielement vacuum tubes, a filament heated to incandescence emits electrons that provide a conducting path.

rated to moderately high driving voltages to the insulating material causes only a very few electrons to break away. In good insulation this very small "leakage" current is generally less than one-millionth of an ampere (microampere). However, if a driver of sufficiently high voltage is applied, e.g., lightning or other high-voltage surges, the electrons will literally be torn away from their molecular bonds, destroying the insulating properties of the material. In the case of organic materials, such as wood, rubber, etc., this electron avalanche through the material generates sufficient heat to cause carbonization.

1-9 Semiconductors

A semiconductor material, such as silicon or germanium, has a conductivity about midway between good conductors and good insulators. Useful semiconductor devices are grown as crystals from a melt of semiconductor material to which extremely small amounts of impurities have been added. Some of the semiconductor devices grown in this manner are diodes, silicon-controlled rectifiers (called thyristors or SCRs), and transistors. A diode has the characteristics of a conductor when the applied voltage is in one direction and those of an insulator when the applied voltage is reversed. This is illustrated in Fig. 1-6 for forward and reverse directions of applied voltage (also called forward and reverse bias). In the forward direction (Fig. 1-6a), the diode offers very little resistance to the current. In the reverse direction (Fig. 1-6b), the diode offers considerable resistance to the current, allowing only a very small leakage current. Hence, if an alternating driving voltage is applied, the diode will act as a rectifier, blocking the current when the voltage is in the reverse direction.

The use of four diodes, in what is called a *bridge type* of full-wave rectifier circuit, provides unidirectional current to the load, even though the generator current is alternating. This is shown in Fig. 1-6c. The identifying colors of the connecting wires represent the NEMA standard.[1] The solid arrows indicate the direction of electron flow when generator terminal L_1 is negative, and the broken arrows indicate the electron flow when L_2 is negative. Note that the direction of current in the load is the same in both cases. Other applications of semiconductor devices are discussed in connection with specific apparatus.

1-10 Series and Parallel Circuits

Figure 1-7a and b illustrates typical series and parallel circuits, along with the associated formulas for determining the overall resistance. Although these formulas are handy, a convenient method for determining

[1] Industrial Control Standards, NEMA IC1-1967.

the overall resistance of a given circuit is to measure it with an ohmmeter, or by the voltmeter-ammeter method. This is particularly advantageous when the circuit is a combined series-parallel circuit or a more complex one, as shown in Fig. 1-7c.

The current through any element of a series circuit is the same as through any other element. Hence, if the current through any one element is measured, the current through the other elements will be known.

The voltage across any element of a parallel circuit is the same as that across any other element.

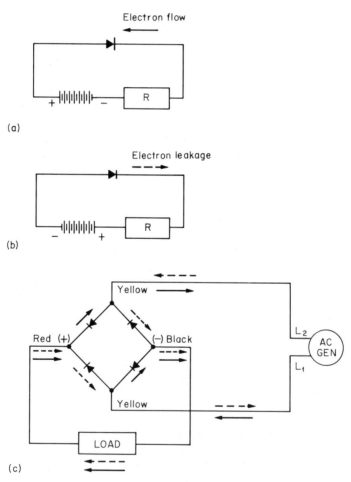

FIG. 1-6 Semiconductor diodes. (a) Biased in forward direction; (b) biased in reverse direction; (c) bridge type of full-wave rectifier.

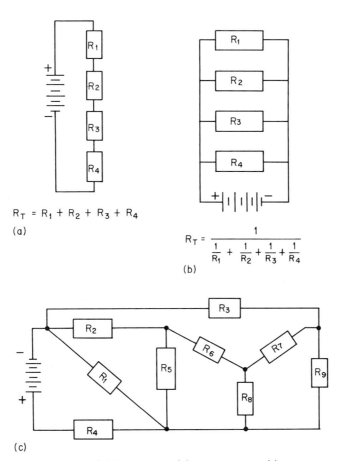

$$R_T = R_1 + R_2 + R_3 + R_4$$

(a)

$$R_T = \frac{1}{\frac{1}{R_1} + \frac{1}{R_2} + \frac{1}{R_3} + \frac{1}{R_4}}$$

(b)

(c)

FIG. 1-7 Electric circuits. (a) Series circuit; (b) parallel circuit; (c) complex circuit.

1-11 Voltage Drops and Kirchhoff's Voltage Law

When a generator or battery supplies current to a lamp, motor, or other load, the resistance of the connecting wires or cables causes the voltage at the load to be less than that at the generator or battery terminals. This is illustrated in Fig. 1-8a. If one "crawls through the conductor" from the negative terminal of the battery through the circuit and back to the positive terminal, adding up the voltage as he goes, the sum of the voltages along the path will equal that of the driver. For the circuit shown in Fig. 1-8a,

$$2 + 120 + 4 = 126$$

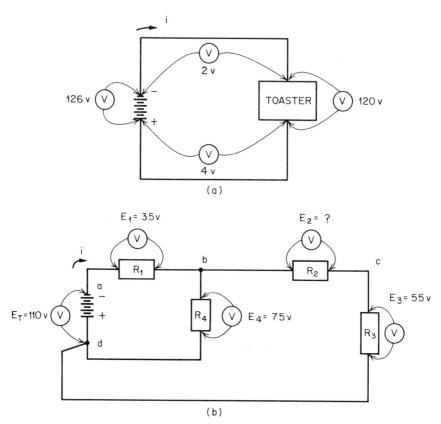

FIG. 1-8 Measurement of voltage drops.

This very simple and extremely useful relationship, called *Kirchhoff's voltage law,* has many applications in analyzing and trouble-shooting electric circuits. Stated in general terms:

The summation of all the voltage drops around any closed loop containing a driver is equal to the voltage of the driver.

• *Example 1–4.* Consider the circuit shown in Fig. 1–8*b*. For this example the voltage drops in the connecting wires are so very small compared with the voltage drops across the blocks of resistance that they are neglected.

For loop *abd,*

$$35 + 75 = 110$$

For loop *abcd,* the voltage across resistor R_2 is not known. Applying Kirchhoff's voltage law to the loop,

$35 + E_2 + 55 = 110$

Solving for E_2,

$E_2 = 110 - 35 - 55$
$E_2 = 20$ volts

Thus, Kirchhoff's voltage law can be used to determine the voltage across any element in a closed loop, provided that the voltage across the other elements is known.

· *Example 1-5.* For the series circuit shown in Fig. 1-7a let $R_1 = 2$ ohms, $R_2 = 3$ ohms, $R_3 = 6$ ohms, $R_4 = 1$ ohm, and $E = 240$ volts. Determine (*a*) the circuit resistance, (*b*) the circuit current, (*c*) the voltage drop across each resistor, (*d*) the total power supplied by the battery, and (*e*) the power drawn by each resistor.

· *Solution*

(*a*) $R_T = R_1 + R_2 + R_3 + R_4$
      $R_T = 2 + 3 + 6 + 1 = 12$ ohms
(*b*) $I = \dfrac{E}{R_T} = \dfrac{240}{12} = 20$ amp
(*c*) $E = IR$
      $E_1 = 20 \times 2 = 40$ volts
      $E_2 = 20 \times 3 = 60$ volts
      $E_3 = 20 \times 6 = 120$ volts
      $E_4 = 20 \times 1 = 20$ volts

Note that the total voltage is $40 + 60 + 120 + 20 = 240$ volts, as should be expected from Kirchhoff's voltage law.

(*d*)   The power may be calculated in three different ways:
      $P = I^2 R_T = (20)^2 \times 12 = 4{,}800$ watts
      $P = EI = 240 \times 20 = 4{,}800$ watts
      $P = \dfrac{E^2}{R_T} = \dfrac{(240)^2}{12} = 4{,}800$ watts
(*e*) $P_1 = I^2 R_1 = (20)^2 \times 2 = 800$ watts
      $P_2 = I^2 R_2 = (20)^2 \times 3 = 1{,}200$ watts
      $P_3 = I^2 R_3 = (20)^2 \times 6 = 2{,}400$ watts
      $P_4 = I^2 R_4 = (20)^2 \times 1 = 400$ watts

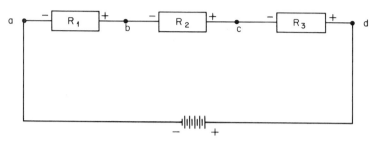

FIG. 1-9 Determining relative polarity.

1-12 Relative Polarity

To prevent the possible bending of the pointer when using a dc voltmeter to take voltage-drop measurements in a series or series-parallel circuit, the relative *polarity* of the terminals under test should be known. The *relative polarity* of the two terminals of a resistor or other apparatus may be determined by an observer "standing" at each terminal in turn and facing the driver terminal to which it is connected. The polarity of the driver terminal that the observer faces is the relative polarity of the terminal in question.

Referring to Fig. 1-9, the relative polarity of the terminals for the three resistors are

For R_1: b is $+$, a is $-$
For R_2: b is $-$, c is $+$
For R_3: d is $+$, c is $-$

Hence, terminal b (or c) is neither $+$ nor $-$; its polarity is relative to the other terminal of the resistor in question.

There are times when the polarity of the supply voltage must be determined before the switch is closed. For example, when charging a battery from a dc supply, the $+$ terminal of the battery must connect to the $+$ terminal of the charging source. A reversed connection ($+$ to $-$) may damage the battery. To determine the polarity of the charging source use a voltmeter set at its highest voltage range. If the voltmeter reads upscale when connected to the source, the $+$ terminal of the voltmeter is connected to the $+$ terminal of the source.

1-13 Summation of Currents and Kirchhoff's Current Law

Kirchhoff's current law states:

The summation of all the currents going away from a junction equals the summation of all the currents going to the junction.

This law is very useful for determining an unknown current in a branch circuit when all other currents at the junction are known.

• *Example 1-6.* Figure 1-10 shows a feeder line supplying three branch circuits. The feeder current and two of the branch currents are known. Using Kirchhoff's current law, the third branch current can be determined:

$$2 + I + 5 = 10$$
$$I = 3 \text{ amp}$$

• *Example 1-7.* For the parallel circuit shown in Fig. 1-7b, let $R_1 = 2$ ohms, $R_2 = 4$ ohms, $R_3 = 10$ ohms, $R_4 = 5$ ohms, and $E = 120$ volts. Determine (*a*) the current to each resistor, (*b*) the total current supplied by the battery, (*c*) the power drawn by each resistor, (*d*) the total power supplied by the battery, and (*e*) the circuit resistance.

• *Solution*

(*a*) In a parallel circuit the voltage across each resistor is the same. Therefore

$$I_1 = \frac{E}{R_1} = \frac{120}{2} = 60 \text{ amp}$$

$$I_2 = \frac{E}{R_2} = \frac{120}{4} = 30 \text{ amp}$$

$$I_3 = \frac{E}{R_3} = \frac{120}{10} = 12 \text{ amp}$$

$$I_4 = \frac{E}{R_4} = \frac{120}{5} = 24 \text{ amp}$$

(*b*) From Kirchhoff's current law

$$I_T = I_1 + I_2 + I_3 + I_4$$
$$I_T = 60 + 30 + 12 + 24 = 126 \text{ amp}$$

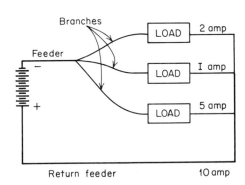

FIG. 1-10 Feeder line supplying branch circuits.

(c) $P_1 = EI_1 = 120 \times 60 = 7{,}200$ watts
      $P_2 = EI_2 = 120 \times 30 = 3{,}600$ watts
      $P_3 = EI_3 = 120 \times 12 = 1{,}440$ watts
      $P_4 = EI_4 = 120 \times 24 = 2{,}880$ watts

(d) $P_T = P_1 + P_2 + P_3 + P_4 = 15{,}120$ watts

    or

    $P_T = EI_T = 120 \times 126 = 15{,}120$ watts

(e)   From Ohm's law

$$R_T = \frac{E}{I_T} = \frac{120}{126} = 0.9525 \text{ ohm}$$

    or

$$R_T = \frac{1}{\frac{1}{2} + \frac{1}{4} + \frac{1}{10} + \frac{1}{5}} = 0.9525 \text{ ohm}$$

1-14 Magnetic Field

A magnetic field is present whenever electric charges are caused to move. Furthermore, as shown in Fig. 1-11*a,* the magnetic field, or *flux* as it is called, completely surrounds the current that causes it and is perpendicular to it. The direction of the magnetic flux around the current may be determined by using the left-hand thumb rule: Grasp the conductor with the left hand so that the thumb points in the direction of electron flow; the fingers will then point in the direction of the flux. The presence of this magnetic field and its perpendicularity relative to the current can easily be detected with a magnetic compass, as shown in Fig. 1-11*b.*

If a long insulated conductor carrying an electric current is wound in the form of a coil, the flux around the wire will be concentrated in a relatively small area. This is shown in Fig. 1-12*a.* The magnetic polarity of the coil may be determined by using the left-hand thumb rule: Grasp the coil with the left hand so that the fingers point or curl in the direction of electron flow; the thumb will then point in the direction of the north pole. The direction of the flux will be from north to south outside the coil, and from south to north inside the coil.

The flux produced by a given amount of current in a coil depends on the number of turns in the coil and the core material on which the coil is wound. Core materials made of iron and steel offer considerably less opposition to the flux than does air or other nonmagnetic materials, such as wood, copper, and aluminum. Hence, iron and steel are used extensively in electrical apparatus that require large magnetic fields.

The following formula may be used to determine the flux produced by the current in a coil when the number of turns and the opposition

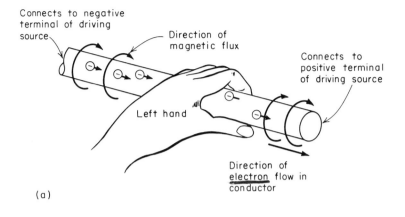

Connects to negative
terminal of driving
source

Direction of
magnetic flux

Connects to
positive terminal
of driving source

Left hand

Direction of
<u>electron</u> flow in
conductor

(a)

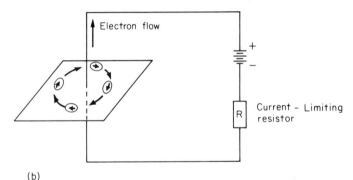

Electron flow

+
−

R   Current - Limiting
    resistor

(b)

FIG. 1-11 Determining the direction of the magnetic field that surrounds an electron current.
(a) Using the left-hand rule; (b) using a magnetic compass.

to the flux, called *reluctance,* are known:

$$\Phi = \frac{IN}{\mathcal{R}} \qquad \text{webers}$$

where Φ = flux, webers (1 weber = 100 million lines of magnetic
    induction)

I = current, amp

N = number of turns of wire in coil

$\mathcal{R}$ = reluctance of core materials that make up flux path, in
    mks units

Representative magnetization curves of flux versus current for a coil
with and without an iron core are shown in Fig. 1-12*b*. Without the
iron core the variation of flux, with increasing current, is a straight line
(linear characteristic). With an iron core, the flux increases linearly for

low values of current and then knees over, increasing only slightly with all further increases in current. The magnetic field produced in the iron by the coil current is caused by the parallel alignment of domains of spinning electrons within the molecular structure of the core material. When all the domains available for alignment are shifted in the direction of the magnetic field, the core is *saturated*. Further increases in current cause only small increases in flux.

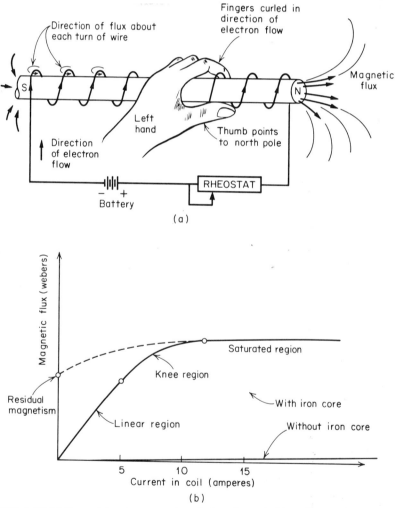

FIG. 1-12 (a) Determining the magnetic polarity of a coil; (b) magnetization curves.

1-15 Permanent Magnets

When the circuit supplying a coil is opened, the current decays to zero. If the coil is wound around an iron core, not all the magnetic domains shift back to their original position. Those which remain in parallel alignment produce the so-called *residual magnetism effect*. See the broken line in Fig. 1-12b. Hard steels and some nickel alloys retain a considerable amount of residual magnetism, and are used for permanent magnets. Soft steels that retain only little residual magnetism are used in such electrical equipment as motors, generators, and transformers. Excessive heat applied to a permanent magnet or striking it severely can cause the paralleled domains to shift to more random positions, thereby decreasing its magnetic field.

The direction of flux about a magnet is from north to south outside the magnet, and from south to north inside the magnet.

1-16 Mechanical Forces Produced by Magnetic Action

The direction of mechanical forces produced by adjacent magnets on each other is usually determined from the simple rule that like poles repel and unlike poles attract. This is shown in Fig. 1-13a and b, where forces of attraction and repulsion may be identified. Unfortunately, the direction of mechanical forces produced on adjacent conductors, as in Fig. 1-13c and d, cannot be so easily identified, because neither north nor south poles are present. However, if the direction of flux generated by the current in each conductor is sketched, examination of the relative directions of the two component fluxes in the region between the conductors (shaded area) will indicate whether the mechanical force produced is one of attraction or repulsion. If the flux contributions generated by the two conductors in the shaded area are in the same direction, both downward or both upward, as in Fig. 1-13d, a force of repulsion is produced; if the flux contributions in the shaded area are in opposite directions, as in Fig. 1-13c, then an attractive force is produced that tends to force the conductors together. For excessively high currents, whether ac or dc, the mechanical forces produced on adjacent conductors can literally bend them out of shape and even tear them from their supporting structures (see Fig. 14-8).

The force of attraction or repulsion produced on two long straight parallel conductors in air (away from magnetic materials) may be determined by substituting in the following formula:

$$F = 4.48\, KI_1I_2(l/d)\, 10^{-7} \qquad \text{lb}$$

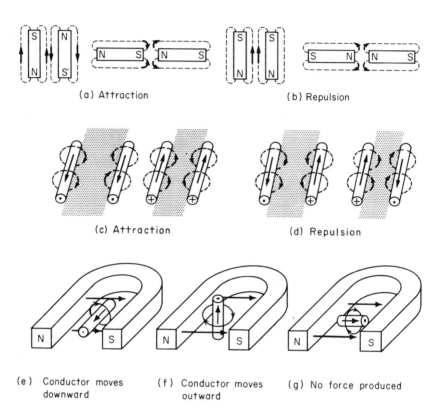

(a) Attraction   (b) Repulsion

(c) Attraction   (d) Repulsion

(e) Conductor moves   (f) Conductor moves   (g) No force produced
    downward   outward

FIG. 1-13 Mechanical forces produced by magnetic action.

where I_1, I_2 = currents in respective conductors, amp

(l/d) = ratio of conductor length to axial distance between conductors

K = shape factor = 1 for round conductors

The force of attraction produced in parallel conductors carrying currents in the same direction, shown in Fig. 1-13c, is the condition that exists in all coils. Thus the coil shown in Fig. 1-12 has all its turns squeezed together. For this reason the armature coils of large generators are supplied with mechanical braces to reduce coil movement under high-current conditions.

Figure 1-13e, f, and g shows a conductor carrying a current and situated in the magnetic field of a horseshoe-shaped magnet. In Fig. 1-13e and f the reaction between the two fields causes a force on the conductor that tends to move it downward in Fig. 1-13*e* and outward in Fig. 1-13*f*. This is the principle of motor action. In Fig. 1-13*e* the respective flux contributions generated by the magnet and the current in the conductor, in the region on top of the conductor, are both in the same direction (toward the right); hence, the force on the conductor is downward. In Fig. 1-13*f* the respective flux contributions generated by the magnet and the current in the conductor, in the region behind the conductor, are both in the same direction (toward the right); hence, the force on the conductor is outward. In Fig. 1-13*g* the respective components of conductor flux and magnet flux are neither in the same direction nor in opposite directions. Because they are perpendicular to each other, no mechanical force is produced.

1-17 Capacitive Property of a Circuit

When two conductors separated by a nonconducting material, such as air, paper, glass, etc., are connected to a dc generator or battery, as shown in Fig. 1-14, the "free electrons" in the conducting materials drift in the direction of the driving voltage. The battery, acting as an "electron pump," transfers some of the "free electrons" from conductor *A* to conductor *B*. The transfer of electrons causes conductor *B* to become increasingly negative, and conductor *A* increasingly positive. The material that lost electrons is said to be *positively charged,* and that which gained elec-

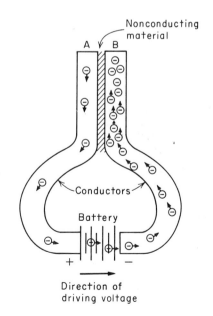

FIG. 1-14 Capacitive property of a circuit.

Direction of
driving voltage

trons is said to be *negatively charged*. As the charging process continues, conductor *B* eventually becomes sufficiently negative to prevent any additional transfer of electrons. When this occurs, the voltage across the conductors is equal and opposite to the driving voltage. Because the rate of movement of these electrons is affected by the resistance of the conducting materials, the charging process will take longer if higher-resistance materials are used.

A lengthening of the conductors will increase their capacity for electron storage. This will permit more of the "free electrons" to be shifted from *A* to *B* before the accumulated charge is sufficiently concentrated to build up an equal and opposite voltage. However, although a greater charge can be accumulated, it will take longer to attain the same voltage. Similarly, shorter conductors will have less capacity for electron storage, and the same density of electrons, and hence the same voltage, will occur with less electron transfer and in less time. This capacitance property, which is present to some extent in all circuits, delays the buildup of voltage across the conductors.

The unit of capacitance is the farad, which is the ratio of accumulated charge divided by the applied voltage:

$$C = \frac{Q}{E} \quad \text{farads}$$

where Q = electric charge, coul (1 coulomb = 6.28×10^{18} electrons)
E = voltage, volts

The farad is a rather large unit, and in most circuit applications the capacitance is on the order of microfarads (one-millionth of a farad, also written μf).

The process of transferring charges from one conductor to the other results in the storage of energy. This energy, in the form of displaced electric charges (static electricity), remains stored for some time after the battery is disconnected. The storage time depends to a good extent on the resistance of the nonconducting material that separates the conductors. The amount of energy stored in the capacitor depends on the capacitance times the square of the voltage developed across it, and may be determined by the following formula:

$$W = \tfrac{1}{2}CE^2 \quad \text{joules} \tag{1-5}$$

where C = capacitance, farads
E = voltage, volts

In those electric-circuit applications where more energy-storage capability is required or where it is desirable to increase the time delay in the buildup of voltage, "blocks" of capacitance, called *capacitors,* are added to the circuit. Some of the many varieties of capacitors are shown in Fig. 1-15. Large capacitors are constructed of two very long strips of aluminum foil separated by insulating paper and rolled into compact units for enclosure in a protective can. A nonconducting liquid is poured into the metal can, completely surrounding the unit. Large capacitors can store a considerable amount of energy a long time. Before cleaning a capacitor, the switch to the capacitor circuit should be opened, and the capacitor discharged by joining the two terminals together with a copper wire called a *jumper* (see Sec. 15-5).

Capacitors are the simplest of all electrical devices. They have no moving parts and draw essentially no power from a battery or generator. Yet, they are essential to the operation of a large variety of electronic and power equipment. In electric-power applications, capacitors are used for motor starting (see Sec. 5-8), power-factor improvement (see Sec. 15-4), energy absorbers to reduce sparking, energy storage for laser power supplies, flash photography, hot-shot wind tunnels, etc.

Figure 1-16a shows a capacitor in series with a battery and a switch. When the switch is closed, the driving voltage causes a transfer of electrons from foil *A* to foil *B*. The rate of transfer of electrons at the instant

FIG. 1-15 Assorted capacitors. (*Cornell Dubilier.*)

the switch is closed depends on the magnitude of the driving voltage and the resistance of the battery and connecting wires. Using Ohm's law,

$$I_0 = \frac{E_{bat}}{R} \quad amp$$

where I_0 is the current at the instant the switch is closed. Although the capacitance has no effect on the initial value of the current, as the charging progresses the buildup of an opposing voltage across the foils reduces the current to zero. The capacitor in the circuit in Fig. 1-16a is, in fact, a discontinuity (break, or open) in an otherwise closed loop and is further evidence that any current must be transitory, decaying to zero as the charging progresses.

Curves showing the buildup of voltage with time, and the decay of current with time, for a circuit containing capacitance and resistance are shown in Fig. 1-16b and c respectively. The part of each curve that extends over a period of time beginning with the closing of the switch (zero time) and ending when the change is almost complete ($5T_c$) is called a *transient*. The equations for the voltage and current curves in Fig. 1-16 are

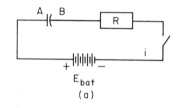

(a)

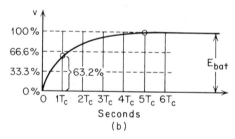

(b)

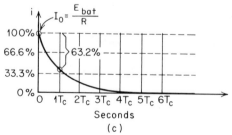

(c)

FIG. 1-16 Voltage and current transients in a capacitive circuit. (a) Circuit diagram; (b) buildup of voltage across the capacitor; (c) charging current to the capacitor.

$$v = E_{\text{bat}}(1 - e^{-t/RC})$$
$$i = I_0 e^{-t/RC} \tag{1-6}$$

where e = 2.718
$\qquad R$ = circuit resistance (of battery, connecting wires, and switch), ohms
$\qquad C$ = circuit capacitance, farads
$\qquad E_{\text{bat}}$ = battery voltage, volts
$\qquad I_0$ = current at instant switch is closed (time = 0), amp
$\qquad t$ = elapsed time, starting when switch is closed, sec

The time delays experienced in the buildup of voltage and the decay of current are related to the amount of resistance and capacitance in the circuit and may be of the order of microseconds, minutes, or even hours. The following formula, derived from Eqs. (1-6), is very useful for approximating the time delay caused by the capacitive effect of a circuit:

$$T_C = RC \qquad \text{sec} \tag{1-7}$$

The time T_c, called the time constant of a capacitive circuit, is the time it takes for the transients to make a 63.2 percent change in value. After an interval equal to five time constants ($5T_c$), the change will be 99.3 percent complete. As indicated by the voltage transient in Fig. 1-16b, the capacitance does not prevent the voltage from changing; it merely delays the change. After an interval equal to five time constants, the voltage across the capacitor will be approximately equal to the battery voltage.

• *Example 1-8.* For the circuit shown in Fig. 1-16a, assume that the capacitance is 10,000 μf, the resistance of the connecting wires and battery is 5 ohms, and the battery voltage is 120 volts. Determine (a) the time constant, (b) the time it will take for the capacitor to attain almost its full charge, (c) the charging current at the instant the switch is closed, and (d) the energy stored in the capacitor 1 hr after closing the switch.

• *Solution*

(a) $\quad T_C = RC = 5(10{,}000 \times 10^{-6})$
$\qquad T_C = 0.05$ sec
(b) $\quad$ The charging will be complete and the voltage across the capacitor will reach its approximate full value of 120 volts in five time constants, or
$\qquad 5T_C = 5 \times 0.05 = 0.25$ sec

(c)   The charging current at the instant the switch is closed is
$$I_0 = \frac{E}{R} = \frac{120}{5} = 24 \text{ amp}$$

(d)   After five time constants (0.25 sec) the voltage is assumed to have attained its approximate full value of 120 volts.   Hence, for all time after 0.25 sec, the voltage will be equal to 120 volts:
$$W = \tfrac{1}{2}CE^2$$
$$W = \tfrac{1}{2}(10,000 \times 10^{-6})(120)^2 = 72 \text{ joules}$$

The capacitive effect is present whether the driving voltage is alternating or direct (ac or dc). However, with an ac generator, the driving voltage is continually reversing, causing the voltage across the capacitor to be in a continuous state of delay. Because the voltage across the capacitor is delayed, the current is considered to be ahead of, or *leading,* the voltage. Figure 1-17a and b illustrates the curves of charging current and voltage for the two directions of battery driving voltage, and Fig. 1-17c shows the charging current and voltage curves when an alternating driving voltage is applied. From time t_0 to t_1, the behavior is similar to that shown in Fig. 1-17a, and from t_2 to t_3 similar to that shown in Fig. 1-17b. The difference in curvature between the curves shown in Fig. 1-17c and their counterparts in Fig. 1-17a and b is caused by the different driving voltages; the driving voltage of the battery is constant, whereas that of the ac generator alternates in a sinusoidal manner.

The lag or lead of alternating current and voltage waves may be determined by an observer "walking" along the time axis from left to right, observing the positive peaks; for the waves shown in Fig. 1-17c, the observer passes the positive peak of the current wave first, and later the positive peak of the voltage wave. Hence, the current wave is said to lead the voltage wave.

1-18 Testing Capacitors

Small capacitors may be tested easily with a megohmmeter or another ohmmeter of the required voltage rating. When a megohmmeter is applied to a good capacitor, the pointer first deflects sharply to zero and then gradually climbs as the capacitor charges. If the pointer deflects to zero and does not climb, the capacitor is shorted. Slight kicks downscale as the pointer climbs indicate leakage current through or across the insulation.

Electrolytic capacitors designed for use in dc circuits only act as capacitors when the applied voltage is in one direction but act as if short-circuited when the leads to the capacitor are reversed. Such capacitors will indicate zero ohms in one direction and will climb to a high value of megohms in the opposite direction. The electrolytic capacitor has one of the two aluminum foils coated with an oxide film that acts as the dielectric or insulator. The foils are separated by thin paper or gauze saturated with a solution such as boric acid and ethylene glycol. Electrolytic capacitors designed for use in ac circuits are actually two dc units mounted back to back and enclosed in one housing. Such units should indicate a high value of megohms in both directions.

Capacitors should be discharged before and after testing; the residual charge in a capacitor will cause an error in the megohmmeter reading and may also damage the instrument.

Large capacitors for power-factor improvement are not easily tested with a megohmmeter; the high capacitive effect requires considerable charging before any appreciable indication is made. Tests on such capacitors require current or power-factor measurements (see Sec. 15-5).

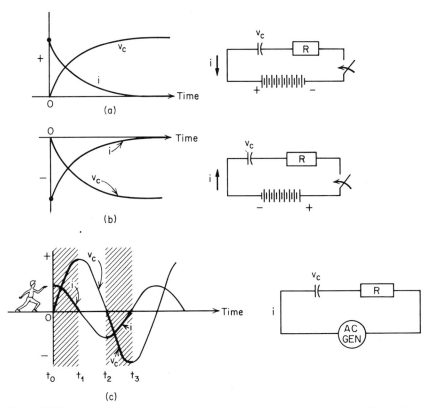

FIG. 1-17 Effect of an alternating driving voltage on the charging current and the voltage across the capacitor. (a), (b) Curves for two directions of battery driving voltage; (c) curves for a sinusoidal driving voltage.

1-19 Inductive Property of a Circuit

Whenever an electric current is made to increase or decrease, for example when starting or stopping the current in a circuit, the magnetic field (flux) that surrounds the conductors increases or decreases respectively. The changing magnetic field causes an inductive effect that delays the change in current. When the current is increasing, the increasing flux generates a voltage within the conductors in opposition to the driving voltage; this delays the buildup of current. When the current is decreasing, the decreasing flux generates a voltage within the conductors that is in the same direction as the driving voltage; this tends to keep the electrons in motion and thus delays the decay of current. Furthermore, the voltage generated by a change in flux is always in a direction to oppose the change; this phenomenon is called *Lenz' law:*

The inductive property, which is present to some extent in all circuits, does not prevent the current from changing; it merely delays the change.

The inductance of the connecting wires and cables in a circuit is generally negligible compared with the inductance of equipment containing coils of wire wound around an iron core. Coiling the wire around an iron core concentrates the inductive effect in a relatively small package, sometimes called a *lumped inductance*. The following formula may be used to determine the inductance of a coil when the number of turns, the current, and the flux produced are known:

$$L = \frac{\Phi N}{I} \quad \text{henries}$$

where L = inductance of coil, henries
I = current, amp
Φ = flux caused by current, webers
N = number of turns of wire in coil

Energy is accumulated in the magnetic field during the buildup of current and is released when the current decays. The amount of energy stored depends on the inductance of the circuit times the square of the circuit current, and may be determined by the following formula:

$$W = \frac{1}{2}LI^2 \quad \text{joules} \tag{1-8}$$

where L = inductance, henries
I = current, amp

When a circuit containing appreciable inductance, such as the field windings of a generator, is opened, the inductive effect prevents the current from dropping to zero right away. The current arcs across the switch, burning the switch blades. Unless adequately protected, such arcing and burning of switches and contacts can cause a serious maintenance problem. The severe arcing and burning is caused by the energy released from the magnetic field when the switch is opened.

Figure 1-18a shows a coil of wire wrapped around an iron core and connected in series with a battery and a switch. When the switch is closed, the inductance of the coil delays the buildup of current. However, because the current is not alternating, the inductive effect only delays the buildup of current and its final value depends solely on the magnitude of the driving voltage and the resistance of the battery, coil, and connecting wires. Thus, using Ohm's law, the final value of current is

$$I_f = \frac{E}{R} \qquad \text{amp}$$

where I_f is the final or steady-state value of current.

A curve showing the buildup of current with time for a circuit containing inductance and resistance is shown in Fig. 1-18b. The part of the curve that extends over a period of time beginning with the closing of the switch (zero time) and ending when the change is almost complete ($5T_c$) is called a *transient*. The equation for the current curve of Fig. 1-18b is

$$i = \frac{E_{\text{bat}}}{R} (1 - e^{-(R/L)t}) \qquad (1\text{-}9)$$

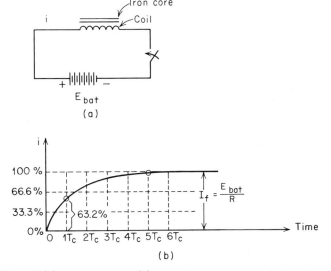

FIG. 1-18 (a) Inductive circuit; (b) current transient in an inductive circuit.

where e = 2.718

     R = circuit resistance (includes resistance of battery, coil, and connecting wires), ohms

     L = inductance of circuit, henries

   E_{bat} = battery voltage, volts

     t = elapsed time, starting when switch is closed, sec

The time delay experienced in the buildup of current is related to the amount of resistance and inductance in the circuit, and may be on the order of microseconds, seconds, or minutes. The following formula, derived from Eq. (1-9), is very useful for approximating the time delay caused by the inductive effect of a circuit:

$$T_c = \frac{L}{R} \quad \text{sec} \tag{1-10}$$

The time T_c, called the *time constant* of an inductive circuit, is the time it takes the transient to make a 63.2 percent change in value. After an interval equal to five time constants $(5T_c)$, the change will be 99.3 percent complete. As indicated in the current transient in Fig. 1-18b, the inductance does not prevent the current from changing; it merely delays the change. After an interval equal to five time constants, the current in the circuit will be approximately equal to its Ohm's law value.

• *Example 1-9.* The shunt-field windings of a particular generator have a resistance of 80 ohms and an inductance of 20 henries. Determine (a) the time it will take the current to reach 63.2 percent of its full value when connected to a 120-volt dc generator, (b) the time it will take the current to reach essentially its final value, (c) the final value of current, and (d) the energy stored in the magnetic field for the current in (c).

• *Solution*

(a) $T_c = \dfrac{L}{R} = \dfrac{20}{80} = 0.25$ sec

(b)   The time it will take to reach essentially its final value is approximately five time constants:

     $5 \times 0.25 = 1.25$ sec

(c)   The approximate final value of current after 1.25 sec can be determined from Ohm's law:

     $I_f = \dfrac{E}{R} = \dfrac{120}{80} = 1.5$ amp

(d) $W = \frac{1}{2}LI^2$

     $W = \frac{1}{2}(20)(1.5)^2 = 22.5$ joules

The inductive effect is present whether the driving voltage is alternating or direct (ac or dc). However, with an alternating-voltage generator, the driving voltage is continually reversing, causing the current to be in a continuous state of delay. Hence, an alternating current in an inductive circuit always lags behind the voltage. Figure 1-19a and b illustrates current curves for the two directions of battery driving voltage, and Fig. 1-19c shows the current and driving voltage when an alternating driving voltage is applied. From time t_0 to t_1, the behavior is similar to that shown in Fig. 1-19a, and from t_2 to t_3 similar to that shown in Fig. 1-19b. The difference in curvature between the curves shown in Fig. 1-19c and their counterparts in Fig. 1-19a and b is caused by different driving voltages; the driving voltage of the battery is constant, whereas that of the ac generator alternates in a sinusoidal manner.

For the waves shown in Fig. 1-19c, an observer "walking" the time axis from left to right passes the positive peak of the voltage wave first,

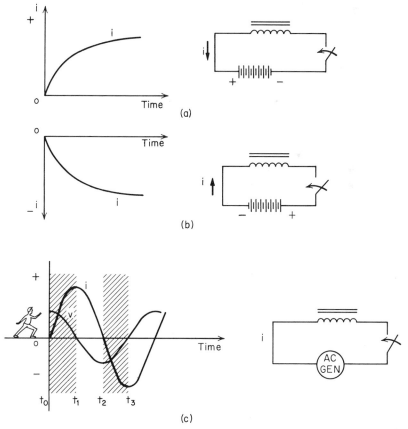

FIG. 1-19 Effect of an alternating driving voltage on the current and voltage to an inductive circuit. (a), (b) Current curves for two directions of battery driving voltage; (c) current and voltage curves for a sinusoidal driving voltage.

and later the positive peak of the current wave. Hence the current wave is said to lag the voltage wave.

1-20 AC Circuit

If an ac generator is connected to any circuit containing nonnegligible amounts of all three circuit properties (inductance, capacitance, and resistance), the current will lag if the overall inductive effect is greater than the overall capacitive effect and will lead if the overall capacitive effect is greater than the overall inductive effect. However, if the inductive and capacitive effects are equal, the lagging action of the inductance will be canceled by the leading action of the capacitance, and the current will not lag or lead the voltage. The current will be *in phase* with the voltage, rising, falling, and reversing in perfect time with the alternations of the driving voltage. The resistance property of the circuit does not cause the current to lag or lead; however, increasing the resistance of lagging or leading circuits, by the introduction of additional resistors, makes the respective currents less lagging or less leading.

In addition to the lagging or leading effects, the presence of inductance or capacitance in a circuit driven by an ac generator results in a smaller current than would otherwise occur if only resistance was present. Therefore, when applying Ohm's law to determine the current in an ac circuit, the effects of inductance and capacitance must be included. This is accomplished by substituting the overall opposition, called *impedance,* in place of resistance in the Ohm's law equation. Thus, for the ac circuit,

$$I = \frac{E}{Z}$$

where I = current, amp
E = driving voltage, volts
Z = impedance, ohms

Ammeters and voltmeters used in sinusoidal ac power systems indicate the rms (root mean square) value of the respective wave, which is equal to 0.707 times the maximum value. The voltage and current ratings on the nameplate of electrical apparatus always indicate rms values unless otherwise specified.

Examples of series and parallel circuits containing lumped values of resistance, inductance, and capacitance are shown in Fig. 1-20. It is assumed that these lumped values of R, L, and C are much greater than those offered by the connecting wires and generator. Formulas for determining the impedance of these circuits are:

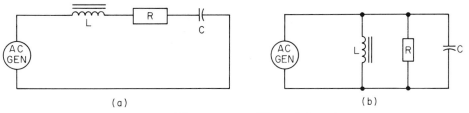

FIG. 1-20 (a) Series circuit; (b) parallel circuit.

For the series circuit,

$$Z = \sqrt{R^2 + (X_L - X_C)^2} \qquad (1\text{-}11a)$$

For the parallel circuit,

$$Z = \sqrt{\frac{1}{(1/R)^2 + (1/X_L - 1/X_C)^2}} \qquad (1\text{-}11b)$$

For both circuits,

$$X_L = 6.28fL \qquad \text{and} \qquad X_C = \frac{1}{6.28fC} \qquad (1\text{-}12)$$

where Z = impedance, ohms
X_L = inductance reactance, ohms
X_C = capacitive reactance, ohms
R = resistance, ohms
L = inductance, henries
C = capacitance, farads
f = frequency of reversal of driving voltage, hertz (1 hertz = 1 cycle per sec)

Note that in those circuits where the effects of R, L, or C are negligible, the respective R, X_L, or X_C may be omitted from the impedance equations without causing any significant error.

Note also that, when connected to a battery or a dc generator, the frequency is zero, causing $X_L = 0$ and $X_C = \infty$ (infinity).

• *Example 1–10.* A 0.2-henry coil, whose resistance is 100 ohms, is connected to a 450-volt 60-hertz generator. (*a*) Determine the impedance of the coil. (*b*) Assuming the impedance of the generator and the connecting wires is negligible, determine the circuit current.

• *Solution.* The coil must be treated as a 0.2-henry inductance in series with a 100-ohm resistance.

(a) $X_L = 6.28\,fL = 6.28(60)(0.2) = 75.36$ ohms

  $Z = \sqrt{R^2 + (X_L - X_C)^2} = \sqrt{(100)^2 + (75.36)^2} = 125$ ohms

(b) $I = \dfrac{E}{Z} = \dfrac{450}{125} = 3.6$ amp

1-21 Circuit Faults

Circuit faults generally fall in one of the following five categories: shorts, flashovers, grounds, opens, and corona.

Short circuit. A short circuit, commonly called a short, is an electrical connection that bypasses part or all of an electric circuit, thus providing an additional route for the current. Figure 1-21*a, b,* and *c* illustrates some examples of undesirable shorts that represent faults in electric circuits. Figure 1-21*a* shows a *solid short,* where a solid metallic connection across the lamp may result in a blown fuse or tripped circuit breaker. Figure 1-21*b* shows a *partial short* between two commutator bars, where an accumulated mixture of carbon dust and oil vapors formed a conducting path; it is called a partial short, or *high-resistance short,* because the resistance of the conducting path is higher than that for a metal-to-metal

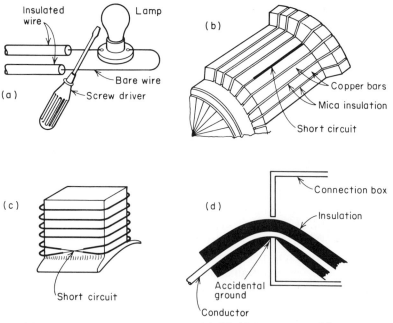

FIG. 1-21 Examples of circuit faults. (a), (b), (c) Short circuits; (d) ground;

contact. Figure 1-21c shows the shorted coil of a field pole; the insulation, worn away through vibration, wind erosion, and excess heat, caused adjacent wires to make contact through the medium of a dirt path.

If electric conduction between adjacent wires of the coil is due to dirt, it is called a high-resistance or partial short; if due to metal-to-metal contact, it is called a *dead short*.

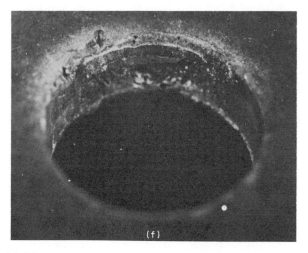

(e), (f) corona burns. (*Mutual Boiler and Machinery Insurance Co.*)

Flashover. A flashover is a violent disruptive discharge around or across the surface of a solid or liquid insulator. Flashovers occur suddenly, involve heavy currents, and generally cause considerable damage. A flashover is always preceded by ionization, which is a process by which the surrounding atmosphere is made into a conductor. This process may be caused by creepage paths developed across dirty and moist insulators; very high temperatures, such as that caused by a nearby arc; switching overvoltages; or lightning. Although the current through creepage paths may not be appreciable, the arcing and sparking produced can ionize the surrounding atmosphere and cause a flashover. Electric arcs are highly mobile and can travel away from the initial striking point, causing extensive damage to nearby equipment. Furthermore, when the arcing occurs in metal-enclosed switchgear, the arc can be expected to spread to ground.

The shortest distance between two bare conductors of opposite polarity or between a conductor and ground, measured along the surface of an insulating material, is called the *creepage distance*. The creepage distance can be increased by the use of petticoat insulators, as shown in Figs. 7-6 and 15-14.

Ground. A ground is an electrical connection between the wiring of an apparatus and its metal framework or enclosure, the wiring and a water pipe, the wiring and the armored sheath of a cable, the wiring and the chassis of a car, the wiring and the hull of a ship, etc. (see Secs. 15-1 and 15-2 concerning the detection of grounds). Figure 1-21*d* illustrates an accidental ground caused by the chafing of the insulation against the sharp edge in the opening of a connection box. Grounds can also occur in electrical machinery when the effects of age, heat, and vibration damage the insulation, permitting the entry of conducting dust; if the dust forms a bridge between the exposed conductor and the frame, the wiring is grounded. Apparatus exposed to atmospheric conditions, even though totally enclosed and "weather-proofed," sometimes develops grounds due to breathing. During the heat of the day the air inside the apparatus expands, forcing some internal air through the seals to the outside. In the evening, when the apparatus cools, a partial vacuum is formed inside the apparatus, and cool moist air is sucked in. The moisture condenses inside the apparatus, and after many heating and cooling cycles, a small pool of water begins to form. Eventually enough water may collect to make an electrical connection between the wiring and the frame, and the equipment is grounded.

Bonding. Bonding (see Sec. 15-3) is the electrical interconnecting of metallic conduit, metal armor or metal sheath of adjacent cables, the framework of electrical apparatus, metallic connection boxes, metallic framework of switchboards, and metallic waterpipes. Bonding and grounding to earth through water pipes or other means reduce shock hazards to personnel and minimize electrostatic noise in electronic equipment.

Open. An open, or open circuit, is a break in the continuity of the circuit. This break can be deliberate or accidental. Deliberate opens are made by the manual opening of a switch or circuit breaker. Accidental opens may be caused by the blowing of a fuse, the melting of a soldered connection, the blowing out of a lamp, a torn wire, etc.

Corona. Corona is a luminous, and sometimes audible, electric discharge caused by ionization of the atmosphere surrounding high-voltage conductors. Its most harmful effect is the production of ozone and nitrogen oxides, which cause chemical deterioration of the organic materials used in insulation. Ozone, a form of oxygen (O_3), is an extremely powerful oxidizing agent and, in addition to damaging insulation, readily oxidizes such metals as copper and iron.

Another adverse effect of corona is severe local heating of insulation resulting in carbonization (called *corona burning*) that produces leakage tracks (paths) in organic materials. The telltale marks of corona discharge on varnished insulation are usually in the form of white or gray dust spots. Each spot, which occurs at a point of high-voltage stress, has a matching spot on an opposite surface.

Figure 1-21e and f shows damage from corona burning in the area where a 13.8-kv potential transformer lead passed through an insulation barrier. Figure 1-21e shows damage to the lead insulation, and Fig. 1-21f shows the effect of corona burning inside the hole and on the surface of the insulating barrier. If such corona tracking is not found and corrected, the buildup of leakage paths can result in a line-to-line or line-to-ground flashover. The corona discharge and surface leakage at the barrier were probably due to a buildup of conductive deposits on the cable sheath and the insulating barrier.

A particularly annoying problem associated with corona is radio and television interference.

• *Example 1-11.* A generator supplies 2,400 watts at 240 volts to an electric heater via a 50-ft-long two-conductor cable. The two conductors are each size 12 tinned copper wire. (*a*) Sketch the circuit; (*b*) determine the current in the cable; (*c*) determine the generator voltage; (*d*) assuming that an accidental solid short occurs across the heater, determine the current in the cable.

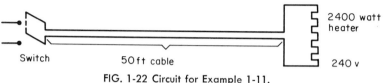

FIG. 1-22 Circuit for Example 1-11.

• *Solution.*

(*a*) The circuit is shown in Fig. 1-22.

(*b*) $P = EI$

$$I = \frac{2,400}{240} = 10 \text{ amp}$$

(*c*) From Table 1-2, the resistance of size 12 tinned copper wire is 1.68 ohms per 1,000 ft. Thus, the resistance per foot is 0.00168 ohm. The resistance of a total of 2×50 or 100 ft of wire is $100 \times 0.00168 = 0.168$ ohm. From Kirchhoff's law,

$$E_{\text{gen}} = 240 + 10 \times 0.168$$
$$E_{\text{gen}} = 241.68$$

(*d*) The short-circuit current in the cable is

$$I = \frac{V}{R} = \frac{241.68}{0.168} = 1430.1 \text{ amp} \quad \textit{Nonsence. SC Amps}$$

depends on generators
impedance!

1-22 Nameplate Data

Nameplate data offer very pertinent information on the limits, operating range, and general characteristics of electrical apparatus. Interpretation of these data and adherence to their specifications are vital to the successful operation and servicing of such equipment. Correspondence with the manufacturer should always be accompanied by the complete nameplate data of the apparatus. Figure 1-23 illustrates a typical nameplate.

Some of the pertinent information entered on the nameplates of electrical apparatus is as follows:

HP   Useful horsepower output of a motor.

RPM   Revolutions per minute at rated horsepower.

CYCLES   Rated operating frequency of the machine.

PH   Number of phases (see Sec. 4-2).

VOLTS   Rated operating voltage.

AMP   Current at rated horsepower and rated voltage.

KW   Rated kilowatts of apparatus.

°C   Normal rise in temperature above the ambient when operating at rated load. Unless otherwise specified, it is based on an ambient temperature of 40°C. The machine temperature should not exceed the nameplate rise plus 40°C.

PF   Power factor (see Sec. 4-4).

SF   Service factor; maximum permissible continuous overload. (At this service factor the machine may operate at reduced efficiency.)

KVA   Kilovolt-amperes of a generator.

FRAME   Code number for the physical dimensions of the frame.

TYPE   Type of machine, e.g., shunt, compound.

SERIAL   Indentifying number.   *sp*

LOCKED KVA CODE   Code letter that corresponds to the locked kilovolt-amperes per horsepower.

ENC   Enclosure: open, drip-proof, explosion-proof, etc.

DUTY   Length of time the machine may operate without overheating.

STYLE   Manufacturer's indentifying number designating the type of construction.   *sp*

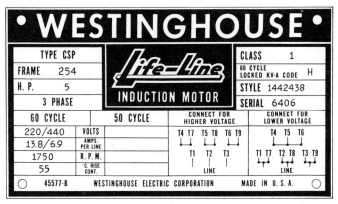

FIG. 1-23 Motor nameplate. (*Westinghouse Electric Corp.*)

Bibliography

Hubert, C. I.: "Operational Electricity," John Wiley & Sons, Inc., New York, 1961.

Jackson, H. W.: "Introduction to Electric Circuits," Prentice-Hall, Inc., Englewood Cliffs, N.J., 1959.

Morecock, E.: "Direct Current Circuits," McGraw-Hill Book Company, New York, 1953.

Schick, K.: "Principles of Electrical Theory," McGraw-Hill Book Company, New York, 1967.

Questions

1-1. What is preventive maintenance as it pertains to electrical equipment?

1-2. What benefits can be derived from a good preventive-maintenance program?

1-3. What is the essential difference between conductors and insulators?

1-4. A flow of electrons occur when a battery and conductors form a closed loop. Where do these electrons come from?

1-5. How do conductors used for energy transmission differ from those used for limiting current? Give examples of each.

1-6. Explain why the passage of current through a conductor generates heat.

1-7. Differentiate between resistivity and resistance.

1-8. What is the function of electrical insulation? Explain why it is (or is not) normal to have some electron leakage through good insulation?

1-9. What is the characteristic behavior of a semiconductor diode? Sketch a circuit that uses one or more diodes, and explain its operation.

1-10. State Kirchhoff's voltage law, and illustrate with a suitable sketch.

1-11. State Kirchhoff's current law, and illustrate with a suitable sketch. Use five branches.

1-12. With appropriate sketches to show the direction of current and flux, determine the direction of the mechanical force developed on each of the two wires of a lamp cord when it is plugged into a wall receptacle.

1-13. What effect does the following have on the amount of flux produced in a coil by a given current? (*a*) Increasing the number of turns of wire; (*b*) adding an iron core; (*c*) adding a wooden core.

1-14. Differentiate between residual magnetism and magnetic saturation.

1-15. What bad effect can excessively high currents have on conductors and supporting structures?

1-16. Describe the construction details of a capacitor. What effect does a capacitor have on the current in an ac circuit?

1-17. What precautions should be observed before and after testing a capacitor? What instruments may be used to test a capacitor?

1-18. What effect does the inductive property of a coil have on the buildup of current through it? Explain.

1-19. When the switch to a generator field circuit is opened, the current arcs across the switch, burning the blades. Explain why this happens.

1-20. What is an electrical transient?

1-21. Two inductive circuits have the same current and the same resistance values but have different time constants. Which circuit will produce the greatest amount of arcing when its switch is opened?

1-22. What effect does inductance have on the current in an ac circuit?

1-23. Describe a short circuit and illustrate it with an example not given in the text.

1-24. Make a three-dimensional sketch of a piece of electrical apparatus (for example, a capacitor enclosed in an insulated can, a coil wound around an iron core, etc.). Indicate the creepage path between conductors and the creepage path between conductors and ground.

1-25. What is a flashover? How may flashovers be prevented?

1-26. (a) Make a three-dimensional sketch showing an accidental ground that should be corrected; (b) make a three-dimensional sketch showing the bonding of electrical apparatus for safety purposes.

1-27. Describe three accidental opens that can occur in electrical apparatus or wiring systems, other than those described in the text.

1-28. What is corona, and what effect does its presence have on electrical insulation and conductors?

Problems

1-1. A coil wound with aluminum wire has a resistance of 25 ohms at a temperature of 20°C. After several hours of operation, the temperature of the coil rises to 50°C. Determine the resistance of the coil at this temperature.

1-2. (a) Determine the circular-mil area of a 200-ft length of annealed copper bar $\frac{1}{2} \times \frac{1}{4}$ in.; (b) determine its resistance at 20°C.

1-3. (a) Determine the resistance of a 6-ft length of 2-in.-diameter annealed copper bar at 20°C; (b) determine its resistance at 90°C.

1-4. A 60-ft length of circular annealed copper bus has inside and outside diameters of 4.0 and 4.5 in. respectively. Determine its resistance at 20°C.

1-5. The copper field windings of a generator have a resistance of 138 ohms at 40°C, and are connected to a 250-volt driver. After 24 hr of continuous operation its temperature rises to 90°C. (a) Determine the resistance of the field circuit for this hot condition; (b) determine the initial and final values of current.

1-6. A 600-ft-long cable containing two rubber insulated size 6 copper conductors is used to connect an electric range to a 240-volt dc generator. The resistance of each conductor is 0.410 ohm per 1,000 ft at 25°C. Determine (a) the resistance of the cable, (b) the total heat-power losses in the cable when the range draws 60 amp, and (c) the voltage across the range.

1-7. The total heat-power loss in a two-conductor extension cord is 10 watts when it carries a current of 30 amp. Determine the resistance of each of the two wires.

1-8. What is the heat-power loss in 3,000 ft of No. 2 two-conductor cable that supplies 92 amp to a 25-hp 230-volt dc motor? Assume tinned copper conductors.

1-9. Determine which combination of current and time will result in the greatest expenditure of heat energy in a resistor: (a) 100 amp for 1 sec, (b) 50 amp for 5 sec, and (c) 25 amp for 10 sec.

1-10. A 240-volt driver is connected to a series circuit consisting of a lamp, a 4-ohm, a 6-ohm, and a 10-ohm resistor. An ammeter clipped to the circuit indicates 2 amp. (a) Sketch the circuit; (b) using Kirchhoff's voltage law, determine the voltage across the lamp; (c) determine the heat-power dissipated by each of the three resistors and by the lamp.

1-11. A circuit consisting of four 120-ohm resistors connected in series draws ½ amp from an ac generator. (a) Sketch the circuit; (b) assuming that an accidental partial short occurs across one resistor, reducing its value to 40 ohms, determine the current for this fault condition.

1-12. Four field coils, each with a resistance of 30 ohms, are connected in series and supplied by a 240-volt dc generator. Calculate (a) the current in each resistor and (b) the voltage across each coil.

1-13. A 36-volt battery is connected to a series circuit consisting of a 2-ohm, a 4-ohm, and a 6-ohm resistor. Determine the voltage across each resistor.

1-14. A 12-volt battery is connected to a feeder that supplies current to three branch circuits composed of a 3-ohm resistor, a 120-ohm resistor, and a 60-ohm resistor respectively. (a) Sketch the circuit; (b) using Kirchhoff's current law, determine the battery current.

1-15. A 3-volt battery is connected to a parallel circuit consisting of the following three resistors: ½ ohm, ⅓ ohm, and ⅕ ohm. (a) Determine the overall circuit resistance; (b) determine the overall circuit current.

1-16. Determine the component currents and the total circuit current for a system composed of a 12-volt battery connected to a parallel arrangement of the following three resistors: 3-ohm, 2-ohm, and 6-ohm.

1-17. A 120-volt generator supplies power to the following parallel-connected lamps: 60-watt, 100-watt, and 150-watt. Each lamp has a voltage of 120 volts. (*a*) Sketch the circuit; (*b*) determine the current drawn by each lamp; (*c*) determine the total current; (*d*) determine the total current if an open occurs in the 150-watt lamp.

1-18. A 240-volt generator supplies 50 amp to a feeder that supplies four branch circuits. The currents in three of the branches are 12, 6, and 25 amp respectively. (*a*) Sketch the circuit; (*b*) determine the current in the fourth branch; (*c*) determine the overall resistance of the system of lamps supplied by the generator; (*d*) determine the power supplied to the system.

1-19. A 120-volt dc generator supplies a total of 6 amp to two resistors connected in parallel. The current drawn by one resistor is 2 amp. (*a*) Sketch the circuit. (*b*) Determine the current drawn by the other resistor. (*c*) What is the resistance of the other resistor?

1-20. Four 600-watt, 120-volt heating elements are required for use in an oven. The only voltage available is 240 volts. (*a*) Make a sketch showing the proper way to connect the elements to obtain 2,400 watts in the oven. (*b*) What is the resistance of each element? (*c*) What is the overall circuit current? (*d*) What is the overall circuit resistance?

1-21. A series circuit containing a 20-μf capacitor in series with a 0.5-meg-ohm resistor is connected to a 45-volt battery. (*a*) Determine the time constant of the circuit; (*b*) determine the approximate time it will take for the voltage across the capacitor to attain approximately 45 volts.

1-22 A 300-μf capacitor is connected to a 200-volt battery through a 40-ohm resistor. (*a*) Sketch the voltage-rise and current-decay curves; (*b*) after a lapse of time equal to one time constant, determine the voltage across the capacitor; (*c*) determine the value of current at that instant.

1-23. A 3-volt battery is connected in series with a 4-μf capacitor, an unknown resistor, and a switch. (*a*) Sketch the circuit. (*b*) When the switch is closed the capacitor voltage rises to 63.2 percent of its final value in 10 sec. Determine the resistance of the unknown resistor. (*c*) Determine the energy stored in the capacitor when it is completely charged.

1-24. A 400-μf capacitor is charged to 1,000 volts. When discharged through a resistor, the voltage drops to 328 volts in 6 sec. (*a*) Sketch the voltage-discharge curve to an approximate scale; (*b*) determine the resistance of the resistor.

1-25. A 16-henry coil has a resistance of 7 ohms and is connected in series to a 24-volt battery and a switch. Assume that the total resistance of the connecting wires, switch, and battery is 1 ohm.

(*a*) Sketch the curve of current buildup; (*b*) determine the final value of current; (*c*) determine the time it will take to reach 63.2 percent of this final value.

1-26. The electrical wiring of a certain machine has an inductance of 0.1 henry and a resistance of 0.5 ohm. How much time will it take for the current to reach approximately its full value when connected to (*a*) a 120-volt battery and (*b*) a 12-volt battery? (*c*) What will be the approximate current in each case?

1-27. The field circuit of a certain dc generator has resistance and inductance values of 20 ohms and 5.4 henries respectively. (*a*) Determine the approximate time in seconds that it will take for the current to reach full value; (*b*) determine the final value of current if the driving voltage is 240 volts; (*c*) determine the energy stored for the value of current in (*b*).

1-28. A parallel circuit consisting of a 0.1-henry "pure" inductance coil and a 50-ohm resistor is connected to a 208-volt 60-hertz driver. Determine (*a*) the inductive reactance of the coil, (*b*) the impedance of the parallel circuit, and (*c*) the current supplied by the generator.

1-29. A parallel circuit consisting of a capacitive reactance of 12 ohms, a resistor of 6 ohms, and an inductive reactance of 4 ohms is connected to a 120-volt 60-hertz generator. (*a*) Sketch the circuit diagram; (*b*) determine the circuit impedance; (*c*) determine the circuit current.

two

insulation testing and maintenance

Part of the insulation of practically all electrical machinery is in the form of organic compounds that contain water as part of their chemical make-up. Excessive temperatures, which tend to dehydrate and oxidize the insulation, cause it to become brittle and disintegrate under vibration and shock. The insulation used in electrical machinery does not last forever. It gradually deteriorates, slowly at low temperatures and more rapidly at higher temperatures. The greater the load, the higher the temperature and the shorter the life of the insulation. Thus the question of how hot a machine may safely be operated can be answered only by how long a life is desired for the machine.

Economic factors, such as initial cost, replacement cost, obsolescence, and maintenance, are of prime importance when determining the years of useful service desired for the electrical insulation. Hence, the permissible horsepower and kilowatt ratings of machines, as indicated on the nameplate, are determined and standardized by the allowable temperature rise dictated by economic considerations. Exceeding the load rating of the machine heats the insulation above the allowable limit and hastens its deterioration, whereas operating below the rating prolongs the insulation's useful life.

Available statistics[1] indicate the life expectancy of insulation to be approximately halved with each 10°C rise in operating temperature. For ex-

[1] General Principles upon Which Temperature Limits Are Based in the Rating of Electric Machines and Equipment, IEEE, 1962.

ample, a machine that is designed for continuous operation at 70°C will have its useful life cut in half when operating at 80°C.

2-1 Classification of Insulation

Electrical insulation is classified[1] by the temperature stability of the materials used in its construction. For the classifications shown in Table 2-1, all classes provide essentially the same life expectancy when operated at the indicated nominal temperatures.

Table 2-1   Temperature Classification of Insulating Materials

Class 90 (O)	Materials or combinations of materials such as cotton, silk, and paper without impregnation. Other materials or combinations of materials may be included in this class if by experience or accepted tests they can be shown to have comparable thermal life at 90°C.*
Class 105 (A)	Materials or combinations of materials such as cotton, silk, and paper which are suitably impregnated or coated or which are immersed in a dielectric liquid. Other materials or combinations may be included in this class if by experience or accepted tests they can be shown to have comparable thermal life at 105°C.*
Class 130 (B)	Materials or combinations of materials such as mica, glass fiber, and asbestos, with suitable bonding substances. Other materials or combinations of materials may be included in this class if by experience or accepted tests they can be shown to have comparable thermal life at 130°C.*
Class 155 (F)	Materials or combinations of materials such as mica, glass fiber, and asbestos, with suitable bonding substances. Other materials or combinations of materials may be included in this class if by experience or accepted tests they can be shown to have comparable thermal life at 155°C.*
Class 180 (H)	Materials or combinations of materials such as silicone elastomer, mica, glass fiber, and asbestos, with suitable bonding substances such as appropriate silicone resins. Other materials or combinations of materials may be included in this class if by experience or accepted tests they can be shown to have comparable thermal life at 180°C.*
Class 220	Materials or combinations of materials that by experience or accepted tests can be shown to have the required thermal life at 220°C.*
Over Class 220 (C)	Materials consisting entirely of mica, porcelain, glass, quartz, and similar inorganic materials. Other materials or combinations of materials may be included in this class if by experience or accepted tests they can be shown to have the required thermal life at temperatures over 220°C.*

* These temperatures are, and have been in most cases over a long period of time, benchmarks descriptive of the various classes of insulating materials, and various accepted test procedures have been or are being developed for use in their identification. They should not be confused with the actual temperatures at which these same classes of insulating materials may be used in various specific types of equipment, or with the temperatures on which specified temperature rise in equipment standards are based.

[1] General Principles upon Which Temperature Limits Are Based in the Rating of Electric Machines and Equipment, IEEE, 1962.

2-2 Permissible Temperature Rise

During the normal operation of electrical machinery its temperature rises above that of the surrounding air. Because the ambient air temperature of machines operating in the United States seldom exceeds 40°C, this value is established as the reference temperature when the permissible rise is determined. Hence the permissible temperature rise[1] of the hottest spot in a winding may be obtained by subtracting 40°C from the hottest allowable temperature. The actual measurement of the temperature rise may be made externally with a thermometer or internally with embedded detectors or resistance measurement.

Because the internal temperature is greater than the external value, the thermometer measurement is always less than that obtained by embedded detectors or resistance measurement. Furthermore, owing to variations in the thickness of insulation, nonuniformity of cooling, inaccessibility of the hottest spot, etc., the observable temperature rise may be less than the actual value. Hence allowances are made for such variation when the permissible observable rise above the 40°C ambient is determined.[1] Unless otherwise indicated, the permissible temperature rise stamped on

[1] General Principles upon Which Temperature Limits Are Based in the Rating of Electric Machines and Equipment, IEEE, 1962.

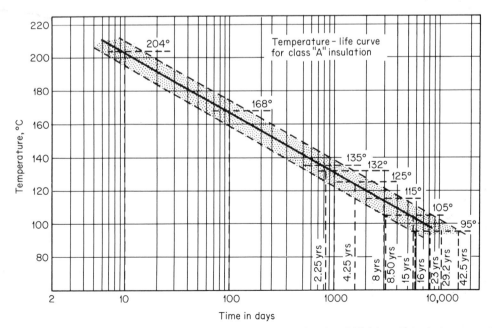

FIG. 2-1 Life expectancy versus operating temperature for class 105 (class A) insulation. (*Allis-Chalmers Mfg. Co.*)

nameplates of electrical machinery is based on thermometer determinations in a 40°C ambient. Hence the maximum total temperature permissible is the nameplate temperature rise in degrees centigrade plus 40°C.

A temperature rise of more than 10 or 15°C above the known normal operating temperature, when operating at normal loads and in a normal ambient, is an almost positive indication that the machine should be cleaned. The accumulation of dirt on the surface of the insulation and in the ventilating ducts reduces the dissipation of heat, raises the temperature, and causes thermal degrading of the insulation. Because the useful life expectancy of electrical machinery is approximately halved with each 10°C increase in operating temperature, good preventive-maintenance procedures require that periodic checks be made on the operating temperature of machines, particularly those operating on a continuous basis.

Figure 2-1 illustrates the temperature-life curve for class A insulation. The temperature indicated is the ambient, plus the temperature rise, plus 15° to obtain the hottest internal value. The sides of the shaded band give the maximum and minimum life expectancy for a given operating temperature, and the straight line through the center of the band indicates the average value. Thus a machine with class A insulation operating at 40°C rise and in an ambient of 40°C has its hottest internal temperature approximately equal to $40 + 40 + 15 = 95°C$. The life curve of Fig. 2-1 indicates that this machine has a life expectancy of 29.2 years when operating at that temperature. Yet, when it operates at a total internal temperature of 105°C, only 10° higher, its life expectancy is reduced to 15 years.

Although the temperature-life curve helps to estimate the life expectancy of a machine, the mode of operation is an important determining factor. Machines operated intermittently have a longer life span than those in continuous use. Vibration, overvoltage, and other adverse operating conditions also decrease the useful life by weakening the insulation.

2-3 Insulation Resistance

Insulation resistance is a measure of opposition offered to the current by the insulating materials. The insulation resistance is affected by moisture and dirt and is therefore a good indicator of deterioration from such causes.

The insulation resistance may be measured by nondestructive tests applied between the conductors and the framework of the apparatus. The resistance value may be read directly from a megohmmeter, or indirectly by calculation, using the voltmeter-ammeter method, or from voltmeter readings alone. When properly made and evaluated, such tests assist in diagnosing impending trouble.

Moisture absorbed in the windings or condensed on the surface of insulation results in a decrease in the measured values of insulation resistance. Hence, for insulation measurements to be significant, the tests should be made immediately after shutdown. This avoids errors due to condensation of moisture on the windings. When the machine temperature is lower than the temperature of the surrounding air, moisture condenses on the machine and is gradually absorbed by the insulation. The insulation-resistance values of dc machines are generally more sensitive to change in humidity than are those of ac windings; this is due to the greater number of leakage paths in the armature and fields of dc machines.

2-4 Temperature Correction of Measured Values of Insulation Resistance

Insulating materials have a negative resistance characteristic; that is, the resistance of the insulation decreases greatly with increasing temperature. Hence, if insulation-resistance readings taken at different operating temperatures are to be compared, the readings should be corrected to a common reference temperature, usually 40°C. The curve in Fig. 2-2a provides temperature-resistance correction data for rotating machinery.[1] To obtain the corrected value, multiply the observed resistance reading by the correction factor corresponding to the observed temperature:

$$R_{40} = K_t \times R_t$$

where R_{40} = insulation resistance corrected to 40°C, megohms
R_t = measured insulation resistance at $t°$C, megohms
K_t = correction factor

• *Example 2-1.* The insulation resistance of a dc motor is observed to be 20 megohms at a temperature of 60°C. What is its value corrected to 40°C?
• *Solution.* The correction factor obtained from the line in Fig. 2-2a is 4.0. Hence

$$R_{40} = 20 \times 4.0 = 80 \text{ megohms}$$

Figure 2-2b illustrates quite vividly the need for correcting insulation-resistance readings to a common reference temperature Curve *A* is an uncorrected resistance curve, and curve *B* is a plot of the adjusted values. The large fluctuations indicated by the unadjusted curve result in an erroneous picture of the actual insulation trend. The curve of the corrected values shows only slight changes in resistance with time.

---

[1] Recommended Guide for Testing Insulation Resistance of Rotating Machinery, IEEE No. 43, 1961.

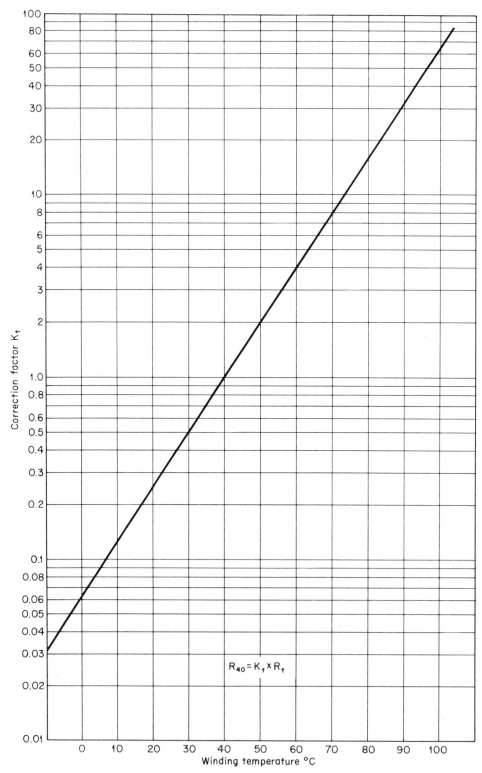

FIG. 2-2 Temperature correction of measured values of insulation resistance. (a) Correction factors; (b) sample record of a class B–insulated ac armature winding.

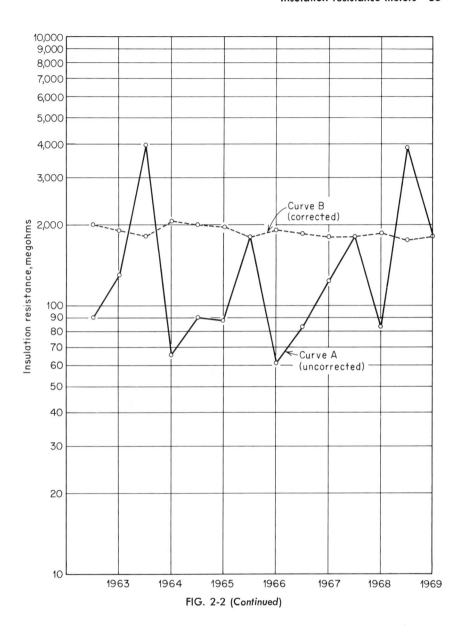

FIG. 2-2 (Continued)

2-5 Insulation-resistance Meters

An insulation-resistance meter consists of a megohmmeter that indicates the resistance directly and a source of voltage, such as a battery, rectifier, or self-contained generator. A battery-operated megohmmeter is illustrated in Fig. 2-3. The generator type of megohmmeter illustrated in Fig. 2-4 is called a *Megger insulation tester* and obtains its energy from a small built-in hand-cranked generator. To make an insulation-resistance test, disconnect the apparatus from the power line, and then connect one terminal of the Megger to the winding and the other to the frame, as shown. Cranking the Megger at its normal speed causes the pointer to move to a position on the scale corresponding to the value of insulation resistance under test. The battery-operated megohmmeter shown in Fig. 2-3 requires no cranking and may be used to measure ohms as well as ac and dc volts.

> **Warning.** If a rotor has brushless excitation, check the manufacturer's instructions before making an insulation-resistance test. The high voltage may destroy the diodes.

Large machines with good insulation have a considerable capacitive effect between the conductors and the frame and during the normal course of operation may acquire a charge of static electricity and retain it for an extended time after shutdown. Because the presence of even a small amount of stored energy causes an error in the megohmmeter reading,

FIG. 2-3 Battery-operated megohmmeter. (*Associated Research, Inc.*)

FIG. 2-4 Megohmmeter test of electrical insulation. *(James G. Biddle Co.)*

the windings of large machines should be grounded for about 15 min immediately prior to the test.

Megohmmeters are available with voltage ratings ranging from 100 to 5,000 volts. The correct selection should be based on the voltage of the equipment to be tested. If the rated voltage of the equipment under test is 100 volts or less, 100- or 250-volt instruments may be used; 440- and 550-volt apparatus may be tested at 500 and 1,000 volts; 2,400-volt apparatus at 1,000 to 2,500 volts; and 4,160-volt apparatus at 1,000 to 5,000 volts.

When a megohmmeter is being operated, the behavior of the pointer should be carefully observed, because much can be learned from its movements. The leakage of current along the surface of dirty insulation is generally indicated by slight kicks downscale; whereas the response of the pointer when testing good insulation is a downward dip followed by a gradual climb to the true resistance value. The initial dip of the pointer toward zero is caused by the capacitance of the windings and is especially noticeable in large machines, cables, and capacitors. However, the charging time is short, seldom more than several seconds. The gradual rise in pointer reading with continued cranking is caused by the dielectric-absorption effect of the insulation. It may take hours or days before the electrification is completed and the pointer ceases to rise.

Warning. All large machines that undergo a long-time insulation test should be discharged after the test is completed. This may be accomplished by grounding the wiring to the framework of the apparatus, and the discharge time should be at least as long as the charging time. When making the ground discharge connection, one end of the grounding wire must first be solidly connected to the framework of the apparatus (ground), and then the other end clipped to the wiring.

The megohmmeter *current-time curve* for relatively good insulation is shown in Fig. 2-5a, and the component currents (leakage, capacitive, and absorption) that result in this behavior are illustrated separately in

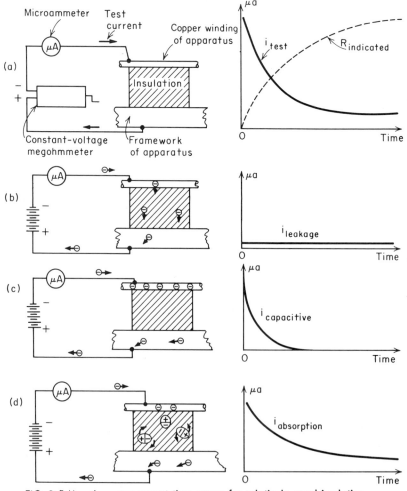

FIG. 2-5 Megohmmeter current-time curves for relatively good insulation.

Fig. 2-5b, c, and d. The leakage component, shown in Fig. 2-5b, passes through or across the surface of the insulation. The magnitude of this leakage current depends on the resistance of the leakage paths and the value of the driving voltage; it is an Ohm's law relationship. For good insulation, only a very small amount of leakage current occurs.

The capacitive component, shown in Fig. 2-5c, is caused by the capacitance between the wiring and the metal frame of the apparatus, and is typical of the charging current to a capacitor (see Sec. 1-17). This component of test current starts high but drops rapidly, reaching almost zero in a very short time (five time constants). Because the duration of this current is very brief, it has very little effect on the indicated values of insulation resistance.

The absorption component, shown in Fig. 2-5d, converts electric energy to stored energy in the form of a molecular strain in the insulating material. Although each molecule is electrically neutral (the positive charge is equal to the negative charge), its positive and negative charges form an *electric dipole*. In the presence of an applied voltage, the positive charges are pulled toward the negative terminal, and the negative charges toward the positive terminal. Thus each electric dipole not already aligned in the direction of the applied voltage experiences a torque that tends to position it parallel to the line of action of the applied voltage. This behavior, called *dielectric absorption,* is a relatively slow process that may take many hours or days to complete. When the applied voltage is removed, and the wiring grounded, the molecules return slowly to their unstressed equilibrium position.

The dielectric-absorption characteristic of electrical insulation makes it both difficult and time-consuming to obtain an absolute measurement of insulation resistance. The insulation resistance indicated by the megohmmeter, at any instant of time, is the ratio of the megohmmeter voltage to the megohmmeter test current (Ohm's law):

$$R_{indicated} = \frac{E}{i_{test}}$$

Assuming a constant megohmmeter voltage, the indicated insulation resistance depends solely on the test current. If the test current is high, the indicated resistance will be low; if the test current is low, the indicated resistance will be high. Because the test current to relatively good insulation starts high and gradually decreases with time, as shown in Fig. 2-5a, the indicated resistance starts low and increases with the continued application of test voltage. Higher leakage currents through and across the surface of relatively poor insulation permit less accumulation of stored energy within the insulating material; this reduces the dielectric-absorption effect, causing both the current and resistance curves to flatten faster.

2-6 Sixty-second Insulation-resistance Test

This is recommended for comparison with previous records. In a 60-sec test, sometimes called a *spot test,* the instrument is applied for 60 sec, and the indication is recorded at the end of that time. If a hand-cranked megohmmeter is used, the reading should be taken while still cranking at rated speed. Readings should be taken at the end of the 60-sec period, even though the pointer is still climbing. Then, if all future tests are made after 60 sec of application of megohmmeter voltage, a good comparison of the trend of the insulation can be made. All readings should be corrected to a common reference temperature, so that a fair comparison may be made. Use the curve in Fig. 2-2a. If the meter pointer reaches the ∞ mark before the end of the 60-sec period, as will happen when testing small machines with good insulation, the test may be terminated and the reading of ∞ recorded. An ∞ reading is not the value of insulation resistance; it is an indication that the insulation resistance is too high to be measured on that particular test instrument. Figure 2-6 shows a typical scale for a Megger insulation tester with a useful ohms range from 10,000 ohms to 200 megohms. An ∞ reading on this scale indicates an insulation resistance in excess of 200 megohms. Similarly a zero reading indicates a resistance of less than 10,000 ohms.

The keeping of accurate records is essential to a logical analysis of insulation condition. The 60-sec insulation-resistance tests should be made regularly and in the same manner with the same test equipment. These tests may be made on a monthly, semiannual, or annual basis as conditions demand, and the resistance, wet- and dry-bulb temperatures, and date should be recorded. The 60-sec megohmmeter readings should be corrected to some base temperature, such as 40°C, and should be plotted on semilog paper for easy determination of the trend. Even though each individual reading may be above the minimum requirements, a consistent downward trend is an indication of impending trouble. Figure 2-7 illustrates a type of test-record card that facilitates the keeping of insulation-resistance records.

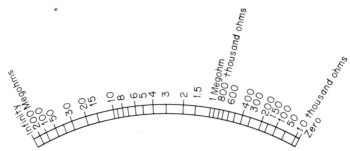

FIG. 2-6 Typical scale of a Megger insulation tester. (*James G. Biddle Co.*)

INSULATION RESISTANCE TEST RECORD

for use with "Megger"* Insulation Testers up to 200 Megohms

APPARATUS_____ NO._____ LOCATION_____

RATING_____ DATE INSTALLED _____

DATE									

INSULATION RESISTANCE

MEGOHMS

INFINITY
200
100
50
30
20
15
10
8
6
5
4
3
2
1.5

OHMS

1 MEGOHM
800000
600000
500000
400000
300000
200000
150000
100000
50000
10000
ZERO

Cat. No. 946   *TRADE MARK REG. U. S. PAT. OFF.   JAMES G. BIDDLE CO., 1316 ARCH ST., PHILA. 7, PA.

DATE	READING	TEMPERATURE			° BASE TEMPERATURE	
		AMBIENT		APPARATUS	Correction Factor	Adjusted Insula Resis.
		DRY BULB	WET BULB			

These two columns are for those who wish to consider dew point and relative humidity in interpreting insulation resistance values.

These three columns are for those who wish to know insulation resistance values after they have been adjusted to a base temperature.

100M 10-53 28576

FIG. 2-7 Insulation-resistance record card.

Sixty-second insulation-resistance tests conducted at two different voltages, preferably in the ratio of 1:5 (500 and 2,500 volts, for example) can provide more conclusive indications of moisture or other contaminants than a single voltage test. Using a multivoltage megohmmeter, a 60-sec test is made at the low-voltage setting and again at the high-voltage setting. A reduction of 25 percent in the measured insulation resistance at the higher voltage, with a 1:5 ratio of test voltages, is a sure indication of contamination. Cracked, aged, or otherwise deteriorated insulation that is fairly clean and dry may pass a one-voltage test but generally fails to pass a two-voltage test; the insulation resistance of such faults drops rapidly with higher voltages. In this test it is the percentage of change in insulation resistance rather than the actual value of resistance that is significant. Therefore, temperature corrections are not required in the two-voltage test.

2-7 Voltmeter Method

This is recommended only when a megohmmeter is not available. A high-resistance dc voltmeter with a range of 500 or 600 volts may be used to determine the insulation resistance, provided that the sensitivity of the meter is 100 ohms per volt or higher. Figure 2-8 illustrates the circuit connections for the test. Voltmeter indications are recorded with the single-pole switch closed and again with it open. The insulation resistance may be determined by substitution in the following formula:

$$R_1 = R_2 \frac{(E_1 - E_2)}{E_2}$$

where R_1 = insulation resistance to be determined
R_2 = resistance of voltmeter (found by multiplying sensitivity in ohms per volt by full-scale voltage of voltmeter)
E_1 = voltmeter indication with switch closed
E_2 = voltmeter indication with switch open

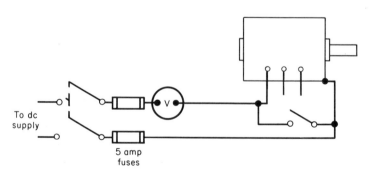

FIG. 2-8 Voltmeter method for measuring insulation resistance.

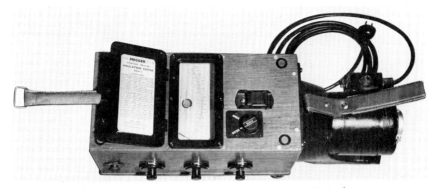

FIG. 2-9 Motor-driven Megger. *(James G. Biddle Co.)*

If the power supply has a grounded neutral, the motor frame must be insulated from ground before making this test. The voltmeter method of testing insulation should be used only as an emergency measure. Self-contained megohmmeters, such as the Megger insulation tester, are preferred.

2-8 Dielectric-absorption Test

The dielectric-absorption test[1] provides considerably more information about the condition of insulation than does the 1-min test. It is particularly useful for diagnostic testing when no previous insulation-resistance records are available. A motor-driven Megger, shown in Fig. 2-9 (battery and rectifier types are also available), is applied to the equipment under test, and readings taken every minute for 10 min are plotted on log-log paper. Typical curves showing the variation of resistance with time for class B–insulated ac armature windings are shown in Fig. 2–10. A steadily rising curve indicates a clean, dry winding, but one that flattens out rapidly indicates dirt or moisture. The test, usually made with 500- to 5,000-volt megohmmeters, is fairly independent of temperature. However, because the temperature must be a few degrees above the dew point to prevent condensation of moisture on the insulation, it is generally best to conduct the test immediately after shutdown.

2-9 Polarization Index

The polarization index[1] provides a quantitative appraisal of the condition of the insulation with respect to moisture and other contaminants. The

---

[1] Recommended Guide for Testing Insulation Resistance of Rotating Machinery, IEEE No. 43, 1961.

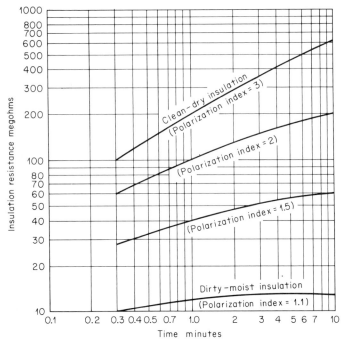

FIG. 2-10 Typical curves showing the variation of insulation resistance with time for class 130 (class B) insulated ac armature windings.

index is obtained by taking the ratio of the 10-min to the 1-min insulation-resistance measurements. The recommended polarization-index values that suggest clean, dry machine windings are, for class 105 (class A) insulation, 1.5 or more; for class 130 (class B) insulation, 2.0 or more. An index of less than 1.0 indicates a need for immediate reconditioning. As a general rule, a machine could be safely returned to service if the polarization index is 1.0 or more and the insulation resistance is above the minimum requirement.

2-10 Insulation-resistance Standards

There is no established rule or method for determining the minimum value of insulation resistance at which a machine may operate without breaking down. However, standard values of insulation resistance have been recommended and are defined as the least value that a winding should have after cleaning, or if an appropriate overpotential test is to be applied. Although it is possible to operate machines with much lower values of insulation resistance, the standard should be used as a guide to safe operation.

Rotating machinery. The recommended minimum insulation resistance for ac and dc machine armature windings and for field windings of ac and dc machines[1] can be determined from the following formula:

$$R_m = kv + 1$$

where R_m = recommended minimum insulation resistance corrected to 40°C, megohms (measured with a 500-volt megohmmeter)
kv = rated machine voltage, kilovolts

Transformers (dry-type). The recommended minimum insulation resistance for dry-type-transformers[2] is 2 megohms per 1,000 volts of nameplate voltage rating, but in no case less than 2 megohms. The test should be conducted for 60 sec with a 500-volt megohmmeter.

• *Example 2-2.* The armature of a 600-kw 240-volt 1,200-rpm dc generator has an indicated insulation resistance to ground of 2 megohms at a temperature of 30°C. The test was made with a 500-volt megohmmeter applied to the armature for 1 min. How does the reading compare with the recommended standard?

• *Solution.* The recommended standard is

$$R_m = kv + 1$$

$$R_m = \frac{240}{1,000} + 1 = 1.24 \text{ megohms}$$

The temperature correction factor for 30°C, as determined from Fig. 2-2a, is 0.5:

$$R_{40} = K_t \times R_t$$
$$R_{40} = 0.5 \times 2 = 1.0 \text{ megohm}$$

The measured insulation resistance corrected to 40°C is below the recommended standard; the machine should be reconditioned.

2-11 Nondestructive High-potential Testing

A high-potential (high-voltage) test is the only means of obtaining positive proof that the insulation of electrical apparatus has sufficient voltage strength to ride out overvoltages caused by switching and lightning surges and to continue to provide useful service. However, this test should be

[1] Recommended Guide for Testing Insulation Resistance of Rotating Machinery, IEEE No. 43, 1961.
[2] Proposed Guide for Operation and Maintenance of Dry-type Transformers with Class B Insulation, USAS C57.94-1958.

applied only if the equipment has a measured insulation resistance, at 40°C, equal to a minimum of 1 megohm per 1,000 volts of nameplate rating, plus 1 megohm; this measurement to be made with a 500-volt megohmmeter. The objective of nondestructive high-potential tests is either diagnostic or go/no-go determinations. These are not breakdown tests, which are destructive in nature and designed to determine the voltage at which actual breakdown would occur.

The major area of application for high-potential testing is in checking equipment that has been scheduled out of service for evaluation and repair. This test must not be made on vital equipment that may, if damaged, disrupt the operation of a plant or pose a safety hazard.

Unless otherwise specified by the manufacturer or testing agency, the following recommended maximum allowable test voltages may be used for a nondestructive high-potential test: ($1.5 \times$ rated line-to-line voltage), for an ac test voltage; $1.7(1.5 \times$ rated line-to-line voltage), for a dc test voltage.[1]

When making connections for a high-potential test, all wiring of the apparatus under test should be connected to the "hot" terminal of the tester, and the framework of the apparatus should be connected to the ground terminal. This is shown in Fig. 2-11.

• *Example 2-3.* A 200-hp 2,300-volt three-phase 60-hertz induction motor is to be given a high-potential test. The maximum allowable test voltage is determined to be

$1.5 \times 2,300 = 3,950$ volts   for an ac test voltage

and

$1.7 \times 1.5 \times 2,300 = 6,715$ volts   for a dc test voltage

---

[1] Guide for Insulation Maintenance for Medium and Small Rotating Electrical Machinery (rated less than 10,000 kva), IEEE, third draft, 1967.
(now IEEE 432)

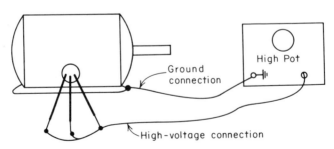

FIG. 2-11 Connections for high-potential testing.

FIG. 2-12 Ac high-potential tester. (*Associated Research, Inc.*)

2-12 AC High-potential Test

An ac high-potential test is generally used for a go/no-go evaluation of electrical insulation. When making this test, the voltage should be raised quickly to the maximum allowable test voltage and held there for 1 min. If breakdown does not occur, the insulation may be considered to have sufficient "voltage strength" for continued service. A breakdown is indicated by a neon light, as on the test equipment shown in Fig. 2-12, by the tripping of an overcurrent relay, or by sparks or arcing. After the 1-min timing period, the voltage should be gradually reduced to a low value and the test set deenergized. This gradual reduction in test voltage prevents a possibly damaging surge voltage. Although an ac high-potential test does provide go/no-go information, it does not indicate the relative quality of the insulation; it does not answer the question "Is the insulation in very good condition or is it approaching the need for reconditioning?"

2-13 DC High-potential Test

A dc high-potential test can provide a considerable amount of data useful for evaluating the condition of electrical insulation. When making this test, an initial voltage step of approximately one-third of the recommended maximum allowable test voltage is applied and maintained constant for 10 min. During this part of the test the leakage current should be plotted against time at the end of each 1-sec timing interval. Figure 2-13*a* illustrates representative current-time curves for the 10-min constant voltage test. Good insulation exhibits a steady decrease in leakage current with time; bad insulation, on the other hand, exhibits a rise rather than a decline in leakage current. Any rise in leakage current during this test is a signal to stop the test. Note that the polarization index can be calculated from these test data by dividing the leakage current after 1 min by that obtained after 10 min:

$$\text{PI} = \frac{I_{\text{leakage}(1)}}{I_{\text{leakage}(10)}}$$

After the initial 10-min test, the dc voltage should be increased in 8 to 10 uniform steps, each 1 min long, from the 30 percent value up to the maximum allowable dc test voltage. Each set of current and voltage measurements taken during this part of the test should be plotted immediately after the end of the particular 1-min interval, even though the current may still be changing. The test should be stopped at the first indication of an upward bend or knee in the curve. A knee in the curve indicates the need for cleaning and drying, and if the test is not stopped, the leakage current may increase to a value that may damage the insulation. Figure 2-13*b* illustrates representative current-voltage curves for the step voltage test. A straight line is indicative of good insulation, whereas a knee indicates the need for reconditioning. A portable dc high-potential test set is shown in Fig. 2-14.

As a safety precaution, when using the step voltage test, the leakage-current relay should be adjusted to an initial setting approximately four times the steady leakage current obtained when 30 percent of the maximum allowable test voltage is applied. As increased increments of test voltage cause the leakage current to approach the trip setting of the overcurrent relay, the trip setting should be gradually inched upwards. It is important that the relay setting not be set too high, or a sudden failure may cause arcing and extensive damage to the insulation. In any event, the test should be stopped when a sudden increase in leakage current is observed.

Repeated diagnostic dc high-potential tests on an apparatus during reconditioning should always be made with the same terminal of the test equipment connected to the apparatus ground. A good rule to follow is always to connect the negative terminal to ground and the positive terminal to the apparatus wiring.

Warning. High-voltage testing, whether ac or dc, represents a potential hazard to life, and every safety precaution recommended by the manufacturer or testing agency must be followed. The voltages used and the available current (although in milliamperes) are lethal. Electrician's rubber gloves must be worn when making these tests, and the wiring of the apparatus must be discharged by grounding for 15 min or more after completing the test. When making the ground discharge connection, one end of the grounding wire must first be solidly connected to the framework of the apparatus, and then the other end connected to the wiring. At very high-test voltages, e.g., 75,000 volts, appreciable voltage recovery may occur when the grounding wire is removed, even though the winding may have been connected to ground for over 30 min; this voltage is caused by the release of energy as the molecules return slowly to their unstressed equilibrium position.

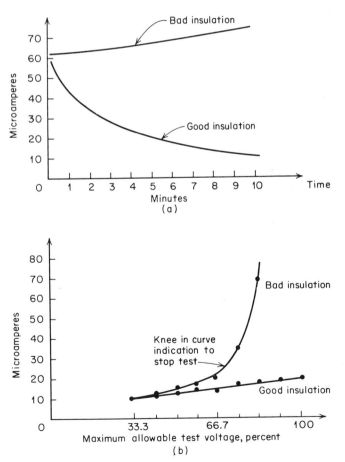

FIG. 2-13 Dc high-potential test curves. (a) Current-time characteristics; (b) current-voltage characteristics.

FIG. 2-14 Dc high-potential test set. (*Associated Research, Inc.*)

2-14 Other Insulation-resistance Tests

Insulation-resistance tests using dielectric absorption, the two-voltage method, and nondestructive high-potential tests are the most common means for determining the general condition of electrical insulation. However, to pinpoint the trouble in very large machines, many other tests have been devised. Some require very elaborate equipment that is too expensive for most shops. These are insulation power-factor tests, slot-discharge tests, surge-comparison tests, corona-probe tests, and interlaminar insulation tests. These tests are highly specialized and are beyond the scope of this book.

2-15 Insulating Liquids

The insulating liquids commonly used for insulating and cooling transformers and for arc interruption in oil circuit breakers are mineral oil and askarel. Askarel is a noninflammable, chemically stable, nonsludging synthetic liquid sold under such trade names as Chlorextol, Inerteen, and Pyranol. These insulating liquids become contaminated and deteriorate during normal service. The absorption of moisture and the formation of acids and sludge reduce its dielectric strength, and the deposit of sludge on coils, in cooling ducts, and other parts of the apparatus reduces the normal heat transfer capabilities of the equipment. Increased temperatures, whether caused by overloads or reduced heat-transfer capabilities, hasten the formation of sludge and acids. Transformers and other apparatus that use a seal of inert gas over the oil are not subject to oxidation, and hence sludging should not occur.

2-16 Testing and Sampling Insulating Liquids

The standard dielectric test for insulating liquids[1] requires the use of an insulated test cup with 1-in. diameter flat electrodes spaced 0.1 in. apart, an adjustable ac high-potential test set, and a pint sample of the liquid to be tested. The test cup, shown in Fig. 2-15a, should be wiped clean with a clean, dry chamois skin and thoroughly rinsed with moisture-free, oil-free, white gasoline. After cleaning, the cup should be refilled with clean white gasoline and given a high-potential test. The test voltage should be increased smoothly at a rate approximating 3 kv per sec until breakdown occurs. Breakdown is indicated by a continuous discharge across the electrodes or the tripping of the breaker. If the gasoline withstands 25 kv or more, the cup is clean enough for test purposes. If it tests less than 25 kv, the cleaning and testing cycle should be repeated.

Assuming that the cup is in suitable condition, it should be emptied of gasoline, rinsed with a portion of the sample to be tested, and then filled to overflowing. Any air bubbles in the test sample should be allowed to escape from the cup before applying the test voltage. Gently rocking the cup and then allowing it to set for 3 to 5 min should get rid of the entrapped air. The test voltage should then be applied and increased smoothly at approximately 3 kv per sec until breakdown occurs. Occasional momentary discharges that do not cause a permanent breakdown should be disregarded. After breakdown occurs, the test cup should be emptied, refilled with a fresh sample, and the test repeated. The average of the breakdown voltages of five tests is taken as the dielectric strength of the sample.

If the dielectric strength of mineral oil tests less than 22 kv, it should be filtered to raise its dielectric strength to 25 kv or higher. If the dielectric strength of askarel tests less than 25 kv, it should be filtered to raise it to 30 kv or higher.

[1] American Society for Testing Materials, Test for Dielectric Strength of Insulating Oils, D-877.

To avoid erroneous test results, it is extremely important that the samples be obtained under controlled conditions and properly cared for until tested. Polyethylene or glass containers with glass, cork, or polyethylene stoppers are preferred, and every effort should be made to use them. Rubber or rubber composition stoppers contaminate the liquid and

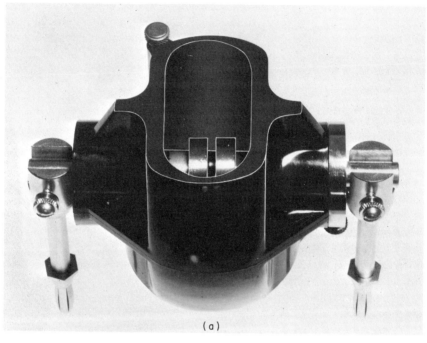

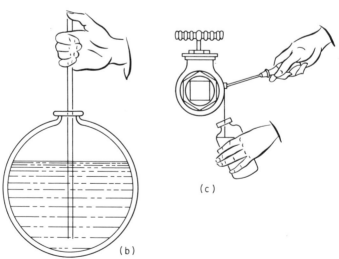

FIG. 2-15 (a) Standard test cup (*Associated Research, Inc.*); (b) sampling oil from the bottom of a drum by means of a "thief"; (c) sampling oil from a valve. (*General Electric Co.*)

must not be used. The containers should be rinsed with oil-free white gasoline, dried, and then washed with strong soap suds, rinsed with distilled water, and baked in an oven for at least 8 hr at 105 to 110°C. The containers should be corked immediately after drying, and remain corked until ready for filling.

Samples may be taken from a valve, or from the drum by means of a glass or brass "thief," as shown in Fig. 2-15. The "thief" should be cleaned in the same manner as prescribed for containers. Samples from outdoor apparatus should be taken on a clear, dry day with adequate protection against windblown dust. To avoid condensation of moisture on the sample, the temperature of the liquid should be no lower than the surrounding air. Before sampling from a valve, it should be cleaned, and then enough liquid allowed to run through it to flush out any moisture or other contaminants trapped in the valve. An 8-hr settling time should be allowed before sampling from a drum. Furthermore, because askarel is heavier than water and mineral oil is lighter than water, the presence of water can be more easily detected if askarel samples are drawn from the top of the tank and oil samples from the bottom. Each sample should be identified with the serial number of the apparatus from which it was drawn, and whether it was taken from the top or the bottom of the drum or from a valve.

Warning. Every effort should be made to avoid getting liquid askarel or concentrated vapors from hot askarel on the skin. It is very irritating, especially to the eyes, nose, and lips. Continued exposure to askarel may cause skin eruptions because of the absorption of the liquid through the pores of the skin. The recommended treatment for such irritations is castor oil for the eyes, and castor oil or cold cream for the nose, lips, and other areas of skin.

2-17 Filtering Insulating Liquids

Mineral oils that test below the minimum acceptable standard should be filtered to remove accumulated sludge and moisture. The oil should be filtered by forcing it through pads of dry filter paper. A filter press for accomplishing this is shown in Fig. 2-16. The moisture is absorbed by the paper, which also filters out the sludge.

Askarel is generally purified using dry fuller's earth in conjunction with filter paper. An oil filter press can be used for this purpose provided that it and all connecting hoses and pipes are thoroughly cleaned of oil.

Warning. Electrostatic charges can be built up when insulating oils are caused to flow in pipes, hoses, tanks, etc. The static voltages generated may be in excess of 50,000 volts. Thus, for safety reasons the tanks, filter press, pipes, and hoses should be grounded during the filtering operation and remain grounded for at least 1 hr after

FIG. 2-16 Filter press for cleaning and drying insulating liquids. *(General Electric Co.)*

filtering is completed. Furthermore, because electron conduction through dielectric liquids is slow, an arc can occur between the free surface of the liquid and adjacent metal surfaces. Hence, provisions must be made for venting any explosive gases that may accumulate in the receiving tank.

2-18 Cleaning Electrical Insulation

The greatest single cause of electrical failures is the breakdown of insulation. Such failures may be brought about by the absorption of moisture, oil, grease, and dust into the windings and by excessive heat, vibration, overvoltage, and aging.

The effect of inadequate maintenance is dramatically illustrated in Fig. 2-17. This 25-hp induction motor, used in a lumber mill, received no attention except an occasional squirt of oil on loosely fitted sleeve bearings. The accumulated mixture of oil and sawdust completely covered the windings, restricting the necessary free movement of cooling air. The resultant overheating caused the motor windings to burn out. Damage of this type is due to negligence and would not happen if an effective preventive-maintenance program were followed.

The frequency of cleaning should be determined by need and not by the calendar alone. Unnecessary cleaning is not only wasteful of time, but it causes excessive wearing away of insulation and may do more harm than good. Electrical apparatus should be cleaned when one or more of the following conditions are noted:

1. Visual determination of accumulated dirt on the windings.
2. The operating temperature of the equipment is 10 to 15°C above its normal operating value.
3. Insulation-evaluation measurements indicate the presence of dirt or moisture.

All equipment to be cleaned must be disconnected from the power source, and the wiring grounded to the frame to eliminate any accumulated static charge.

When cleaning large motors, such as the main propulsion motors of electric-drive ships, it is often necessary to climb inside the machine. All one's pockets should be emptied prior to such entry, because a screw, coin, or other small article may fall into the machine and cause serious damage when starting up.

FIG. 2-17 Motor damage caused by lack of maintenance. (*Mutual Boiler and Machinery Insurance Co.*)

The removal of loose, dry dust is accomplished best with a vacuum cleaner. A brush and a thin nonconducting nozzle make excellent attachments for cleaning commutator risers and other hard-to-get-at places. After vacuuming, oil-free, dry compressed air at about 30 to 40 psi should be used to loosen and clear out any remaining dust. The use of safety goggles and a dust mask permits close work and encourages a thorough cleaning job. When using compressed air for cleaning, an adequate exhaust must be provided; otherwise it merely rearranges the dust, blowing it off one part of the machine and depositing it on another or on some other apparatus.

Encrusted dirt that blocks air passages should be removed carefully with a hardwood or fiber scraper. A metal scraper will damage the insulation and should not be used.

Insulation that is coated with an oily film should be cleaned by wiping or rubbing with a piece of cheesecloth or lintless rag barely moistened with an appropriate commercial solvent. Excessive use of solvents will damage the protective coating provided by the varnish; use only enough to remove the oily scum, and then wipe dry with a clean cloth. Insulations containing silicones are very sensitive to commercial solvents. Such insulation requires special treatment and is discussed later in the chapter. Petroleum distillates, such as Stoddard solvent (commercial standard CS 3-40, U.S. Bureau of Standards), mineral spirits, and cleaner's naphtha are generally adequate for most cleaning situations. However, these solvents are inflammable and should not be used in the vicinity of open flames. Adequate ventilation must be provided, because high concentrations of these vapors have toxic effects when inhaled.

The so-called "safety" solvents are mixtures of petroleum distillates with a chlorinated solvent. The chlorinated solvent reduces the fire hazard but increases the toxicity. Unfortunately as the chlorinated solvent evaporates, the mixture becomes more inflammable.

Apparatus covered with an exceptionally heavy oily scum may require a stronger cleaning action than that provided by the petroleum distillates. In such cases, chlorinated solvents such as methyl chloroform, trichloroethylene, and perchloroethylene may be used. Although noninflammable, all chlorinated solvents are considerably more toxic than the petroleum distillates. Of the three chlorinated solvents, methyl chloroform is the least toxic but eats away aluminum. Carbon tetrachloride (carbon tet, CCL_4) is extremely toxic and should not be used for cleaning electrical insulation. Violent illness and death from suffocation have resulted from its use.[1] It also has a detrimental effect on the surface film of commutators and slip rings and deteriorates rubber.

[1] B. L. Hardin, Jr., M.D. Carbon Tetrachloride Poisoning: A Review, *Industrial Medicine and Surgery*, 23(3):93–105, March, 1954.

Cleaning by spraying. Spraying the solvent through an atomizer held close to the work is an excellent way of loosening and removing the grime from hard-to-get-at places, such as the ventilating ducts or recesses in armature and stator windings. The disadvantage of spraying is that any carbon dust or other conducting dust set free by the solvent may be redeposited in hairline cracks. If this occurs, the insulation resistance after cleaning may have a lower value than before the cleaning process was started. Spraying also causes some of the solvent to puddle in inaccessible places, where it softens the insulation and does irreparable damage. If emergency conditions require the direct application of liquid solvent, it should be used as a spray, not as a solid stream. Confined spaces should be ventilated by forced draft, and the sprayer should be grounded to prevent the buildup of high-static voltages. It is recommended that machines cleaned by spraying be revarnished (see Sec. 2-22) before being returned to service.

The immersion and running of electrical machinery in solvents is not recommended. Although the machine comes out looking shiny and new, the redepositing of conducting dust in unseen crevices may result in shortening the life of the equipment. Immersion of commutators is particularly bad, because carbon dust can be carried into inaccessible places. Furthermore, soaking in solvent degrades the insulation, making revarnishing necessary.

Insulations containing silicone, such as class 180 (class H), should not be cleaned with the standard commercial solvents. Such solvents attack the silicone varnish, and should not be used. The recommended practice for cleaning silicones is to use water containing a nonconducting detergent. It is best to consult the manufacturer before attempting to clean this class of insulation.

Cleaning with water. The cleaning of electrical machinery with fresh water and a detergent, such as Dreft, is often done successfully in repair shops, and is the recommended procedure for cleaning silicone-treated insulation. The cleaning should be done as rapidly as possible, the machine rinsed with fresh warm water, excess moisture wiped off with a clean, dry cloth, and the apparatus baked dry in an oven. The procedure outlined for submerged or flooded equipment should be followed.

Cleaning with abrasives. Very hard dirt deposits that are not removable by solvents or detergents may be cleaned by air blasting with ground corn cobs or the more abrasive ground walnut shells. This must be recognized as a drastic measure, because the insulation can be very easily worn away. To prevent damage, the operator must be close enough to the work to see what he is doing. Hence it is imperative that he wear close-fitting goggles and a dust mask.

2-19 Submerged or Flooded Equipment

Submerged or flooded equipment should be washed with fresh warm water, no hotter than 80°C (176°F), to remove all traces of salt and silt. The water pressure at the windings should not exceed 25 psi. The washing operations should continue until salinity tests of the wash water show the insulation to be free of salt. Contaminants, such as tar, grease, oil, or wax, should be removed by the use of nonconducting detergents (emulsion cleaners) added to the wash water. After a thorough rinsing with fresh water, the water should be removed as quickly as possible by wiping and blowing with low-velocity compressed air. In some cases it may be necessary to remove the commutator clamping ring to drain any accumulated water. The machine should then be baked dry with externally applied heat. However, even with the best treatment and careful tests, a submerged machine sometimes fails when put back into service. It is therefore recommended that reconditioned flooded equipment be limited for use in nonvital operations that would not endanger the plant or personnel if the equipment failed. After four months of continuous service with no failure, reconditioned equipment can be assumed reliable for unrestricted use.

2-20 Drying Electrical Insulation

The most favored method of drying insulation is through the application of external heat. This is conveniently done with a regular baking oven, as shown in Fig. 2-18, or by means of an improvised oven of tarpaulin surrounding the machine. A vent at the top of the tarpaulin permits the exit of moisture-laden air. Small machines are often successfully dried by placing them on top of a boiler. When external heat is used for baking, the temperature of the windings should not exceed 90°C as measured by thermometers taped to the coils. The best source of heat for makeshift ovens is an electric heater or radiant-heat lamps; steam heaters or hot-air furnaces also do an adequate job.

Sixty-second insulation-resistance measurements should be recorded every 4 hr during the drying-out process. A typical drying curve for a dc motor armature is shown in Fig. 2-19. During the first part of the drying operation the increase in temperature causes a decrease in the indicated values of insulation resistance. Then, with a constant drying temperature, the resistance increases as the moisture is driven out. When the insulation is dried and allowed to cool, the resistance increases to a high value. The plotted values of insulation resistance are not corrected for temperature, because such correction would serve no useful purpose in this case.

Baking with internal heat. Windings that have an insulation resistance greater than 50,000 ohms when cold may be baked with internal heat. Current is fed into the windings at low voltage, and the heat developed internally (I^2R losses) serves to drive off the moisture. The required current may be obtained from a dc electric welder, as shown in Fig. 2-20, or may be generated within the machine itself. In the latter case, the armature is short-circuited, and while the machine is running at less than half speed, the field current is gradually increased until an ammeter placed in the armature circuit indicates rated current. The reduced speed of the machine prevents damage to damp or wet insulation.

Because the internal temperature of the machine will be higher than the external value, the total temperature of the windings as measured by thermometers taped to the coils should not exceed 80°C, and it should take at least 6 hr to attain that value. In addition, the gradual increase in current to the rated value should be spread out over the entire 6-hr period. Rapid heating of the winding will form steam pockets that may rupture and permanently damage the insulation. Alcohol-bulb

FIG. 2-18 Baking oven in a large repair shop. (*Westinghouse Electric Corp.*)

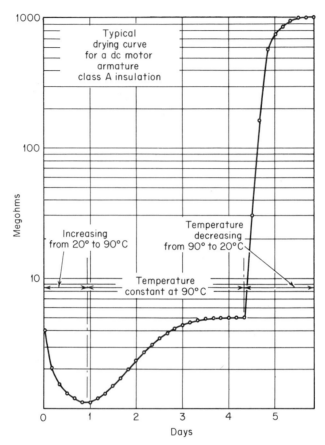

FIG. 2-19 Typical drying curve for a dc motor armature. (*James G. Biddle* Co.)

thermometers should be used with ac machines, because a mercury ther-mometer may be affected by induced voltages, and read higher than the actual value.

When internal heat developed by generator action is used to dry out the armature of a dc machine, the series field must be disconnected. Other-wise it may cause a rapid buildup of voltage and destroy the insulation.

Induction motors may be dried by blocking the rotor and applying 25 percent of its rated voltage to the stator.

The field of an alternator or synchronous motor should be dried with direct current through the windings. If alternating current is used, transformer action will overheat and possibly damage the squirrel-cage or damper windings. The brushes should not be used to conduct the cur-rent to the rotor, or the rings will be blackened and pitted where they

FIG. 2-20 Improvised oven made with a tarpaulin. Internal heat supplied by an electric welder. (*Westinghouse Electric Corp.*)

contact the brush, and the brushes will be damaged by overheating; copper bands should be clamped around the rings for this purpose.

Electrical determination of internal winding temperatures during the drying period. This method of temperature determination may be used on almost any type of winding and is especially adaptable to the cylindrical-rotor type of field for large high-speed ac machines. Such rotors contain a considerable amount of copper and iron, and the external temperature of the rotor lags considerably behind the internal value. Hence thermometer readings, which can only be observed externally, do not give a true indication of the internal temperature. Temperature determinations should therefore be made by resistance measurements of the winding. Figure 2-21 illustrates the circuit connections required for drying out the rotor of an ac generator, and the instruments required for temperature determinations.

The resistance of the winding is determined by Ohm's law from the observed indications of the voltmeter and ammeter, that is,

$$R_1 = \frac{E_1}{I_1}$$

A thermometer reading of the winding temperature, immediately prior to the application of voltage, is used as the cold reference temperature; and the initial values of voltage and current applied to the winding are used to calculate the resistance at that temperature. Then, as heating

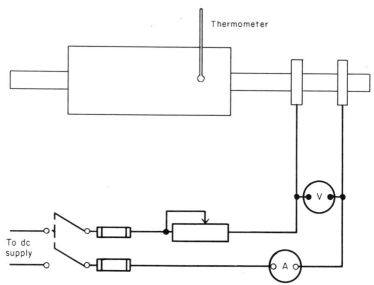

FIG. 2-21 Electrical connections for drying out the rotor of an ac generator, and the instruments required for determining the internal temperature.

progresses, the temperature at any instant can be determined by substitution of the indicated values of current and voltage in the following formula:[1]

$$t_2 = \frac{E_2(234.4 + t_1)}{I_2 R_1} - 234.4$$

where R_1 = "cold" value of resistance as determined by voltmeter and ammeter indications at instant current starts to flow

E_2 = voltmeter indication at instant for which temperature is to be determined

I_2 = ammeter indication at instant for which temperature is to be determined

t_1 = reference or "cold" temperature of winding immediately prior to application of voltage, measured by thermometer, °C

t_2 = internal temperature at which V_2 and I_2 are measured, °C

It is assumed that, prior to the application of voltage, the winding has been allowed to rest long enough for the internal temperature to become equal to the external value—over 24 hr for large cylindrical types of rotors.

---

[1] Master Test Code for Temperature Measurement of Electric Apparatus, IEEE No. 119, 1966.

2-21 Care of Electrical Apparatus During Periods of Inactivity

The greatest injury done to electrical apparatus during periods of idleness is the absorption of moisture that condenses on the insulation. Condensation of moisture on electrical apparatus results when the surface temperature falls below the dew point. Hence, if the temperature within the machine is kept several degrees higher than the ambient, condensation will not take place.

Space heaters should be used to keep the insulation dry during periods of inactivity. Electric lamps may be placed in the housing of machines to prevent the condensation of moisture, and a tarpaulin covering may be used to conserve heat and reduce the amount of energy required. Propulsion generators and motors for shipboard use have built-in heaters for this purpose. Controllers for deck winches are similarly equipped.

If the machine is to be left idle for a considerable period, the brushes should be raised off the commutator and collector rings to prevent electrolytic action from pitting or wearing flat spots on the metal surfaces. This is especially true aboard ship, where an atmosphere of moist salt air prevails.

2-22 Revarnishing Electrical Insulation

The revarnishing of electrical insulation should be done only when absolutely necessary. The buildup of many layers of varnish leads to surface crazing and the resultant absorption of dirt and moisture. The windings should be thoroughly cleaned and dried before the application of varnish. The best method of revarnishing is to dip the unit in varnish, allow for proper impregnation, drain, and bake in an oven. Armatures too large to fit into a tank may be rolled slowly in a pan of varnish and then baked.

Aboard ship, or in shops where baking ovens are not available, or where machines are too large, air-drying varnish may be applied with a sprayer. However, it should be recognized that spraying can never reach more than 60 to 70 percent of the surfaces, and although it is better than doing nothing, the lack of varnish in some vital areas may cause failure.

The commutator, slip rings, shaft, and bearings should be protected against the spray with several layers of heavy paper. The operator should wear a gas mask for protection against the noxious vapors. Generally two light coats are recommended, with the second application made when the first coat is no longer tacky. The drying time between coats is generally much less than 24 hr and depends on the surrounding temperature. If necessary, the drying may be accelerated by the application of a moderate amount of external heat, provided that the temperature of the apparatus does not exceed 80°C.

The vapors given off by the varnish are highly explosive. Hence, to reduce the risk of fire, the spray gun as well as the apparatus to be varnished should be grounded to a common point. This prevents the buildup of a static charge that may touch off an explosion. A CO_2 type of fire extinguisher should be conveniently located in the spray area.

Bibliography

General Principles upon Which Temperature Limits Are Based in the Rating of Electric Machines and Equipment, IEEE No. 1, 1962.
Guiding Principles for Dielectric Tests, IEEE No. 51, 1955.
Miller, H. N.: "Non-destructive High-potential Testing," Hayden Publishing Company, New York, 1964.
Moses, G. L.: "Electrical Insulation," McGraw-Hill Book Company, New York, 1951.
Proposed Guides for Insulation Maintenance of Large A-C Rotating Equipment, IEEE No. 56, 1958.
Recommended Practice for Testing Insulation Resistance of Rotating Machinery, IEEE No. 43, 1961.
U.S. Department of the Navy: "Bureau of Ships Manual," chap. 60, 1965.

Questions

2-1. What factors affect the useful life of electrical insulation?

2-2. If a machine is designed for continuous operation at a total temperature of 90°C, what effect will operating at 70°C have on its life expectancy?

2-3. In what way does excessive heat deteriorate insulation?

2-4. What is the maximum allowable temperature for the hottest spot in a machine wound with class H insulation?

2-5. What class of insulation has no temperature limitations? Why is this so?

2-6. What is meant by "temperature rise" as stamped on the nameplate of electrical machinery? Of what use is this information to the maintenance program?

2-7. What is the reference temperature generally used by the manufacturer when determining the permissible rise?

2-8. If a motor with a 50°C permissible rise is found to operate at a temperature of 90°C while in a 20°C ambient, what conclusions would you draw? Explain your answer.

2-9. What is insulation resistance? What instruments are used to measure it, and how are they used?

2-10. What effect does absorbed moisture have on the measured values of insulation resistance?

2-11. What effect does the temperature of a machine have on the measured values of insulation resistance?

2-12. What is a Megger insulation tester? How is it used?

2-13. Can electrical equipment be damaged as a result of a megohmmeter test? Explain.

2-14. An insulation-resistance test of an ac motor indicates ∞ on the megohmmeter scale. Does this mean that the insulation resistance is infinite? Explain.

2-15. Why should a large machine be grounded for several minutes before an insulation-resistance test is made?

2-16. What voltage megohmmeters should be used when testing the insulation resistance of a 50-volt generator; a 450-volt motor; a 2,300-volt transformer?

2-17. When testing insulation resistance with a megohmmeter, what would cause the pointer to make a few downscale kicks while gradually climbing upscale?

2-18. What is dielectric absorption?

2-19. What three components of test voltage are present when meggering insulation?

2-20. What effect does capacitance have on the indicated values of insulation resistance? Explain.

2-21. What is the "60-sec" insulation-resistance test? Of what use is it in a preventive-maintenance program?

2-22. What may be concluded from the plotted data of a dielectric-absorption test if it shows a steady increase in resistance with the continued application of test voltage?

2-23. Define polarization index. What does an index of less than 1.0 indicate?

2-24. What is the most conclusive insulation-resistance test for checking the presence of absorbed moisture?

2-25. A 10,000-hp motor has undergone a long-time insulation-resistance test. State the correct procedure for discharging the apparatus.

2-26. What is the objective of nondestructive high-potential testing? Should this test be made on equipment vital to the operation of a plant or process? Explain.

2-27. What are the advantages and disadvantages of an ac high-potential test? What precautions should be observed when making this test?

2-28. What are the advantages and disadvantages of a dc high-potential test? What precautions should be observed when making this test?

2-29. What factors can cause a reduction in the dielectric strength of insulating liquids?

2-30. (*a*) Describe the standard dielectric test for insulating liquids. (*b*) What precautions should be observed when obtaining samples of the liquid? (*c*) How is breakdown indicated?

2-31. Do occasional momentary discharges through the insulating liquid indicate a breakdown?

2-32. (*a*) How askarel purified? (*b*) In what way is the purification method for askarel different from that used for mineral oil?

2-33. Although askarel is noninflammable, all equipment, including hoses, pipes, etc., should be grounded during the purification process. Why is this necessary?

2-34. What is the recommended treatment for skin irritations caused by askarel?

2-35. Is it advisable to clean electrical insulation on an annual basis? Explain.

2-36. What observable conditions would indicate that the insulation of electrical apparatus requires cleaning?

2-37. (*a*) Describe the recommended procedure for cleaning apparatus that is covered with loose, dry dust. (*b*) How should encrusted dirt be removed?

2-38. What is the recommended procedure for cleaning insulation that is coated with an oily film?

2-39. State briefly the advantages and disadvantages of petroleum distillates, "safety" solvents, and carbon tetrachloride when used to clean electrical insulation.

2-40. How should insulation containing silicone be cleaned?

2-41. Why should the cleaning of insulation by spraying with a solvent be avoided?

2-42. What cleaning techniques may produce a nice-looking machine but result in a lower insulation resistance than before the cleaning was started? Explain.

2-43. Explain why the immersion and running of electrical machinery in solvents is not recommended.

2-44. Is air blasting with abrasives a generally recommended procedure for cleaning insulation? Explain.

2-45. What restrictions should be placed on the use of reconditioned flooded equipment?

2-46. Outline the procedure to be followed for cleaning and drying submerged or flooded equipment.

2-47. Describe an improvised method that may be used for drying electrical machinery in your plant or school.

2-48. Under what conditions may internal heat be used to dry electrical insulation? How is it applied, and what precautions should be observed?

2-49. Describe the method used to determine the internal temperature of a winding during the drying period.
2-50. Why does a thermometer not give a true indication of the temperature of a cylindrical-rotor field winding?
2-51. What can be done to prevent moisture from condensing on the insulation and windings of electrical apparatus during extended periods of inactivity?
2-52. When a machine is to be left idle for a considerable period, the brushes should be raised off the commutator and collector rings. Why is this necessary?
2-53. Describe two methods for revarnishing electrical insulation. What precautions should be observed?
2-54. What method of revarnishing insulation should be used when a baking oven is not available?
2-55. Should the insulation of a machine be revarnished every time it is cleaned? Explain.

Problems

2-1. The insulation resistance of a dc generator is 30 megohms at a temperature of 70°C. What is its value corrected to 40°C?
2-2. A series of periodic insulation-resistance tests made over a 1-year period provide the tabulated data for a 100-hp 2,300-volt induction motor as shown in the following table. The insulation is class A. Plot a curve of the observed and the corrected values of insulation resistance. What does the curve indicate?

Month	Observed resistance, megohms	Temperature, °C
January	12	35
February	9	38
March	8	43
April	9	40
May	5	48
June	3	53
July	2	62
August	3	54
September	3.2	53
October	6	46
November	7	43
December	10	40

2-3. A 750-volt dc voltmeter with a sensitivity of 1,000 ohms per volt is used to measure the insulation resistance of an alternator. The voltmeter indication with the switch closed is 500 volts and with the switch open is 200 volts. What is the resistance of the insulation?

2-4. What is the recommended minimum insulation-resistance standard for a 120-volt dc armature of 90 kw?

2-5. What is the recommended insulation-resistance standard for a 2,300-volt 500-kva armature of a synchronous generator?

2-6. The armature of a 4,160-volt 6,000-kva 3,600-rpm alternator has a measured insulation resistance of 2 megohms at a temperature of 80°C. The insulation is class B, and the test was made with a 500-volt megohmmeter applied for 1 min. How does the measured value compare with the recommended standard?

2-7. The following apparatus were given a 60-sec insulation-resistance test with a 500-volt megohmmeter. The test was made immediately after shutdown.

Machine	Reading
a. 50-hp 450-volt three-phase induction motor	1.5 megohms, 73°C
b. 600-hp 2,300-volt three-phase synchronous motor (stator)	2.0 megohms, 49°C
c. 5-kw 240-volt shunt generator (armature)	750,000 ohms, 30°C
d. 50-kva, 450/110-volt dry-type transformer	5 megohms, 55°C

Which machines meet the minimum acceptable standards recommended by the IEEE?

2-8. A motor-driven megohmmeter is used to test the insulation resistance of a class A–insulated ac armature. The readings observed after 1 min and 10 min are 12 megohms and 13 megohms respectively. (a) What is the polarization index? (b) Would you recommend continued operation of the machine in its present condition? Explain.

2-9. A class B–insulated armature winding has a measured insulation resistance of 200 megohms after 1 min of application of test voltage. After 10 min of continued application of test voltage the reading is 150 megohms. (a) Calculate the polarization index. (b) What does this test indicate about the condition of the insulation?

2-10. The results of a 10-min high-potential test made on a large class A–insulated induction motor show the leakage current to be 3.6 ma after 1 min and 1.5 ma after 10 min. (a) What is the polarization index? (b) What does this indicate?

three

batteries

a group (two or more) ↑

A battery is ~~an~~ electrochemical cell*s* that ~~is used to~~ store chemical energy for conversion later to electric energy. Of the many different types of batteries available, this chapter describes only the principal types in commercial use and presents their characteristics and charging methods.

The two basic types of cells are primary and secondary. A primary cell is chemically irreversible. Such cells cannot be recharged and must be discarded when no longer usable; their active material is consumed during discharge. The secondary cell, known as a *storage cell,* is chemically reversible. Its active materials are not consumed. Such cells may be recharged by passing current through the battery in the reverse direction.

Electrochemical cells are rated in voltage and ampere-hour capacity. The latter represents the number of ampere-hours that can be delivered under specified conditions of temperature, rate of discharge, and final voltage.

3-1 Dry Cell

The dry cell, shown in Fig. 3-1, is the most common type of primary cell. It has a carbon rod for the anode, or positive terminal, and a zinc container for the cathode, or negative terminal. The zinc container has an inner lining of absorbent material, and the space between the electrodes is filled with a mixture of crushed coke, manganese dioxide, and graphite saturated with a solution of salammoniac and zinc chloride. The no. 6 dry cell has an open-circuit voltage of about 1.6 volts and a rating of approximately 30 amp-hr. The life of a dry cell is limited. It gradually discharges by internal

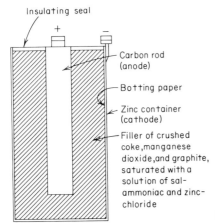

Insulating seal

+   −

Carbon rod
(anode)

Botting paper

Zinc container
(cathode)

Filler of crushed
coke,manganese
dioxide,and graphite,
saturated with a
solution of sal-
ammoniac and zinc-
chloride

FIG. 3-1 Cutaway view of a dry cell.

chemical action, called *local action,* even though no load is connected to its terminals. For this reason, some manufacturers stamp a service date on the outer covering of each cell. Dry cells should be stored in a cool, dry place.

3-2 Lead-Acid Battery

The lead-acid battery, illustrated in Fig. 3-2, has positive plates of lead peroxide and negative plates of sponge lead. The plates are kept apart by separators made of microporous rubber, fiber glass, perforated plastic or hard rubber, or resin-impregnated cellulose. The electrolyte is a solution of sulfuric acid and water. During discharge, the positive and negative plates are chemically changed to lead sulfate. The charging process restores the plates to their original chemical makeup of sponge lead and lead dioxide. The open-circuit voltage of a fully charged cell is about 2.05 volts and varies slightly with the temperature and specific gravity of the electrolyte.

Lead-acid batteries should never be left in a discharged condition for any great length of time, or the lead sulfate may harden and become nonporous, preventing the battery from accepting a full charge. Because the specific gravity of the electrolyte changes with the state of charge or discharge, the battery condition may be determined with a hydrometer, as shown in Fig. 3-3. Enough electrolyte should be drawn into the glass barrel to make the hydrometer float. Hydrometer readings should always be taken before the addition of distilled water. If the level of the electrolyte is low, water should be added, and the battery should be charged before a hydrometer reading is taken.

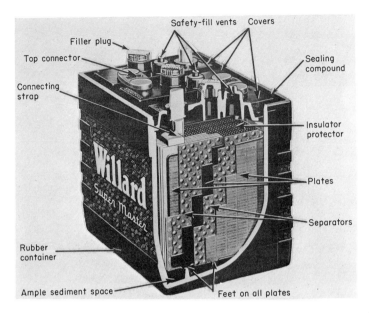

FIG. 3-2 Cutaway view of a lead-acid battery. (*Willard Storage Battery Co.*)

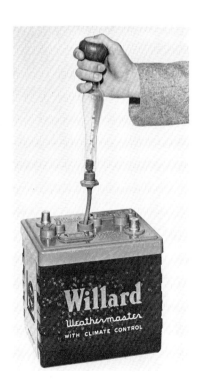

FIG. 3-3 Hydrometer test of a lead-acid battery.

The temperature of the electrolyte has considerable effect on its hydrometric values. High temperatures cause the volume of the electrolyte to increase, resulting in lower values of specific gravity, whereas low temperatures cause the volume of the electrolyte to decrease, resulting in higher values of specific gravity. Hence, if the specific gravity readings are to be of any value in determining the condition of a battery, corrections to some reference temperature must be made. Table 3-1 lists the additions or subtractions which should be made to the observed hydrometer readings, in order to obtain the specific gravity of the electrolyte for a standard reference temperature of 80°F. Only alcohol thermometers should be used; mercury thermometers are dangerous, because accidental breakage will cause severe sparking and explosions.

Table 3-1   Temperature Corrections of Hydrometer Readings for an 80° F Standard

Observed temperature, °F	Correction, points
130	+20
120	+16
110	+12
100	+ 8
90	+ 4
80	No correction
70	− 4
60	− 8
50	−12
40	−16
30	−20
20	−24
10	−28
0	−32
−10	−36
−20	−40

SOURCE: Data supplied by Association of American Battery Manufacturers.

• *Example.* If the temperature and specific gravity readings of a battery are 120°F and 1.240, respectively, what is its specific gravity corrected to 80°F?

• *Solution.* From Table 3-1, the correction from a temperature of 120°F is 16 points added to the observed hydrometer reading. Hence, 1.240 + 0.016 = 1.256. Using the corrected values of specific gravity, the condition of the battery may be determined from Table 3-2.

When lead-acid batteries are operated in extremely low temperatures, it is imperative that they be kept in a fully charged condition at all times. Low temperatures reduce the useful capacity of a battery, and at 0°F

**Table 3-2   Condition of Lead-Acid Battery
as Indicated by Specific Gravity of Electrolyte**

Specific gravity corrected to 80°F	Amount of charge, percent
1.285	100
1.245	75
1.225	66
1.205	50
1.165	25
1.125	Discharged

SOURCE: Data supplied by Association of American Battery Manufacturers.

its capacity is reduced to one-half its value at 80°F. Furthermore, the freezing point of the electrolyte is affected by the condition of charge. A fully charged battery has a freezing point of −90°F, whereas a dead battery freezes at +22°F. Freezing of the electrolyte damages the plates and ruptures the case, thus rendering the battery useless. The approximate freezing points of the electrolyte for different conditions of charge are given in Table 3-3.

**Table 3-3   Freezing Point of Sulfuric Acid
Electrolyte**

Condition of charge, percent	Approximate freezing point, °F
100	−90
75	−40
66	−25
50	−15
25	+ 5
Discharged	+15
Overdischarged	+22

SOURCE: Data supplied by Association of American Battery Manufacturers.

The lowering of electrolyte level in normal usage is caused by evaporation of water, and the restoration to the correct level should be made only with distilled water. However, if electrolyte is lost because of spilling or bubbling over, it should be replaced with sulfuric acid of the correct specific gravity. Concentrated sulfuric acid (specific gravity 1.835) should be mixed with distilled water in the proportions outlined in Table 3-4. The acid should be poured slowly into the water. Water should never be poured into acid; it will splash violently and may cause serious burns. Spilled acid may be neutralized with a solution of baking soda or a dilute ammonia solution. Acid splashed into the eyes should be washed immediately with large quantities of water, and medical treatment should be administered by a doctor. The electrolyte should be mixed only in glass, hard rubber, lead, or glazed earthenware containers.

Table 3-4   Proportions of Acid and Water
for Battery Electrolyte

Specific gravity desired	Ratio (by volume), distilled water to acid (sp gr 1.835)
1.050	16:1
1.100	10:1
1.150	$6\frac{1}{2}$:1
1.200	$4\frac{1}{2}$:1
1.250	$3\frac{1}{2}$:1
1.300	$2\frac{1}{2}$:1
1.350	2:1
1.400	$1\frac{3}{4}$:1

SOURCE: Data supplied by Association of American Battery Manufacturers.

When charging lead-acid batteries, the initial charging current should not exceed its ampere-hour rating. For example, a 100-amp-hr battery should not be charged at a rate greater than 100 amp. This high rate should taper off as the battery accepts the charge. When nearing full charge, the battery gasses freely, and it is necessary to reduce the current to a low finishing rate. This will prevent excessive gassing and excessive temperatures from damaging the plates by loosening the active material. The temperature of the electrolyte should not be allowed to exceed 125°F (51°C). The battery has attained full charge when the specific gravity of all cells reaches the correct value and no longer increases over a period of 3 to 4 hr. The source of power for charging a lead-acid battery should be 2.5 volts per cell.

3-3 Nickel-Iron Battery

The nickel-iron battery uses for its electrolyte a solution of potassium hydroxide (caustic potash) to which a small amount of lithium hydroxide has been added. The active material for the negative plate, or grid, is iron oxide pressed into perforated steel pockets. The positive plate consists of steel tubes containing nickel hydrate and nickel flakes in layers. When charged, the active material in the negative plate reduces to iron, and the active material in the positive plate is changed to a higher oxide of nickel. A cutaway view of a nickel-iron cell is shown in Fig. 3-4. The plates and electrolyte are housed in a nickel-plated welded steel container. The filler cap is used for adding water or changing electrolyte.

The valve permits generated gases to be released but blocks the entry of air; this is necessary because potassium hydroxide absorbs carbon dioxide from the air, reducing the effectiveness of the electrolyte.

The nickel-iron battery has an average voltage of 1.2 volts per cell during normal discharge and has a higher watthour capacity per pound than the lead-acid battery. The battery may be completely discharged, short-circuited, and left in this condition for indefinite periods without injury. It may be accidentally overcharged, charged accidentally in the reverse direction, and accidentally short-circuited without harmful effects. Alkaline batteries are not injured by freezing, and an electrolyte with

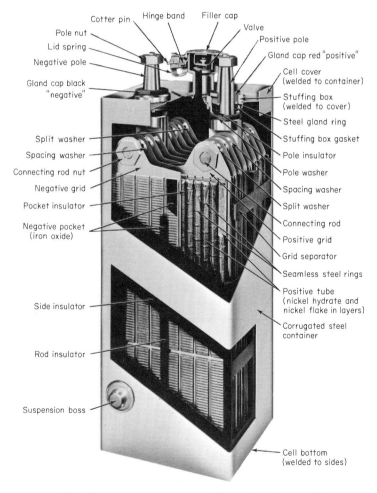

FIG. 3-4 Nickel-iron storage cell. *(Thomas A. Edison, Inc.)*

a specific gravity of 1.200 at 60°F freezes solid at −87°F. The specific gravity of the electrolyte does not change appreciably between charge and discharge and has a value of 1.200 at 60°F (15.5°C). Hence, hydrometer readings cannot be used to determine the condition of the battery. The state of charge must be determined with a voltmeter, during charge or discharge.

When charging or discharging a nickel-iron battery, the electrolyte temperature must not exceed 115°F (46°C). Although the battery gasses freely throughout the entire charging period, it has no adverse effect, because the active material is enclosed in perforated steel tubes and pockets. The battery may be fully charged at its normal rate. No special finishing rate is required. The source of power for alkaline batteries should have a minimum value of 1.85 volts per cell. When a trickle charger is used to keep a battery in a full-charge condition, it is better to overcharge than undercharge.

Although the specific gravity of a nickel-iron cell is unaffected by the state of charge or discharge, the specific gravity gradually drops during the life of the battery. When it decreases to 1.160 at 60°F, it should be renewed. Specific gravity readings should be taken only when the electrolyte is at the correct level and the battery is fully charged. Before renewing the electrolyte, the battery should be completely discharged and then short-circuited for several hours. The old solution should be poured out and the cell filled immediately with fresh electrolyte. The battery should not be rinsed with water nor left unfilled. After renewing the electrolyte, the battery should be charged at the normal rate for 15 hr. A few drops of pure paraffin oil in each cell will prevent atmospheric carbon dioxide from contaminating the electrolyte.

3-4 Nickel-Cadmium Battery

The nickel-cadmium cell uses a solution of potassium hydroxide in distilled water for the electrolyte at a specific gravity of 1.200 at 60°F. The active material of the positive plate consists of nickel hydroxide and specially treated graphite. The active material of the negative plate consists of a mixture of oxides of cadmium and iron. The battery has an average voltage of 1.2 volts per cell during normal discharge rates and a voltage of 1.3 volts at full charge. Because the specific gravity of the electrolyte is essentially constant, the condition of the battery must be determined with a voltmeter during charge or discharge. Figure 3-5 illustrates the relationship of cell voltage to state of charge for a nickel-cadmium battery. As shown by the curves, if the cell is on charge, and charging at its normal rate, a voltage of 1.7 volts indicates the cell to be 75 percent charged. Similarly, if the cell is supplying a load and discharging at its

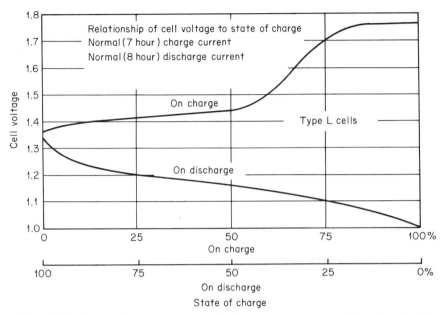

FIG. 3-5 Relationship of cell voltage to state of charge in a nickel-cadmium type L cell. (*Nickel Cadmium Battery Corp.*)

normal rate, a cell voltage of 1.2 volts will indicate a 75 percent charge.

When charging a nickel-cadmium battery, the temperature of the electrolyte should not be allowed to exceed 115°F (46°C). The battery does not commence to gas until after the first $4\frac{1}{2}$ hr when charging at the 7-hr rate. No finishing rate is needed. The source of power for charging should have a minimum value of 1.85 volts per cell. Because the battery cannot be damaged by overcharging, it is always advisable to overcharge and thereby ensure a fully charged battery. Furthermore, a reasonable amount of overcharging has a beneficial effect on the battery. Like the nickel-iron cell, the specific gravity of the nickel-cadmium battery is unaffected by the state of charge or discharge, but gradually decreases during normal usage. When the gravity falls below 1.170 at 60°F, the electrolyte should be renewed. Operation of the battery below this gravity causes a rapid reduction in its life. A few drops of pure paraffin oil in each cell will prevent atmospheric carbon dioxide from contaminating the electrolyte.

3-5 Mixing Alkali Electrolyte

Electrolyte lost because of spilling or bubbling over should be replaced with a caustic potash solution of 1.190 to 1.200 specific gravity (60°F).

The caustic potash is generally supplied as a solid in airtight cans and should be mixed 1 part by weight of KOH to 2 parts by weight of distilled water. The solution should be mixed only in glass, iron, or earthenware containers, using a glass or iron stirring rod. The iron container and stirring rod must not be galvanized or have soldered joints. Great care should be used when handling electrolyte. The solution is very caustic and dissolves skin as well as other organic matter. It also attacks copper, brass, lead, aluminum, and zinc. Spilled electrolyte should be washed immediately with a large quantity of water and then neutralized with a 4 percent solution of boric acid. Goggles, rubber gloves, and rubber apron should be worn when changing the electrolyte.

3-6 Battery Charging

During the charging process, current is forced into the battery in the opposite direction to that which occurs during discharge. Hence, the negative terminal of the battery must connect to the negative terminal of the charging source. When charging from an ac source, a rectifier, motor-generator set, or other source of direct current must be used. When charging two or more batteries in series, the positive of one must connect to the negative of the other. The rate and method of charge depends on the physical condition of the battery, its state of charge, and its application.

Figure 3-6 illustrates some representative charging circuits. Figure 3-6a is a simple half-wave rectifier circuit that uses a diode to block the negative half-cycle of current. Figure 3-6b is an adjustable-voltage full-wave rectifier; voltage adjustment is obtained through a tap-changing transformer or a slide-wire rheostat. Figure 3-6c is a circuit that uses an available dc source.

Trickle charging. Because of minor impurities in the ingredients of the active material, a certain amount of local discharging occurs at the plates. This is called local action, and gradually discharges the battery. If a very small charging current is used to offset the local action, the battery can be kept in a fully charged condition at all times. This is called trickle charging.

Safety precautions. The gases given off during the charging process are an explosive mixture of hydrogen and oxygen. Hence adequate ventilation must be had, and no smoking or carrying of open flames should be allowed in a battery room or compartment. Only spirit thermometers should be used when measuring battery temperatures. The mercury from a broken mercury thermometer will cause sparking and explosions.

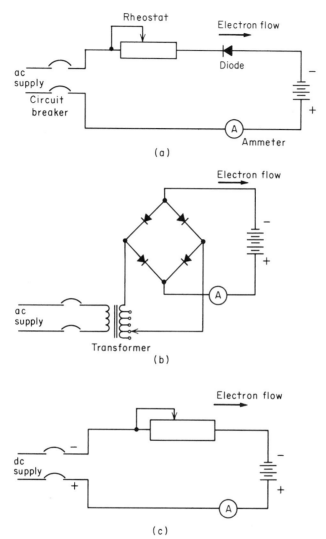

FIG. 3-6 (a) Simple half-wave rectifier charging circuit; (b) full-wave adjustable-voltage rectifier circuit; (c) charging circuit using a dc supply.

3-7 Battery Maintenance

Although storage batteries require relatively little maintenance, they must not be neglected. The terminals should be kept clean, and the electrolyte maintained at the correct level. Pure petroleum jelly applied to the cable connections of both acid and alkaline batteries will avoid corrosion. Dirty battery connections may be cleaned by removing them and brushing with a wire brush. The cell cases of alkaline batteries should be kept clean, and the tops greased with pure petroleum jelly.

Bibliography

Vinal, G. W.: "Primary Batteries," John Wiley & Sons, Inc., New York, 1950.
————: "Storage Batteries," John Wiley & Sons, Inc., New York, 1955.

Questions

3-1. What is the basic difference between primary and secondary cells? How are battery cells rated?

3-2. What causes a dry cell to discharge gradually, even though no load is connected to its terminals?

3-3. How may dirty and corroded battery terminals be cleaned?

3-4. Sketch a circuit for a battery-charging system. What safety precautions should be observed when charging?

3-5. What are the principal components of a lead-acid battery?

3-6. How is the electrolyte of a lead-acid battery tested? What precautions should be observed to ensure an accurate indication?

3-7. What effect does temperature have on the specific gravity of the electrolyte?

3-8. If the specific gravity of a lead-acid battery is 1.209, measured at 130°F, what is its condition expressed as a percentage of full charge?

3-9. What effect do extremely low temperatures have on the useful capacity of a lead-acid battery?

3-10. Can the electrolyte of a lead-acid battery freeze?

3-11. What causes the electrolyte level of a battery to lower during normal usage? Discount accidental spillage.

3-12. State the correct procedure for mixing acid and water when making battery electrolyte. What precautions should be observed?

3-13. How much distilled water must be added to concentrated sulfuric acid in order to make a mixture having a specific gravity of 1.300?

3-14. What are the two principal types of alkaline batteries in use throughout the United States?

3-15. How must an alkaline battery be tested?

3-16. Why is it that a hydrometer cannot be used for testing the condition of an alkaline battery?

3-17. State the correct procedure for mixing electrolyte for an alkaline battery. What precautions should be observed?

four

ac generator

The generation of an ac driving voltage for applications requiring large amounts of electric energy is accomplished by rotating electromechanical apparatus, such as turbine-driven generators, diesel-driven generators, waterwheel generators, etc.

4-1 Generation of an Alternating Voltage

In its simplest form, an ac generator, also called *alternator,* consists of a set of magnet poles, a coil of wire, called an *armature coil,* and a prime mover, as shown in Fig. 4-1. As the magnet poles rotate, the flux through the window of the armature coil is made to vary with time. A graph of this variation for one revolution of the prime mover is shown in Fig. 4-2*a*. The direction and relative magnitude of flux through the window of the coil for different positions of the magnet is indicated by arrows; the greater the flux, the longer the arrow. The apparent sinusoidal variation of the flux through the window, as the magnet rotates, is obtained by proper shaping of the field poles.

The change in flux through the window, as the magnet rotates (increasing, decreasing, and reversing), generates an alternating voltage within the coil. The magnitude of this magnetically induced voltage is proportional to the number of turns of wire in the coil and the rate of change of flux through the window. The mathematical relationship that expresses this behavior is called *Faraday's law:*

$$v = -N\left(\frac{d\phi}{dt}\right) \qquad \text{volts} \qquad (4\text{-}1)$$

where v = generated voltage

N = number of turns of wire in coil

$\left(\dfrac{d\phi}{dt}\right)$ = rate of change of flux through window of coil

The minus sign $(-)$ indicates that the voltage generated will be in a direction to oppose any change in flux through the window of the coil (Lenz's law).

The magnitude of the generated voltage does not depend on the amount of flux passing through the window of the coil but on the rate at which the flux through the window is changing. Referring to Fig. 4-2a, as the magnet rotates, at 90° and 270°, the flux through the window attains its maximum amount and is about to decrease; at these brief in-

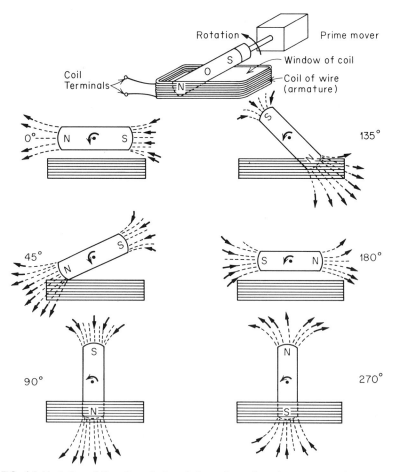

FIG. 4-1 Variation of flux through the window of a coil as the magnet poles rotate.

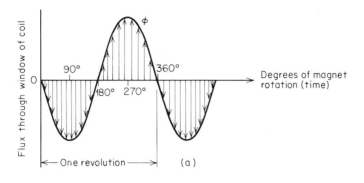

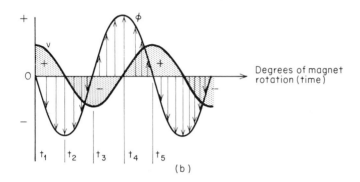

FIG. 4-2 (a) Graph showing variation of flux through the window of the coil for one revolution of the prime mover; (b) voltage and flux curves.

Time interval	Flux through window	Direction of generated voltage as determined from Lenz's law
t_1 to t_2	Increasing negatively	Positive
t_2 to t_3	Decreasing negatively	Negative
t_3 to t_4	Increasing positively	Negative
t_4 to t_5	Decreasing positively	Positive

stants of time the flux is neither increasing nor decreasing, there is no change in flux though the window, and no voltage is generated. At 0°, 180°, and 360° the flux through the window is changing at its greatest rate; hence at the instants of time corresponding to these instantaneous positions of the revolving magnet, even though the flux through the window is zero, the voltage generated in the coil has its maximum value.

The direction of the generated voltage depends on the direction of the flux through the window and whether it is increasing or decreasing. This is shown in Fig. 4-2b; the accompanying table illustrates the use of Lenz's law for determining the direction (polarity) of the generated voltage.

Increasing the speed of rotation of the magnet increases the rate of change of flux through the window, and a higher voltage is induced. The same increase in voltage may be accomplished without changing the speed, by using a stronger magnet; because the magnet is rotating at the same speed, a greater flux causes a greater rate of change through the window and hence a higher voltage. Still another means of obtaining a higher voltage without changing the strength of the magnet or changing the speed is to increase the number of turns of wire in the armature coil. Practical generators use dc electromagnets whose strength is adjusted by varing the current to the magnet coils. Direct current is used to provide poles of fixed polarity. The small generator used to supply the current to the magnets is called an *exciter generator* or *exciter*. The exciter builds up its own voltage from residual magnetism.

The voltage generated in an ac machine can be conveniently expressed in terms of speed, flux, and a machine constant:

$$E = n\Phi K_G \qquad \text{volts}$$

where n = rpm
Φ = flux per pole (webers)[1]
K_G = generator constant

The generator constant includes such design factors as the number of conductors, arrangement of armature winding, and the number of poles. If the generated voltage for a given speed and flux is known, the voltage for some other speed and flux can be determined by substituting into the following equation:

$$\frac{E_2}{E_1} = \frac{n_2 \Phi_2}{n_1 \Phi_1}$$

The subscript 1 denotes the original combination of voltage, speed, and flux; the subscript 2 denotes the new combination.

• *Example 4-1.* A single-phase generator running at 3,600 rpm generates 450 volts. If the speed is reduced to 3,400 rpm and the field current is not changed, determine the new voltage.

• *Solution*

$$\frac{E_2}{450} = \frac{(3,400)\Phi_2}{(3,600)\Phi_1}$$

[1] 10^8 maxwells (lines of force) equals one weber.

Because Φ_2 and Φ_1 are equal,

$$E_2 = \frac{3{,}400 \times 450}{3{,}600} = 425 \text{ volts}$$

The voltage output of an ac generator rises, falls, and reverses with time in a sinusoidal manner, as shown in Fig. 4-2b. Examination of this voltage wave shows it to consist of repetitive cycles, each composed of a positive and a negative alternation. The number of cycles that occur in 1 sec, called the *frequency*, depends on the rpm of the magnets and the total number of magnet poles in the field structure. Expressed mathematically,

$$f = \frac{pn}{120} \tag{4-2}$$

where f = cycles per second or hertz
$\quad\quad\quad p$ = total number of magnet poles
$\quad\quad\quad n$ = rpm of rotating member

Because the generation of a voltage depends on a changing flux through the window of a coil, it does not matter whether the armature coils are stationary and the magnet poles rotate, or the magnets are stationary and the armature coils rotate. Elementary circuit diagrams and one-line diagrams for rotating-armature and rotating-field generators are shown in Fig. 4-3. In each case, it is the armature that supplies current to the load. Two, three, or more output terminals may be provided, depending on the type of armature winding and output connections desired. A turbine is used to drive the exciter generator and the alternator armature (or dc field in the case of a revolving-field stationary-armature alternator). The one-line diagram is a shorthand drawing that shows the interconnection of circuit components. The return wires that complete the circuit are not shown. One-line diagrams are particularly useful in complex systems as an aid to understanding the mechanics of plant operation.

4-2 Three-phase Generator

Except for very small generators of the type used for home emergency lighting, most ac generators are three-phase machines. A three-phase generator has three separate but identical armature windings that are acted on by one system of rotating magnets. Each winding, called a *phase*, consists of a set of armature coils. Figure 4-4 shows the construction details for a six-pole three-phase rotating-field generator. The armature coils are tied securely to a steel bracing ring to prevent the large mechanical forces caused by the armature current from pulling the coils out of

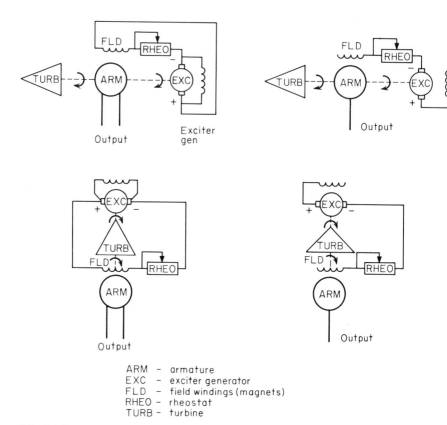

ARM   –   armature
EXC   –   exciter generator
FLD   –   field windings (magnets)
RHEO – rheostat
TURB – turbine

FIG. 4-3 Elementary circuit diagrams and corresponding one-line diagrams for rotating armature and rotating field machines.

shape. The magnet coils are connected in series and terminate at two slip rings. Carbon brushes, not shown, are pressed against the slip rings and provide the connecting link between the magnet coils and the small dc generator called an exciter. The exciter is mounted on the same shaft as the rotating magnet member. The ventilating duct at the top of the stator and the fans on the rotor provide forced-draft cooling. The field frame for the exciter is not shown. Figure 4-5 shows the armature and field members of a two-pole three-phase generator. Because the field structure is in the form of a cylinder, it is called a *cylindrical rotor*. The field structure of the machine in Fig. 4-4 is called a *salient-pole rotor*.

Although the three phases are acted on by the same rotating magnets and therefore have the same maximum values of voltage, the spacing and connections of the coils cause the three voltages to reach their respective positive maximums 120° apart. The order, or sequence, in which the voltage waves of the three phases reach their maximum positive values is called the *phase sequence*. Thus, for the waves shown in Fig. 4-6*a,* the phase sequence is

$v_a, v_b, v_c, v_a, v_b, v_c, \ldots$

To determine the phase sequence, the observer should "walk" to the right along the horizontal axis, observing the sequence of positive peaks (the peaks above the axis) as he travels.

Figure 4-6*b* and *c* illustrates the current and voltage relationships for wye-connected and delta-connected generators, each supplying the same line current and the same line voltage to identical loads. The neutral, or common connection, shown with a broken line in the wye arrangement, is not always provided. The voltage and current relationships illustrated in the figures are summarized in the following table:

Connection	Current	Voltage
Wye (Y)	$I_{line} = I_{phase}$	$V_{line} = 1.73V_{phase}$
Delta (Δ)	$I_{line} = 1.73I_{phase}$	$V_{line} = V_{phase}$

Figure 4-7 shows the construction details and the corresponding elementary circuit diagram for a brushless generator system. This system differs from its counterpart in Fig. 4-4 in that it has no commutator, slip rings, or brushes. The generator field, exciter armature, and rectifier assembly are all part of the rotating member. The brushless exciter consists of a small rotating three-phase exciter armature whose output current is rectified, changed from ac to dc, by means of semiconductor diodes and fed to the alternator field. The heat sink absorbs the heat generated by the flow of electrons through the diodes.

The circuit shown in Fig. 4-7*b* is a simplified circuit for manual control of the generator voltage and illustrates the connecting link between the brushless alternator and the brushless exciter. The exciter field obtains its direct current from the alternator through a step-down transformer, rheostat, and rectifier. The rheostat provides manual control of the generator voltage. Although not shown, the complete brushless excitation system includes an automatic voltage regulator and means for "flashing" the exciter field. Flashing is used to restore residual magnetism after a prolonged idle period and for initial start-up.

4-3 Voltage and Speed Regulation

To prevent the dimming of lights and the malfunctioning of motors and other electrical equipment, the output voltage and frequency of an ac

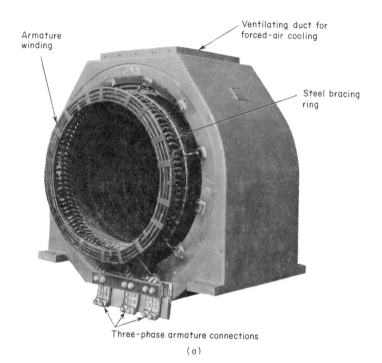

Armature winding

Ventilating duct for forced-air cooling

Steel bracing ring

Three-phase armature connections

(a)

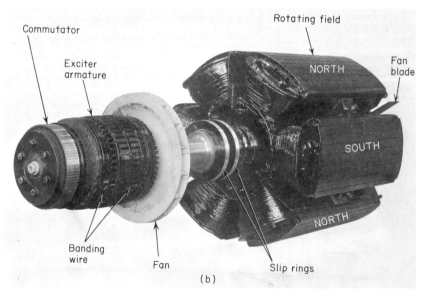

Commutator

Exciter armature

Rotating field

NORTH

Fan blade

SOUTH

NORTH

Banding wire

Fan

Slip rings

(b)

FIG. 4-4 Three-phase ac generator with a six-pole field. (a) Stationary armature; (b) rotating field and exciter armature. (*General Electric* Co.)

(a)

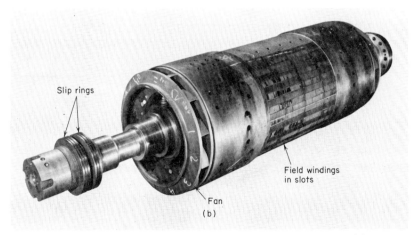

Slip rings

Field windings
in slots

Fan

(b)

FIG. 4-5 Three-phase ac generator with a two-pole cylindrical field rotor. (a) Stationary armature called the stator; (b) rotating field called the rotor. (*Westinghouse Electric Corp.*)

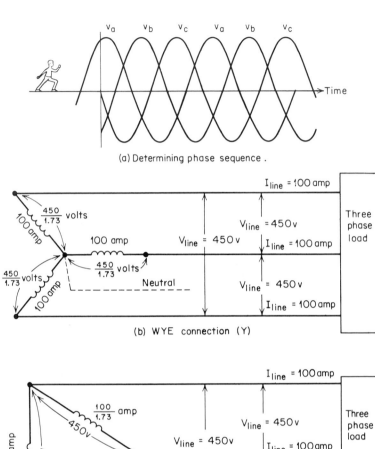

(a) Determining phase sequence.

(b) WYE connection (Y)

(c) DELTA connection (Δ)

FIG. 4-6 Current and voltage relationships for wye and delta connections.

generator must be held relatively constant as electrical loads are switched on and off. Unfortunately, when a load is connected to a generator, two undesirable reactions take place. First, the prime mover slows down, causing a lowering of both the frequency and the generated voltage. Second, the passage of current through the impedance of the armature winding causes a further drop in output voltage.

The reduction in prime-mover speed is caused by the development of a countertorque. When a load is connected to the generator terminals,

the current in the armature conductors sets up a magnetic field that interacts with the field of the rotating magnets. The mechanical force produced
by this action is in opposition to the driving torque of the prime mover.
This magnetic action is discussed in Sec. 1-16 and illustrated in Fig. 1-13e.

An elementary circuit diagram and its one-line diagram counterpart

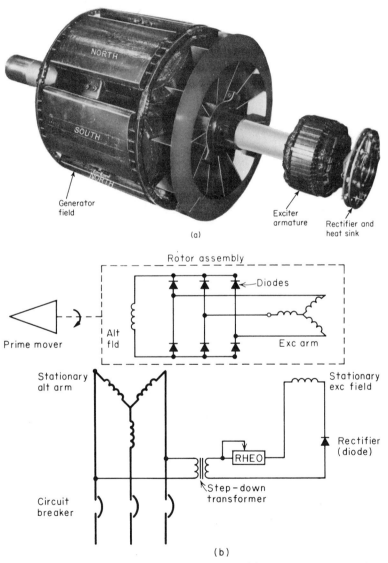

FIG. 4-7 Brushless generator system. (a) Rotor assembly; (b) elementary diagram showing
a simplified manual control circuit. (*General Electric* Co.)

for a generator and distribution system are shown in Fig. 4-8. The ac generator and exciter are driven by a single prime mover. The generator feeds a distribution bus, which in turn supplies various connected loads. A current transformer (CT) is used to provide a reduced but proportional current to the ammeter. Changes in prime-mover speed caused by the application of a load are corrected by an automatic speed governor that uses electronic or mechanical sensors to detect a change in speed and then automatically increases or decreases the energy input to the prime

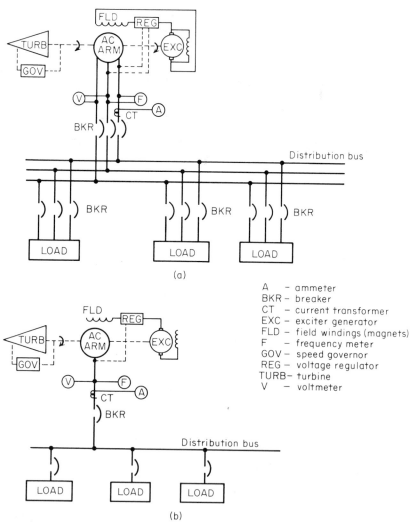

A   – ammeter
BKR – breaker
CT   – current transformer
EXC – exciter generator
FLD – field windings (magnets)
F   – frequency meter
GOV – speed governor
REG – voltage regulator
TURB– turbine
V   – voltmeter

FIG. 4-8 Circuit diagram and one-line counterpart for an ac generator and distribution system.

mover. Similarly, an automatic voltage regulator uses electrical sensors to detect a change in voltage and then automatically adjusts the resistance in the field circuit to raise or lower the voltage.

4-4 Watts, Vars, KVA, and Power Factor

When an ac generator supplies energy to a distribution system consisting of resistors, lamps, motors, etc., some of the energy does useful work, but some remains locked in the system, seesawing between the generator and the load.

This is illustrated in the energy-flow diagrams in Fig. 4-9. Heating

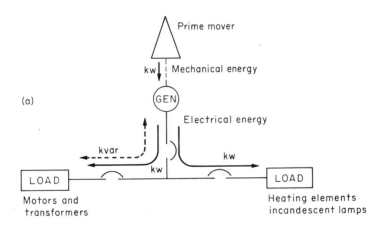

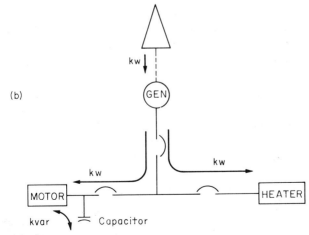

FIG. 4-9 Energy-flow diagrams. (a) Without a capacitor; (b) with a capacitor that draws the same magnitude of kvars as does the motor.

elements, such as electric ranges, toasters, space heaters, incandescent lamps, etc., convert electric energy to heat energy. The flow of this energy is unidirectional in that it passes from the prime mover to the load via the generator and does not return to the generator. The rate of transfer of this energy as it flows unidirectionally from the generator to the load is called *active power* and is measured with a wattmeter in watts or kilowatts (kw). This active power is sometimes called *real power* or *true power*.

Whenever an element possessing appreciable inductance is connected to an ac generator, the oscillations of current cause the magnetic field about the element to build up and decay alternately; the energy accumulated in the magnetic field during the buildup of current is returned to the generator when the field decays. This cyclic exchange of energy between the inductive element and the generator continues undiminished as long as the element is connected to the generator. The rate at which this energy is exchanged is called *reactive power* and is measured with a varmeter in vars or kilovars (kvar). The term "var" comes from the first letters of the words "volt-amperes reactive." Reactive power is sometimes called *wattless power*. Because the current to the inductance lags the voltage across it, the vars are called *lagging vars*.

An ac motor converts part of the electric energy it receives to mechanical work (active power) and a smaller part to heat (active power); but the energy used to establish the magnetic field of the motor seesaws between the generator and the motor, doing no work at all (reactive power).

If a capacitor is connected to an ac generator, the oscillations of the driving voltage cause the electric charge in the capacitor to build up and decay alternately. The energy accumulated in the capacitor during the buildup of voltage is returned to the generator when the voltage decays. This cyclic exchange of energy between the capacitor and the generator continues undiminished as long as the capacitor is connected to the generator. The rate at which this energy is exchanged is reactive power. However, because the current to a capacitor leads the voltage, these vars are called *leading vars*.

If a circuit includes inductive elements that draw lagging vars and capacitive elements that draw leading vars, the total vars supplied by the generator is the difference between the two. The element with the lagging vars gives up its energy during the same period of time that the element with the leading vars is accepting energy. Hence, some exchange of energy takes place between the inductance and the capacitance, as shown in Fig. 4-9b, and fewer vars are drawn from the generator. Thus, the total vars supplied by the generator is the difference between the lagging vars and the leading vars.

The combination of active and reactive components of power (kw;

kvar) drawn by a load is called *apparent power* and is expressed in volt-amperes (va) or kilovolt-amperes (kva). The apparent power can be determined from the measured values of its active and reactive components by substituting in the following formula:

$$\text{kva} = \sqrt{(\text{kw})^2 + (\text{kvar})^2} \tag{4-3}$$

The kva supplied by a generator, operating at an essentially constant voltage and frequency, is determined solely by the magnitude and type of the connected load; the generator delivers only the amount of watts and vars "demanded" by the load. Thus a resistor load "demands" only watts, and a pure inductive load "demands" only vars.

The ratio of the active power to the apparent power drawn by a load is called the *power factor:*

$$pf = \frac{\text{kw}}{\text{kva}} \tag{4-4}$$

The power factor of a circuit or system is a measure of its effectiveness in utilizing the apparent power it draws from the generator (see Sec. 15-4). The power factor of an element or system can be measured with a power-factor meter and may have values ranging between 1.0 and zero. At the high end of the scale are such elements as resistors, incandescent lamps, electric ranges, and similar heating equipment that draw only active power. For these elements and circuits the kva and kw drawn from the generator are one and the same, and the power factor is 1.0.

At the other extreme, a capacitor draws essentially reactive power from the generator. (The very small heat losses in a ceramic capacitor, for example, do not provide a significant wattmeter indication, and may be neglected). Thus, for the capacitor, the kva drawn from the generator is essentially kvar. Hence, the power factor of a capacitor is almost zero:

$$pf = \frac{\text{kw}}{\text{kva}} = \frac{0}{\text{kva}} = 0$$

Other useful formulas for the determination of kw and kva are listed below. For a single-phase generator:

$$\text{kva} = \frac{V_{\text{line}} I_{\text{line}}}{1,000}$$

$$\text{kw} = V_{\text{line}} I_{\text{line}} pf / (1,000)$$

For a three-phase generator (wye or delta):

$$\text{kva} = 1.73 V_{\text{line}} I_{\text{line}} / (1,000)$$
$$\text{kw} = 1.73 V_{\text{line}} I_{\text{line}} pf / (1,000)$$

4-5 Procedure for Single-generator Operation

Before starting the prime mover, make sure the generator breaker is open, and then close the disconnect links, or disconnect switch, generally located in the rear of the switchboard. The purpose of the "disconnects" is to allow the complete isolation of the breaker from the bus for purposes of test, maintenance, and repair. The links, shown in Fig. 4-10, must be bolted to the connection bars with an insulated socket wrench.

> **Warning.** The circuit breaker must always be opened before opening or closing the disconnect switch or links.

The following procedure may be used as a guide for single-generator operation:

1. Make sure that the generator breaker is open.
2. Bolt the disconnect links to the connection bars. (Use an insulated wrench.)
3. Switch the voltage regulator to "automatic."
4. Start the turbine or engine, and bring it up to speed.
5. Using the governor-control switch, adjust the frequency to the rated nameplate frequency.
6. Adjust the automatic voltage regulator to obtain rated voltage.
7. Close the circuit breaker manually or by turning the breaker switch to "close."

4-6 Procedure for Parallel Operation

When the load on an ac system exceeds the amount that can be supplied by a single generator, additional generators must be connected to the system to supply the required energy. The incoming machines must be paralleled in a manner that enables each machine to supply its proper proportion of the active power (kw) and reactive power (kvar) to the common load:

1. Make sure that the breaker for the incoming machine is open.
2. Bolt the disconnect links of the incoming machine to the connection bars. (Use an insulated wrench.)
3. Switch the voltage regulator to "automatic."
4. Start the turbine or engine, and bring it up to speed.
5. Using the governor-control switch, adjust the frequency of the incoming machine to approximately $\frac{1}{10}$ cycle higher than the bus frequency.
6. Adjust the automatic voltage regulator so that the voltage of the incoming machine is approximately equal to or slightly higher than the bus voltage.

FIG. 4-10 Bus disconnecting link. (*General Electric* Co.)

7. Switch the synchroscope to the incoming machine, and adjust the frequency until the synchroscope pointer revolves slowly in the "fast" direction.

8. Close the circuit breaker at the instant the synchroscope pointer passes into the zero (12 o'clock) position. It is good practice to start closing the breaker one or two degrees before the 0° position, thus assuring that the actual closing occurs at about 0°.

9. Turn the synchroscope switch to "off."

10. Turn the governor switch of the incoming machine to "raise," and that of the machine on the bus to "lower" until the kw load is divided equally between machines. If the machines are not identical, the kw load should be divided between the machines in proportion to their kw ratings.

    If both governor controls are close enough, they may be operated simultaneously. However, if they are too far apart, the transfer should be made in small increments. Turn the governor control of the incoming machine to "raise," and hold it there until a small amount of load is transferred; then turn the governor control of the other machine to "lower" until another equal increment is transferred. Repeat until the desired load transfer is accomplished. While making this transfer, keep an eye on the frequency meter to avoid excessive changes in frequency.

11. If the power factors or kvar readings of the two generators are not equal, adjust the automatic regulator control, turning that of the incoming machine in the "raise voltage" direction, and the other in the "lower voltage" direction until they are equalized.

In factories, hotels, or large ships, which require two or three ac generators in parallel to supply the connected load, it is good operating practice to balance both the kw and the kvar loads in proportion to the kw ratings of the machines. By doing so, transient overcurrents and short circuits in the distribution system are shared by the generators, reducing the probability of one generator tripping. The tripping of one generator due to overcurrent will throw the entire load on the remaining unit or units, causing them to trip and blacking out the plant.

4-7 Removing an Alternator from the Bus

To remove an alternator from the bus, transfer all the load to the remaining machine by adjusting the governor controls of both. Then trip the circuit breaker of the outgoing machine, and adjust the voltage of the remaining alternator to the desired value. The transfer of load by tripping the breaker of the outgoing machine, instead of through governor adjustment, should be avoided, because it exerts a severe strain on the mechanical parts of the machine that takes the load. From a safety standpoint, if a generator is to be out of service for an extended period of time, the disconnect switch or links should be opened (after generator shutdown) and kept open until the generator is ready to go back on the line. This prevents damage to the machine in the event that the breaker is inadvertently closed.

4-8 Motorization of Alternators

If the energy input to the prime mover of an alternator is decreased to a point where the energy admitted is insufficient to keep the machine running in synchronism with the bus, the other machines on the bus will feed into the machine and run it as a motor at the same speed and in the same direction as before. This is known as *motorization*. Although motorization is not harmful to the alternator, it does result in an energy loss in the form of an additional load on the bus. A reverse-power relay, also called a power-directional relay or wattmeter relay, shown in Fig.

FIG. 4-11 Power-directional relay. (*General Electric* Co.)

4-11, is provided to trip the machine off the bus if this occurs. A motorized generator can be detected by the reversed reading of its wattmeter and, if detected early enough, may be corrected by admitting more energy to its prime mover.

If a generator pulls out of synchronism with the other machines on the bus, exceedingly heavy currents will circulate between the armature of the affected machine and the bus. Although this fault current has a magnitude closely approaching the short-circuit value, it occurs in periodic surges. The effect is identical to that of a synchronous motor that pulls out of synchronism; as each rotor pole slips past a pole of the rotating flux set up in the stator by the bus voltage, a large pulse of current and a torque reversal are produced (see Sec. 5-9). The high voltage induced in the field windings by the extremely high-current pulsations may cause a flashover at the collector rings, and the violent vibration caused by the torque reversals may cause severe mechanical damage. Hence, to prevent serious damage, an out-of-synchronism machine should be immediately removed from the bus.

4-9 Synchronizing Methods and Problems

There is no advantage in adjusting the incoming machine to have the exact frequency of the bus, for if paralleled in such a manner, the machine would merely float on the line, and a slight decrease in its input energy would immediately result in motorization. Hence, for added protection, the frequency of the incoming machine prior to paralleling should be adjusted to about $\frac{1}{10}$ cycle higher than the bus frequency.

Two generators cannot operate at different frequencies when they are in parallel. Hence, if the frequency of the incoming machine before paralleling is lower than that of the bus, the machines on the bus will drive the incoming machine at the higher electrical speed as soon as it is paralleled. The energy to motorize the incoming machine comes from the other generators on the bus. If these machines are heavily loaded, the additional kilowatts required to motorize the incoming machine may be sufficient to trip them off the bus and black out the plant. If the frequency of the incoming machine before paralleling is higher than that of the bus, it will take some of the load from the other machines at the instant it is paralleled. The absorption of load by the incoming machine causes its frequency to drop.

A synchroscope, shown in Fig. 4-12a, always indicates the condition of the incoming machine with respect to the bus. If the frequency of the incoming machine is higher than the bus frequency, the synchroscope pointer will revolve in the direction marked "fast." If the frequency of the incoming machine is lower than the bus frequency, the pointer will revolve in the direction marked "slow." If the pointer stops at a position other than $0°$, it is an indication that the incoming machine is at the same frequency as the bus but out of phase by an error angle indicated by the position of the pointer. To correct this, adjust the governor control of the incoming machine to admit more energy to its prime mover. This will result in a slightly higher frequency and will cause the synchroscope pointer to revolve slowly in the fast direction. The machines should be paralleled when the pointer reaches the $0°$ position while traveling in the fast direction. When paralleled, the synchroscope pointer will not revolve but will stay at the $0°$ position. Because there is a slight lag in the operation of the breaker or switching mechanisms, it is good practice to start the breaker-closing operation $1°$ or $2°$ before the $0°$ position, thus assuring that the actual closing occurs at about $0°$. A breaker with a 15-cycle closing time takes $\frac{15}{60}$ or $\frac{1}{4}$ sec for the closing mechanism to complete its operation. A 30-cycle closing device takes $\frac{30}{60}$ or $\frac{1}{2}$ sec for closing to occur. (Assume a 60-hertz system.)

If the breaker is closed at an angle other than $0°$ (12 o'clock), a crosscurrent will occur between the generators. The severity of this crosscurrent depends on how far out of synchronism the machines are when the breaker is closed. The greater the error angle, the higher the crosscurrent. The worst condition corresponds to an error angle of $180°$ and has the same effect as a short circuit on both machines; the resulting tremendous forces on the generator winding and other conductors could be disastrous. Figure 4-12b shows how the error angle changes with time when the voltage of the incoming machine has a higher frequency than

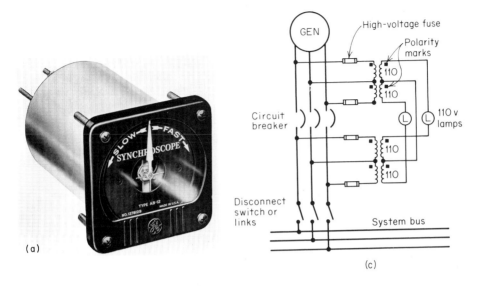

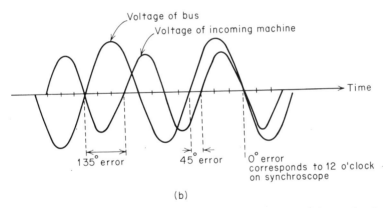

FIG. 4-12 (a) Synchroscope; (b) relationship between voltage of bus and voltage of incoming machine; (c) synchronizing-lamp circuit.

that of the bus. Under no circumstances, emergency or otherwise, should the breaker be closed when the error angle exceeds 15°.

Synchronizing lamps provide a means for checking the synchroscope, and may be used as a secondary method for determining the correct instant for paralleling if the synchroscope is defective. Synchronizing-lamp connections are shown in Fig. 4-12c. If the frequency of the incoming machine is different from that of the bus, as occurs when bringing the incoming machine up to speed, both lamps will go alternately bright and dark in unison. The significant correspondence between the synchroscope and the synchronizing lamps is the 0°–lamps-dark relationship. When the lamps are dark, the synchroscope pointer must be at 0°, or the synchroscope is defective; the lamps are always correct, assuming that they are not burned out. When using lamps for synchronizing, the duration of the dark period should be timed, and the breaker closed midway through the dark period. Timing is important, because the dark period may extend over 8° degrees or more of error angle. Lamps used for this purpose are generally of the nonfrosted type. Both lamps must have the same voltage and wattage ratings. The voltage rating must be as specified by the manufacturer.

Because in all but a few applications, the generator voltage is greater than 250 volts rms, potential transformers (see Sec. 7-10) are connected between the generator and the synchroscope (or lamps) to reduce this voltage to the rated synchroscope (or lamp) value of 120 to 240 volts. Accidental reversing of the connections in the primary or secondary of the transformer will cause the synchroscope and the synchronizing lamps to give false indications; the synchroscope will indicate 0° and the lamps will be dark when the generator is 180° out of phase with the bus. Hence, all polarity connections to repaired or replaced potential transformers must be carefully checked to avoid this very serious but easily made error.

4-10 Instrumentation and Control of AC Generators

Figure 4-13 is a functional diagram for a two-generator system, showing the minimum instruments, switches, and adjustable controls that an operating engineer must be acquainted with when operating ac generators singly or in parallel. The connecting lines indicate what each control device operates, and where each meter gets its signal. Operating controls and connections are drawn with heavy lines; meters and instrumentation connections are drawn with light lines. Panels 1 and 3 are identical generator panels and contain the instruments and controls for the particular machine. Panel 2 contains the instruments and controls necessary for paralleling. As should be expected, the wattmeter and power-factor meters each require both current and voltage signals, and the synchroscope requires voltage signals from each generator to determine the error angle.

4-11 Load-sharing Problems of Alternators in Parallel

The transfer of kw between machines is accomplished by increasing the energy input to one machine and decreasing it by an equal increment from

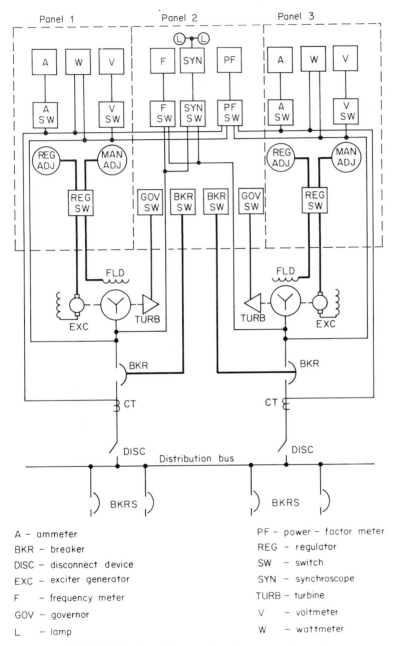

FIG. 4-13 Functional diagram for a two-generator system.

A – ammeter
BKR – breaker
DISC – disconnect device
EXC – exciter generator
F   – frequency meter
GOV – governor
L   – lamp

PF – power – factor meter
REG – regulator
SW – switch
SYN – synchroscope
TURB – turbine
V   – voltmeter
W   – wattmeter

the other machine. This becomes increasingly important as the number of paralleled machines in a system is reduced; the minimum is two. Hence, unless the system of machines is very large, such as a network of power plants, the amount of energy added to one prime mover must be subtracted from the others by adjustment of governor controls, in order to maintain the system in equilibrium at its rated speed and frequency. Means for adjusting the no-load speed setting of the governors is provided on the generator switchboard. On some installations, the governor controls of two machines are on one panel. This permits simultaneous adjustment of both controls, raising one and lowering the other, until the desired transfer is accomplished. Changing the field excitation of an alternator, when in parallel with others, does not appreciably alter the division of load between machines. Load transfer must be done by governor adjustment. The machine that is to accept more load should have its governor adjusted so that more energy is admitted to its prime mover. The machine that is to lose some load should have its governor adjusted so that less energy enters its prime mover.

The division of kw between ac generators operating in parallel, as load is switched on or off the bus, depends solely on the load-frequency (load-speed) characteristics of the governors. It is independent of the generator characteristics. Figure 4-14a illustrates the typical characteristic for a prime mover operating in parallel and governed by mechanical governors. Note how the speed and frequency decrease with increased load. Because this type of governor has a relatively slow action, a speed droop of 1 to 3 percent is essential for stable operation; if two ac generators are in parallel and both characteristics are horizontal (zero speed droop), one machine will tend to hog the load and drive the other as a motor. However, the sophisticated circuitry and rapid response of electronic governors enable machines so governed to operate in parallel without speed droop.

A governor with a 3 percent speed droop causes a 3 percent difference in speed and frequency between its no-load and full-load values. Thus, if a turbogenerator with a 3 percent speed droop is operating at its rated 1,200 kw and 60 hertz, when the load is removed, the frequency will rise to

$$60 + 0.03 \times 60 = 61.8 \text{ hertz}$$

The intersection of the droop characteristic with the frequency axis is the no-load frequency of the alternator for the given no-load speed setting of the governor. Turning the governor-control switch to "raise" or "lower" raises or lowers the no-load speed setting of the governor respectively but does not change the droop. This is shown by the dotted lines in Fig. 4-14a.

For a given no-load speed setting of the governor, there is a fixed relationship between the frequency and the kw load. With the no-load speed setting at 61.8 hertz (Fig. 4-14a), a load of 600 kw will cause the frequency to drop to 60.9 hertz, and a load of 1,200 kw will cause it to drop to 60 hertz.

Figure 4-14b shows two alternators with identical droop characteristics and the same no-load speed settings operating in parallel. Note how increased bus load, indicated by the lowered system frequency, is divided equally between the two machines.

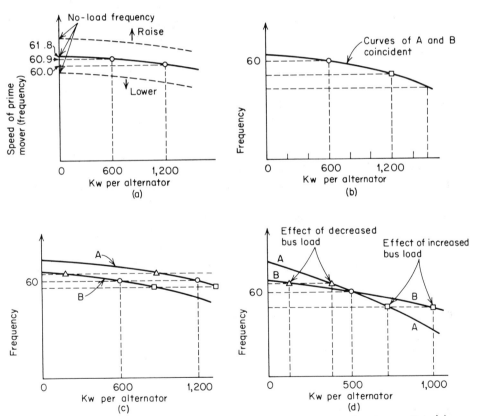

FIG. 4-14 Governor characteristics for two generator sets operating in parallel. (a) Typical governor load-speed characteristic; (b) identical droop characteristics with the same no-load speed; (c) identical droop characteristics with different no-load speeds; (d) radically different droop characteristics.

Figure 4-14c shows two alternators with identical droop characteristics but with different no-load speed settings. Note how increased bus load is divided almost equally between machines.

Figure 4-14d shows the effect of radically different governor droop characteristics on the division of kw load. Starting with an initial balanced condition of 500 kw per alternator and a bus frequency of 60 hertz, the lowered frequency that is caused by increased bus load will cause machine B to take more of the additional load than machine A. If, as loading continues, the kw load on machine B causes its breaker to trip, all the bus load will fall on machine A and it too will trip, blacking out the plant. Referring again to Fig. 4-14d, if the bus load is decreased, the system frequency will increase. For this condition machine B will lose a greater amount of the overall reduction in bus load than will machine A. A large reduction in bus load can cause machine B to lose all its load and then be motorized by machine A. If this occurs, a power-directional relay, shown in Fig. 4-11, will trip machine B off the bus.

If, after balancing the bus kw between two turbogenerators with supposedly identical governor droops, the switching of load on or off the bus causes an unbalance condition, it is an indication that the governor droops of the two machines are not the same. One or both require droop adjustment, which should be done by experienced personnel.

There are some applications where the parallel operation of machines with dissimilar governor droops is desirable. For example, if one machine is adjusted to have zero droop, and the others in parallel with it to have a 2 or 3 percent droop, the droop machines may be loaded for optimum operating efficiency. The machine with droop can carry only one value of load at a specific frequency for a given no-load speed setting of its governor. Hence all fluctuations in the bus load will be absorbed by the machine with zero droop; it will take all the additional oncoming load and will also lose any load that may be disconnected from the bus. The load should be transferred by adjusting only the governor control of the machine with droop. Adjusting the governor control of the zero-droop machine will change the frequency of the system.

4-12 Power-factor Problems of Alternators in Parallel

Individual alternators supplying power to a distribution system have the same power factor as the connected load. However, when operating in parallel, the alternators may have entirely different power factors. When machines are in parallel, their terminal voltages are the same. Yet the acceptance of load by a machine causes its internal voltage to decrease below the bus voltage. Hence the machine is said to be underexcited for its particular load. The machine that gave up the load has its internal voltage increased above the bus voltage, and it is said to be overexcited for its particular load.

If the bus load has a lagging power factor, which is the general case, the internal-voltage difference brought about by a transfer of load will cause the machine that absorbed some load to take less than its proportional share of the kvars. Thus, the machine that accepted some load has a less-lagging power factor than the bus and may even be leading, whereas the machine that lost some load has a more-lagging power factor than the bus. Although the active power may be equally divided between the machines, the equal division of reactive power requires adjustment of the internal voltage of each machine. To do this, the dc field excitation of the machine that took some load should be increased, and that of the other machine decreased, until the power-factor meter indication of each alternator reads the same and lagging. A power-factor meter is shown in Fig. 4-15. Some installations may use varmeters for the same purpose. (A varmeter measures reactive volt-amperes.) The excitation of each machine is adjusted until both machines have the same amount of kvars. The field excitation of each machine is adjusted with its respective field rheostat.

If the generator panels are not equipped with a power-factor meter or varmeter, the balancing of kvars may be made through observation of the ac ammeters. The field excitation of the machine that took some load should be increased, and that of the other machine decreased, until the ac ammeters of both machines indicate the same. If carried too far, it will result in an unbalance in the other direction.

Machines that are equipped with automatic voltage regulators do not require manual adjustment of their field rheostats to balance the kvars. A compensating device in the regulator automatically adjusts the kvar distribution and hence the power factor of each machine. If they are

FIG. 4-15 Power-factor meter. (*General Electric* Co.)

not equal, the voltage-adjusting control of the voltage regulators may be used to provide equalization.

4-13 Phase Sequence and Phase-sequence Indicators

In addition to satisfying the requirements of voltage, frequency, and phase-angle displacement, the incoming machine, if polyphase, must have the same phase sequence as the bus. Phase sequence, sometimes referred to as *phase rotation,* is the order in which the voltage waves pass through zero going in the positive direction. To parallel successfully, only the corresponding phases of the two machines may be connected together. The generator cables for phases *A, B,* and *C* of the incoming machine should connect to the corresponding *A', B',* and *C'* phases of the bus. If any two cables are interchanged, for example, connecting *A* to *B',* *B* to *A',* and *C* to *C',* and paralleling is attempted, an excessively high current will circulate between the incoming machine and the other machines on the bus, causing the breaker to trip. When an incoming machine does not have its corresponding phases arranged for proper connection to the bus, it is said to have *opposite* phase sequence. The phase sequence of an alternator need be checked only when a new or rebuilt machine is installed or tied into another power system. If successfully paralleled before and if no changes were made in the generator connections or direction of rotation, it may be safely assumed that the phase sequence is correct.

The phase sequence of an incoming machine relative to that of the bus may be determined by the synchronizing-lamp circuit shown in Fig. 4-12c. If the generator has the same phase sequence as the bus, both lamps will go bright and dark in unison. If the generator and the bus have opposite phase sequences, the lamps will go bright and dark one after the other.

A neon-light type of phase-sequence indicator is shown in Fig. 4-16a, and its method of use in Fig. 4-16b. To make the test, the three leads of the phase-sequence indicator are connected to the three top terminals of the breaker. Depending on the phase sequence, lamp *ABC* or lamp *CBA* will glow. After noting which lamp glows, the three test leads should be shifted vertically from the top terminals of the breaker to the corresponding bottom terminals. The same lamp will glow if the phase sequence of the generator is the same as that of the bus. If the other lamp glows, the phase sequence of the generator is opposite that of the bus. To correct the phase sequence of the incoming machine, shut down the generator and reverse any two of the three lines marked *A, B,* and *C.*

A three-phase induction motor may also be used as a phase-sequence indicator. When using it for this purpose, mark the three leads, connect

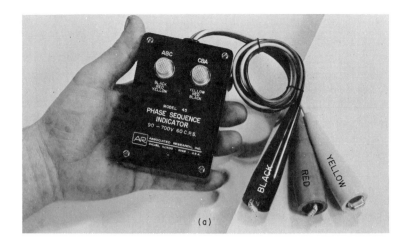

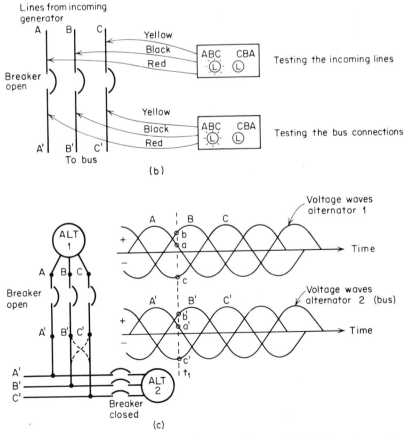

FIG. 4-16 (a) Phase-sequence indicator (*Associated Research, Inc.*); (b) test connections; (c) phase relationships between the voltage waves of the two alternators.

them to the bus terminals of the circuit breaker, and note the direction of rotation. Then transfer the three motor leads to the corresponding generator terminals of the breaker, and again note the direction of rotation. If the motor rotation is the same as before, the incoming machine has the same phase sequence as the bus.

Figure 4-16c illustrates the instantaneous voltage conditions corresponding to correct phase sequence. As indicated by the two sets of curves, at any given instant of time t_1 the voltages b, a, and c of alternator 1 will be equal to and in the same direction as the corresponding voltages b', a', and c' of alternator 2. Thus, if the breaker is closed, there will be no interchange of energy between machines; both machines will feed energy to the common load. However, if any two of the three cables from the bus to the breaker are interchanged, for example, B' and C' shown with broken lines, then the voltages corresponding to the top and the bottom of each breaker pole will be respectively b and c', a and a', c and b'. As indicated by the two sets of curves, at time t_1 a positive voltage b will connect to the same breaker pole as a negative voltage c', and a negative voltage c will connect to the same breaker pole as a positive voltage b'. If the breaker is closed for this condition, there will be a violent interchange of energy between machines, and the circuit breaker overload device will trip the breaker.

Aside from the impossibility of operating ac generators in parallel with opposite phase sequence, if the three-phase power lines supplied to a factory, or to a ship for shore power, inadvertently had two of the three incoming power cables reversed, every three-phase motor in the plant would run in the reverse direction. This can have disastrous consequences.

4-14 Voltage Failure of AC Generators

Voltage failure of an ac generator may be caused by an open in the field circuit, an open in the field rheostat, or failure of the exciter generator. A voltage check of the exciter generator will determine if it is at fault. To test the rheostat, short-circuit its terminals and observe the alternator voltage. A buildup of voltage indicates an open rheostat. If the exciter is operating at rated voltage and the alternator fails to build up when the rheostat is short-circuited, the trouble is in the field circuit or the wires connecting the exciter to the field.

The loss of field excitation to a generator operating in parallel with others causes it to lose load and overspeed. Unless tripped off the bus in a short time, the machine may be damaged by excessively high temperature; high armature current caused by the high voltage differential between the armature and the bus, and the high currents induced in the field iron and field windings by the armature current, will cause rapid heating of the apparatus. However, if the loss in excitation is identified as caused by accidental tripping of the voltage-regulator switch or field breaker, the immediate reclosing of the respective switch or breaker should restore service.

4-15 Operating Outside of the Nameplate Rating

Most generators can be operated with a voltage variation of 5 percent above or below rated voltage and still safely carry rated kva at rated power factor. However, the manufacturer should be consulted if deviations greater than 5 percent are anticipated.

If the generator is to be operated at reduced frequency, a proportional decrease from rated voltage and rated kva is required. Although a reduction in operating frequency of up to 5 percent of rated is not harmful, the manufacturer should be consulted.

Under emergency conditions it sometimes becomes necessary to operate a generator in excess of its rated kva. Because the increased load causes increased heating of the armature and field windings, additional cooling must be provided to prevent damage to the machine. In the case of self-contained air or gas recirculating systems, an increase in the flow of cooling water or other coolant will help keep the temperature down. However, under no circumstances should the temperature of the air or gas be allowed to reach or go below the dew point; temperatures below the dew point cause condensation of any entrapped moisture on the windings of the machine.

Bibliography

Guide for Operation and Maintenance of Turbine-generators, IEEE No. 67, Nov. 1963.

Hubert, C. I.: "Operational Electricity," John Wiley & Sons, Inc., New York, 1961.

Questions

4-1. Explain why a voltage is induced in a coil when the coil is rotating within a magnetic field.

4-2. What adjustments may be made to an operating alternator to change the generated voltage?

4-3. What adjustment may be made to an operating alternator to change its frequency?

4-4. Explain why a generator with a rotating field is just as effective in generating a voltage as is a generator with a rotating armature.

4-5. How does a three-phase generator differ from a single-phase machine? Does a 60-hertz 1,200-rpm three-phase generator have the same total number of field poles as a single-phase 60-hertz 1,200-rpm generator? Explain.

4-6. Make two sketches, using sine waves, to illustrate the two possible phase sequences that may be obtained from a three-phase generator. If the direction of prime-mover rotation cannot be reversed, how may the phase sequence of the voltage supplied to a load be reversed?

4-7. Explain the operation of a brushless excitation system used with some ac generators.

4-8. When a relatively heavy load is connected to an alternator, the frequency and voltage output of the machine are reduced. Explain why this happens. What equipment is provided to minimize this effect, and how does it work?

4-9. Differentiate among active power, reactive power, and apparent power.

4-10. What is meant by the power factor of a circuit? How may the power factor of a circuit be determined without a power-factor meter?

4-11. Explain why a capacitor may be used to reduce the apparent power supplied by a generator to a lagging power-factor load. Make a one-line diagram showing the load, the generator, and the location of the capacitor.

4-12. What is the purpose of generator disconnect links when a generator has a circuit breaker? Where are the links located, and how are they opened and closed? What cautions must be observed?

4-13. Outline the steps required for paralleling an ac generator with another one already on the bus. Assume that the machine on the bus is operating at 450 volts, 60 hertz, and has a 300-kw load at 0.8 *pf*. Include the correct procedure for dividing the active and reactive components of the bus load equally.

4-14. Describe the procedure for dividing the kw load between two alternators in parallel when the governor controls are too far apart to be operated simultaneously.

4-15. Why is it good practice to start the breaker-closing operation 1° or 2° before the 0° position?

4-16. Why is it good practice to balance both the kw and kvar load in proportion to the kw ratings of the machines?

4-17. Assume that two identical generators are operating in parallel and sharing a small bus load. Outline the correct procedure for removing one of the generators from the bus.

4-18. What is motorization? Is it harmful? How may it be detected and corrected?

4-19. What damage may occur if an out-of-synchronism machine is not immediately removed from the bus?

4-20. Is there any advantage in adjusting the frequency of an incoming machine to equal the bus frequency before paralleling? Explain.

4-21. Interpret each of the following synchroscope indications: (a) rotating slowly clockwise, (b) rotating slowly counterclockwise, and (c) stationary at 90°.

4-22. What is the maximum error angle permissible when paralleling alternators? What bad effect does a large error angle have on the machines? Explain.

4-23. If a synchroscope indicates that conditions are right for paralleling but the synchronizing lamps indicate otherwise, which is correct? Explain.

4-24. State the correct procedure for paralleling when using synchronizing lamps. Are there any restrictions on the type, voltage, and wattage ratings of lamps when used as synchronizing indicators?

4-25. What effect do different governor droop characteristics have on the division of oncoming load between generators? Explain with the aid of a sketch.

4-26. Two alternators A and B are in parallel, taking equal shares of the bus load. The governor of prime mover A has zero speed droop, and the governor of B has a 2 percent speed droop. If an additional 100-kw load is connected to the bus, what percentage of the additional load will be taken by each machine?

4-27. Alternator A is connected to the bus and is supplying the total bus load of 400 kw at a lagging power factor of 0.8. Alternator B is paralleled with alternator A, and the bus load is divided equally between machines using the governor controls. Is the power factor of each machine 0.8? Explain.

4-28. Assume that a damaged ac generator has new armature coils installed, the terminals connected to the switchboard, and the machine is apparently ready to be paralleled with the bus. State in detail the procedure that you would follow to determine whether or not the phase sequence is correct.

4-29. What effect does an opposite phase-sequence power supply have on the operation of three-phase motors?

4-30. What will happen if an alternator with the wrong phase sequence is paralleled with the bus?

4-31. State the possible reasons why an ac generator fails to build up voltage.

4-32. What will happen if an alternator in parallel with others loses its field excitation?

4-33. What are the limitations, if any, on (a) operating above or below rated voltage, (b) operating below rated frequency, and (c) operating above rated kva?

Problems

4-1. A 60-hertz 450-volt three-phase generator operates at 1,800 rpm. Assume that a defect in the excitation system that supplies the field current causes the output voltage to drop to 400 volts. (a) At what speed must the generator operate to obtain 450 volts? (b) What effect does this new speed have on the frequency?

4-2. Determine the rated operating speed for a 450-volt three-phase six-pole 60-hertz 1,200-kva generator.

4-3. Determine the operating speed for a three-phase twelve-pole 450-volt 60-hertz 600-kw alternator.

4-4. Determine the frequency of a four-pole 600-kw three-phase generator operating at 1,750 rpm.

4-5. A 450-volt two-pole 60-hertz three-phase generator is operating at a reduced speed of 3,000 rpm. Determine the frequency at this lower speed.

4-6. Determine the number of poles in an alternator that generates 4,160 volts and 60 hertz, at 600 rpm.

4-7. If an ac generator develops 200 volts at 60 hertz, what will be its voltage and frequency when both the speed and the strength of the magnetic field are doubled?

4-8. Calculate the phase current of a delta-connected generator that supplies 320 amp of line current at 4,160 volts to a three-phase induction motor. Sketch the circuit.

4-9. What is the phase current of a wye-connected generator that supplies 50 amp to a three-phase resistor load? Sketch the circuit.

4-10. A three-phase generator has a voltage rating of 260 volts per phase. What is the line voltage when it is (a) wye-connected and (b) delta-connected? Sketch the circuits.

4-11. The instrumentation of a generator system indicates an output of 350 kw at 0.7 power factor from a 450-volt 60-hertz generator. Determine the apparent power supplied by the generator.

4-12. Referring to Fig. 4-9a, assume that the heating elements draw 12 kw and the motor draws 10 kva at 0.8 pf. Determine (a) the kw drawn by the motor, (b) the kvars drawn by the motor, (c) the total kw supplied by the generator, (d) the total kvars supplied by the generator, and (e) the total apparent power supplied by the generator.

4-13. Referring to Fig. 4-9b, assume that the heater draws 4 kw and the motor draws 14 kva at 0.6 pf. Determine (a) the total active power supplied by the generator and (b) the required kvar rating of the capacitor so that the generator does not have to supply kilovars to the motor.

4-14. A lighting load draws 30 amp from a single-phase 120-volt 60-hertz generator. Sketch the circuit, and determine the kva supplied.

4-15. A single-phase load draws 4 kw from a 120-volt line. A clip-on ammeter indicates the current to be 50 amp. Sketch the circuit, and determine (a) the apparent power and (b) the power factor.

4-16. A three-phase induction motor draws 20 amp from a 2,400-volt three-phase 60-hertz generator. (a) Sketch the circuit; (b) determine the kva supplied.

4-17. A three-phase 60-hertz motor draws 25 amp at 0.75 pf from a 460-volt system. Sketch the circuit, and determine (a) the apparent power drawn by the motor, (b) the active power, and (c) the reactive power.

4-18. The instrumentation for a given three-phase load indicates power 24,500 watts, voltage 550 volts, current 28.6 amp. Determine (a) the apparent power delivered to the load, (b) the reactive power, and (c) the power factor.

4-19. In Prob. 4-18, assume that a three-phase capacitor bank rated at 5 kvar is connected in parallel with the load. (a) Make a one-line diagram for the system; (b) determine the total kva supplied by the generator; (c) determine the new power factor; (d) determine the generator current.

4-20. Repeat Prob. 4-19, using a three-phase capacitor bank rated at 12.5 kvar.

4-21. A 600-kw 60-hertz diesel generator with a 2 percent speed droop is operating at rated load and rated frequency. Determine the frequency when the load is removed.

five

ac motor

5-1 Three-phase Squirrel-cage Induction Motor

A cutaway view of a three-phase squirrel-cage induction motor is shown in Fig. 5-1a. The stator coils are inserted in the slots and connected in a wye or delta arrangement similar to that for a three-phase generator. The squirrel-cage rotor, shown in Fig. 5-1b, consists of a laminated steel core with molten aluminum cast into the slots. The aluminum end rings are cast along with the rotor conductors to form an aluminum cage.

The windings of the three phases of the stator are spaced and connected in a manner that causes the development of a rotating magnetic field when the stator is connected to a three-phase supply voltage. As illustrated in Figs. 5-1a and 5-2a, the coils of the three phases are staggered in an overlapping arrangement so that the magnetic field contribution of each, for the development of a given pole, is nonconcentric. Furthermore, because the current in each phase of the three-phase supply attains its respective maximum value at a different instant of time, the centerline of magnetic flux shifts from C to B to A, assuming this to be the phase sequence of the applied voltage. The shifting magnetic field has the same effect on the squirrel-cage rotor as that produced by a magnet sweeping around the rotor, as shown in Fig. 5-2b.

The rotating magnetic field set up by the three-phase current in the stator passes through the many windows formed by the bars in the squirrel-cage rotor. This behavior is simulated in Fig. 5-2c, where the moving magnets represent the "rotating poles" set

Oil ring

Oil reservoir

Stator windings   Rotor

(a)

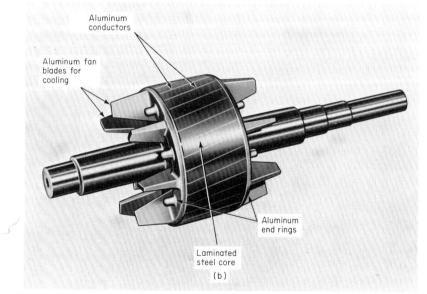

Aluminum
conductors

Aluminum fan
blades for
cooling

Aluminum
end rings

Laminated
steel core

(b)

FIG. 5-1 (a) Cutaway view of a squirrel-cage induction motor (*Westinghouse Electric Corp.*);
(b) squirrel-cage rotor. (*Allis-Chalmers Mfg. Co.*)

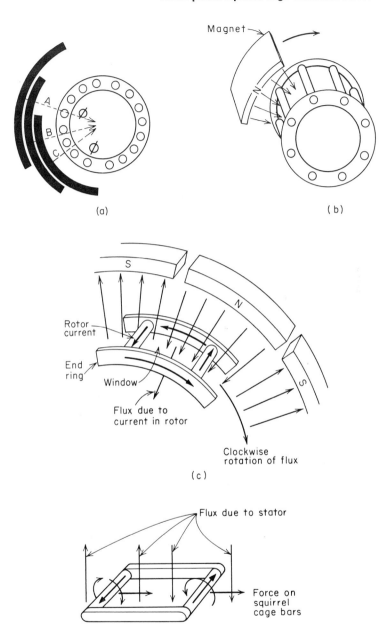

FIG. 5-2 Induction-motor action. (a) Coils of three phases that make up one pole; (b) rotating flux; (c) electron flow in rotor conductors due to the change in flux through the window; (d) force on squirrel-cage bars.

up in the stator and a representative window of the stationary squirrel-cage rotor is in the process of being swept by the clockwise rotation of these "rotating poles" (stator flux). At the instant shown, there are three magnetic lines directed downward through the window, and one line directed upward, for a net flux of two lines in the downward direction. As the flux rotation proceeds, the net downward flux through the window is reduced to zero and then increases in the upward direction. For the

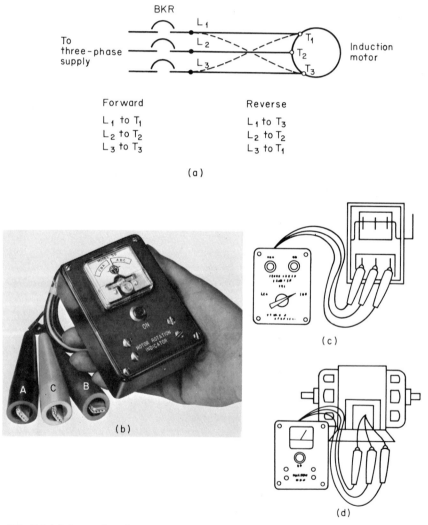

FIG. 5-3 (a) Connections for reversing a three-phase induction motor; (b) motor-rotation indicator; (c) checking phase sequence of supply voltage; (d) determining direction of rotation. (*Associated Research, Inc.*)

instant shown, the net flux is in the downward direction but becoming less and less. The decreasing flux through the window induces a current in the associated squirrel-cage bars in a direction to delay the change in the flux (Lenz's law). The result, as shown in Fig. 5-2c, is a counterclockwise (CCW) induced current that produces an additional downward contribution of flux.

The interaction of the magnetic field caused by the induced current in the squirrel-cage bars with the magnetic field of the stator produces a mechanical force on the bars (see Sec. 1-16). This force, for the instant shown in Fig. 5-2c and d, causes the rotor to experience a torque in the direction of rotation of the stator flux.

5-2 Reversing a Three-phase Induction Motor

Reversing the phase sequence of the applied voltage reverses the direction of flux shift (A to B to C), causing the rotor to reverse direction. This is accomplished by interchanging any two of the three-phase line leads to the stator, as shown in Fig. 5-3a.

In certain motor applications, rotating the shaft in the wrong direction, for even a few degrees, may cause serious damage to the driven equipment. For such applications, the direction of motor rotation must be accurately predicted before voltage is applied. A motor-rotation indicator, shown in Fig. 5-3b, can quickly predict the direction of rotation of a three-phase motor without applying voltage to the machine. Before checking the motor, determine the phase sequence of the applied voltage with a phase-sequence indicator, as shown in Fig. 5-3c; interchange any two leads of the instrument, if necessary, so that it indicates sequence ABC. Mark the breaker terminals with the corresponding ABC markings on the instrument leads. Then, connect the motor-rotation indicator to the motor leads, as shown in Fig. 5-3d. Hold the "on" button down, and rotate the motor shaft about one-quarter turn in the desired direction of rotation. Note whether the first deflection is to ABC or to CBA. If ABC shows first, mark the motor leads to correspond with the terminal markings of the motor-rotation indicator. If CBA shows first, interchange leads A and C of the motor-rotation indicator, and then correspondingly mark the motor leads. Connect the motor to the starter (A to A, B to B, C to C), and it will run in the desired direction.

5-3 Speed Adjustment of an Induction Motor

The speed of rotation of the squirrel-cage rotor depends on the speed of the rotating flux. Assuming that the torque load on the shaft is not excessive, increasing the speed of the rotating flux provides an almost proportional increase in rotor speed. The speed of the rotating stator flux (also called synchronous speed) is affected by the frequency of the three-phase supply voltage and the number of magnetic poles in the stator winding. This relationship is expressed by

$$\text{rpm of rotating flux (synchronous speed)} = \frac{120}{p}$$

where f = frequency, cycles per sec (hertz)
p = number of poles in stator winding

The rotor will always turn at a slower speed than the synchronous speed of the rotating flux, and it is this difference in speed, called the *slip speed* or *slip,* which continues to develop motor action; the rate of change of flux through the windows of the rotor is proportional to the slip speed.

Referring to Fig. 5-2*a*, increasing the frequency shortens the time for the flux to shift from the centerline of coil C to the centerline of coil A, resulting in a proportionately faster sweep around the rotor. The effect is a higher rotor speed.

The number of poles in a stator winding can be determined by the span of a stator coil. If a coil spans a distance equal to one-fourth of the circumference, it is a four-pole winding; if it spans one-sixth of the circumference, it is a six-pole winding; etc. This is illustrated in Fig. 5-4 for eight-pole and four-pole windings. As indicated, the centerline of flux shifts 30° (A to C) for the eight-pole winding and 60° (A to C) for the four-pole winding. Because of the greater angle swept, in the same period of time (assuming that the frequency is the same), the four-pole winding has a higher synchronous speed than does the eight-pole winding.

The speed of a squirrel-cage induction motor operating from a fixed frequency system can be changed only by changing the number of poles in the stator. A machine so designed is called a *multispeed motor.* If a two-speed motor with a speed ratio of 2:1 is desired from a *single stator winding,* additional motor terminals are provided to enable reconnecting for a lower speed. The lower speed is obtained by reconnecting the stator so that all poles have the same polarity (all north or all south). As a consequence of this connection, opposite polarity poles, called *consequent poles,* form between the poles established by the stator windings, thus doubling the number of poles.

For speed ratios other than 2:1, the stator must have two or more independent windings, each with a different number of poles and only

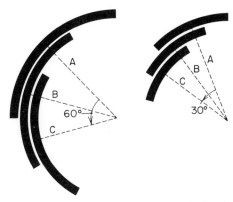

FIG. 5-4 Coil span for eight-pole and
four-pole stators.

4 pole stator (each
coil spans 1/4 of the
circumference)

8 pole stator (each
coil spans 1/8 of the
circumference)

one energized at a time. Figure 5-5 illustrates some of the NEMA standard
terminal markings and connections for multispeed induction motors.

5-4 Representative Performance Data for Squirrel-cage Motors

An ac induction motor of almost any horsepower may be started by connecting it across the line without fear of damage to the motor. The rapid
acceleration of induction motors quickly reduces the high starting current,
called *locked-rotor current,* before the windings overheat. However, such
factors as the large voltage drop in the distribution system caused by
the heavy flow of starting current, or a sudden shock load on the driven
member when starting, or the kilovolt-ampere-demand limit set by an
electric utility company often makes it desirable to use a current-limiting
device when starting. The circuits for such starting devices are discussed
in Chap. 13. Representative values of locked-rotor current for three-phase
220-volt 60-hertz constant-speed squirrel-cage motors are given in Appendix 4.

The torque-speed characteristics of NEMA[1] standard squirrel-cage
induction motors and some representative applications are given in Fig.
5-6. Two very significant points on the curves are the rated breakdown
and the rated locked-rotor values of torque. The rated breakdown
torque is the maximum torque that the machine can develop as a result
of increased shaft loads without causing an appreciable drop in speed.
The rated locked-rotor torque (starting torque) is the torque produced

[1] National Electrical Manufacturers Association.

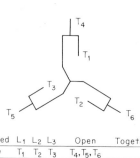

Speed	L₁	L₂	L₃	Open	Together
Low	T₁	T₂	T₃	T₄,T₅,T₆	
High	T₆	T₄	T₅		T₁,T₂,T₃

Two-speed one-winding
variable torque motors
(a)

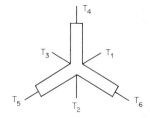

Speed	L₁	L₂	L₃	Open	Together
Low	T₁	T₂	T₃	T₄,T₅,T₆	
High	T₆	T₄	T₅		T₁,T₂,T₃,

Two-speed one-winding
constant torque motors
(b)

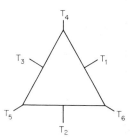

Speed	L₁	L₂	L₃	Open	Together
Low	T₁	T₂	T₃		T₄,T₅,T₆,
High	T₆	T₄	T₅	T₁,T₂,T₃	

Two-speed one-winding
constant horsepower motors
(c)

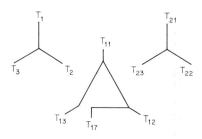

Speed	L₁	L₂	L₃	Open
Low	T₁	T₂	T₃	T₁₁,T₁₂,T₁₃,T₁₇,T₂₁,T₂₂,T₂₃
2d	T₁₁	T₁₂	T₁₃,T₁₇	T₁,T₂,T₃,T₂₁,T₂₂,T₂₃
High	T₂₁	T₂₂	T₂₃	T₁,T₂,T₃,T₁₁,T₁₂,T₁₃,T₁₇

Three-speed three-winding motors
(d)

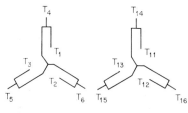

Speed	L₁	L₂	L₃	Open	Together
Low	T₁	T₂	T₃	T₄,T₅,T₆	
2d	T₁₁	T₁₂	T₁₃	T₁₄,T₁₅,T₁₆	
3d	T₆	T₄	T₅		T₁,T₂,T₃
High	T₁₆	T₁₄	T₁₅		T₁₁,T₁₂,T₁₃

Four-speed two-winding motors
(e)

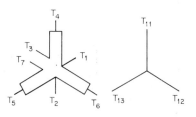

Speed	L₁	L₂	L₃	Open	Together
Low	T₁	T₂	T₃,T₇	T₄,T₅,T₆	
2d	T₆	T₄	T₅		T₁,T₂,T₃,T₇
High	T₁₁	T₁₂	T₁₃		

Three-speed two-winding motors
(f)

FIG. 5-5 Standard connections for multispeed induction motors.

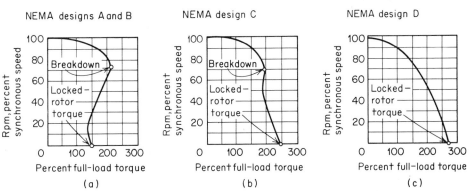

FIG. 5-6 Torque-speed characteristics of NEMA-design squirrel-cage induction motors. (a) NEMA designs A and B; applications are pumps, lathes, drill presses, grinders, machine tools, conveyors (started unloaded), compressors (started unloaded). (b) NEMA design C; applications are electric stairways, pulverizers, compressors (started loaded), conveyors (started loaded). (c) NEMA design D; applications are hoists, elevators, punch presses, and machines with large fly wheels. (*Westinghouse Electric Corp.*)

by the motor when the shaft is blocked and rated voltage is applied to the stator.

Efficiency and power-factor values for $\frac{4}{4}$, $\frac{3}{4}$, and $\frac{1}{2}$ load are given in Table 5-1 for NEMA design B motors.

5-5 Wound-rotor Induction Motor

A wound-rotor induction motor, shown in Fig. 5-7, is an adjustable-speed machine that uses a wound rotor (called the *secondary circuit*) in place of a squirrel cage. However, the stator construction is the same as for the squirrel-cage motor.

The principle of operation is similar to that discussed in Sec. 5-1; the stator develops a rotating magnetic field that sweeps past the rotor, and the interaction of the magnetic field of the stator with that produced by the current in the rotor results in the development of motor torque.

Because the rotor windings do not form closed loops, as does the squirrel cage, the rotor circuit must be completed through a rheostat, as shown in Fig. 5-7b, or by short-circuiting the rotor terminals. The three-phase rheostat is composed of three rheostats connected in wye, and a common lever is used to adjust all three arms simultaneously. The rheostat serves to change the torque-speed characteristic of the machine. When starting, all rheostat resistance should be inserted in the rotor circuit. This provides a relatively high starting torque and a reduced

Table 5-1 Effect of Loading on the Efficiency and Power Factor of Design B Motors

Open and enclosed types
Three-phase, 60-hertz, 208–220–440–550-volt

Hp	Rpm (syn- chronous)	Efficiency			Power factor			Full-load current,* amp per phase, 220 volts, 3 phase
		$\frac{4}{4}$ Load	$\frac{3}{4}$ Load	$\frac{1}{2}$ Load	$\frac{4}{4}$ Load	$\frac{3}{4}$ Load	$\frac{1}{2}$ Load	
$\frac{1}{2}$	900	66	60	53	54	45	37	2.74
$\frac{3}{4}$	1,200	70	68	64	66	55	43	3.18
$\frac{3}{4}$	900	68	63	55	57	49	38	3.8
1	1,800	76	74	68	71.5	62	48	3.6
1	1,200	71	70	64	67	57	45	4.1
1	900	70	66	57	60	51	39	4.66
$1\frac{1}{2}$	3,600	79	76	69	84	78	67	4.48
$1\frac{1}{2}$	1,800	79	76.5	72	72.5	66	54	5.14
$1\frac{1}{2}$	1,200	76.5	76	71	70	61	49	5.5
$1\frac{1}{2}$	900	74	73	67	63	52	40	6.3
2	3,600	81.5	78	73	84.5	78.5	68	5.7
2	1,800	80	78	73	76	68	54.5	6.44
2	1,200	77	76	73	71	62	50	7.16
2	900	75	74	69	65	55	43	8.04
3	3,600	82.5	82	80	85	79	69	8.4
3	1,800	81	81.5	77.5	79	73	61	9.2
3	1,200	80	79	75	75	66	51	9.8
3	900	78	76	70.5	65	56	44	11.6
5	3,600	83.5	83.5	81	85	79	69	13.8
5	1,800	85	85	83	81	75	64	14.2
5	1,200	82	81.5	80	77	73	60	15.5
5	900	81	80.5	79	71	63	50	17
$7\frac{1}{2}$	3,600	85	85	82	87	82	75	19.9
$7\frac{1}{2}$	1,800	84	83.5	81	85.5	80	71	20.4
$7\frac{1}{2}$	1,200	83.5	83	80	80	74	62	22
$7\frac{1}{2}$	900	82.5	82	79	71	63	50	25
10	3,600	86	86	84	88	84	77	26
10	1,800	85	85	84	87	83	76	26.5
10	1,200	84	84	83	81.5	77	68	28.6
10	900	83.5	83.5	81	80	75	63	29.4
15	3,600	86	86	84	90	87	81	38
15	1,800	86	86	85	87	84	76	39.4

Table 5-1   *Effect of Loading on the Efficiency and Power Factor of Design B Motors (continued)*

Hp	Rpm (synchronous)	Efficiency			Power factor			Full-load current,* amp per phase, 220 volts, 3 phase
		$\frac{4}{4}$ Load	$\frac{3}{4}$ Load	$\frac{1}{2}$ Load	$\frac{4}{4}$ Load	$\frac{3}{4}$ Load	$\frac{1}{2}$ Load	
15	1,200	87	87	86	82	77	67	41.2
15	900	84	84	82	81	75	64	43.2
20	3,600	86	86	85	91	89	84	50
20	1,800	87.5	87.5	86.5	87	84.5	79	51.6
20	1,200	87	87	86	82	77	67	55
20	900	86	86	85	81	75	64	56.4
25	3,600	87	86	85	90	87.5	83.5	62.6
25	1,800	88.5	88.5	87.5	87	84.5	79	63.6
25	1,200	88	88.5	87	86	83.5	72	64.8
25	900	87	87	86	82	76	66	68.8
30	3,600	89	88	86	90	87.5	83.5	73.4
30	1,800	89	89	88	88	85	80	75
30	1,200	88.5	89	88	87.5	85.0	73.5	76
30	900	88	88	87	82	78	70	81.4
40	3,600	89	88	86.5	90	87.5	83.5	98
40	1,800	89	89	88	88.5	86	81	99.6
40	1,200	89	89	88	88	84.5	78	100
40	900	88.5	89	88	82	79	71	108
50	3,600	89	88	85.5	89	87	83.5	124
50	1,800	89.5	89.5	88	89	87	82	123
50	1,200	89.5	89.5	88	88	84.5	78	124.4
50	900	88.5	88.5	87	83	79.5	71	134
60	3,600	90	89	87	90	89	85	145
60	1,800	90	90	89.5	89	87	82	147
60	1,200	89.5	89.5	88.5	87.5	84.5	76	150
60	900	88.5	88.5	87	83	79.5	71	160
75	3,600	90.5	90	88	89.5	88.5	84.5	181
75	1,800	90	90	89	89	87	82	184
75	1,200	90	90	89	87.5	84.5	76	187
75	900	89	89	88	86	84.5	74	192

NOTE 1: These values are not to be used for guarantees; consult motor manufacturer.
NOTE 2: For two-pole, totally enclosed fan-cooled motors, efficiency should be reduced 1 percent at $\frac{4}{4}$ load, 2 percent at $\frac{3}{4}$ load, and 3 percent at $\frac{1}{2}$ load.
* Full-load current is not to be used for selecting heater elements; see nameplate rating of motor involved.
SOURCE: Allis-Chalmers Mfg. Co.

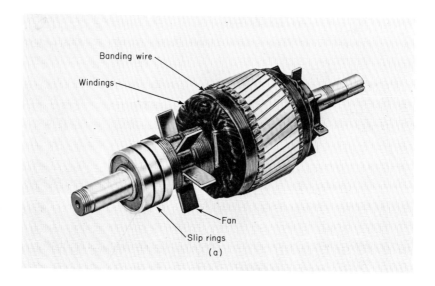

Banding wire

Windings

Fan

Slip rings

(a)

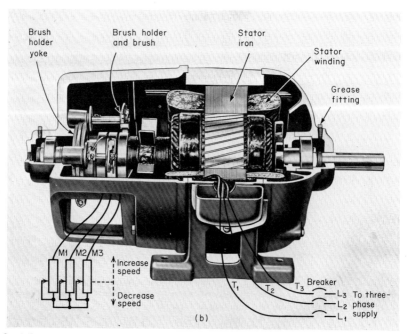

Brush holder yoke

Brush holder and brush

Stator iron

Stator winding

Grease fitting

M1 M2 M3

Increase speed

Decrease speed

T_1 T_2 T_3

Breaker

L_3
L_2
L_1

To three-phase supply

(b)

FIG. 5-7 Wound-rotor induction motor. (a) Rotor; (b) cutaway view of motor and connections to rheostat and three-phase supply. (*Louis Allis Co.*)

speed. Reducing the rheostat resistance, assuming the same shaft load, causes an increase in motor speed. Speed reduction through rheostat control can be obtained only if the machine is loaded; with no load on the machine, the no-load speed is changed very little. The speed of the machine may also be changed by varying the frequency or by changing the number of stator poles.

If the motor is to be used without a rheostat, the three rotor terminals M_1, M_2, and M_3 should be connected together.

To reverse the motor, interchange any two of the stator connections; interchanging the rotor connections does not reverse rotation, nor does it have any effect on the motor performance.

5-6 Effect of Off-standard Voltage and Frequency on Induction-motor Characteristics

Table 5-2 lists the general effects of off-standard voltage and frequency on induction-motor characteristics. Note particularly the adverse effect of low voltage on the starting and running torque. Since the torque varies as the voltage squared, a 10 percent decrease in voltage causes a 19 percent decrease in torque. Thus, too low a supply voltage may prevent a heavily loaded motor from starting or, if running, may cause it to stop. Additionally, the full-load current and the temperature rise increase with decreased supply voltage. The operating speed is also changed by variations in voltage. A 10 percent decrease in voltage causes the speed to drop by approximately 1.5 percent.

The following formula provides a convenient method for determining the induction-motor torque that can be developed when operating at off-standard voltages:

$$T_0 = T_R \left(\frac{E_0}{E_R}\right)^2$$

where T_R = rated (standard) torque
E_R = rated (standard) voltage
T_0 = off-standard torque
E_0 = off-standard voltage

The rated operating characteristics of three-phase motors are based on the assumption that the three line voltages are equal. Unbalanced line voltages cause unbalanced currents, resulting in excessive heating. A relatively small voltage unbalance of approximately 3 percent may cause a considerable increase in temperature, sufficient to damage the insulation.

Table 5-2 General Effect of Voltage and Frequency Variation on Induction Motor Characteristics

		Starting and maximum running torque	Synchronous speed	Percent slip	Full-load speed	Efficiency			Power factor			Full-load current	Starting current	Temperature rise, full load	Maximum overload capacity	Magnetic noise, no load in particular
						Full load	¾ load	½ load	Full load	¾ load	½ load					
Voltage variation	120 percent voltage	Increase 44 percent	No change	Decrease 30 percent	Increase 1.5 percent	Small increase	Decrease ½-2 points	Decrease 7-20 points	Decrease 5-15 points	Decrease 10-30 points	Decrease 15-40 points	Decrease 11 percent	Increase 25 percent	Decrease 5-6°C	Increase 44 percent	Noticeable increase
	110 percent voltage	Increase 21 percent	No change	Decrease 17 percent	Increase 1 percent	Increase ½-1 point	Practically no change	Decrease 1-2 points	Decrease 3 points	Decrease 4 points	Decrease 5-6 points	Decrease 7 percent	Increase 10-12 percent	Decrease 3-4°C	Increase 21 percent	Increase slightly
	Function of voltage	$(Voltage)^2$	Constant	$\dfrac{1}{(Voltage)^2}$									Voltage	....	$(Voltage)^2$	
	90 percent voltage	Decrease 19 percent	No change	Increase 23 percent	Decrease 1½ percent	Decrease 2 points	Practically no change	Increase 1-2 points	Increase 1 point	Increase 2-3 points	Increase 4-5 points	Increase 11 percent	Decrease 10-12 percent	Increase 6-7°C	Decrease 19 percent	Decrease slightly
Frequency variation	105 percent frequency	Decrease 10 percent	Increase 5 percent	Practically no change	Increase 5 percent	Slight increase	Slight increase	Slight increase	Slight increase	Slight increase	Slight increase	Decrease slightly	Decrease 5-6 percent	Decrease slightly	Decrease slightly	Decrease slightly
	Function of frequency	$\dfrac{1}{(Frequency)^2}$	Frequency		Nearly direct	....							$\dfrac{1}{Frequency}$			
	95 percent frequency	Increase 11 percent	Decrease 5 percent	Practically no change	Decrease 5 percent	Slight decrease	Slight decrease	Slight decrease	Slight decrease	Slight decrease	Slight decrease	Increase slightly	Increase 5-6 percent	Increase slightly	Increase slightly	Increase slightly

NOTE: This table shows general effects, which will vary somewhat for specific ratings.

SOURCE: From D. G. Fink and J. M. Carroll, "Standard Handbook for Electrical Engineers," Sec. 18-84, McGraw-Hill Book Company, New York, 1968, with permission.

5-7 Single-phase Operation of Three-phase Induction Motors

Accidental operation of a three-phase motor on single phase (one supply line open) causes the current in the energized windings approximately to double. Although less torque is developed and severe and possibly damaging vibration may occur, the motor will continue to run if it is not heavily loaded. However, once stopped, a three-phase motor will not start on single phase, even if lightly loaded; it will make growling noises, heat, and smoke.

Because the heat-power loss varies as the square of the current, the doubled current in the energized windings generates heat at approximately four times the normal rate. Unless thermal-overload (see Sec. 14-9) or phase-failure devices disconnect the machine from the power line, the excessively high temperature will burn out the windings.

5-8 Split-phase Induction Motors

A split-phase induction motor is a squirrel-cage motor that operates from a single-phase supply (two-wire ac). The rotating member is a squirrel-cage rotor, and the stator has two separate windings. Figure 5-8 illustrates the winding layout for a four-pole split-phase motor, and Fig. 5-9 shows some of the more commonly used circuit configurations. The circuit diagrams show the USA standard terminal markings, and connections for counterclockwise (CCW) rotation, as observed when facing the end

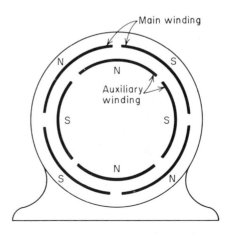

FIG. 5-8 Winding layout for a single-phase motor.

opposite the drive shaft.[1] The rotating magnetic field, necessary to produce induction-motor action, is caused by the current in the main winding and that in the auxiliary winding, each attaining its respective maximum value at different instants of time. The capacitor-start and the two-value capacitor motors use capacitors to cause the current in the auxiliary winding to lead that in the main winding. The resistance-start motor uses additional resistance in its auxiliary winding (it is wound with a smaller-diameter copper wire) so that its current is less lagging than the current in the main winding. For an understanding of the behavior of capacitance and inductance, see Secs. 1-17 to 1-19. Thus, for the stator shown in Fig. 5.8, the flux rotates counterclockwise (CCW). The rotating field sweeping the squirrel-cage rotor produces motor action in the same manner discussed in Sec. 5-1 for the three-phase motor.

The rotor-mounted centrifugal mechanism opens a switch, mounted in the end shield, at approximately 75 percent rated speed, cutting out the auxiliary winding. External relays are also used for this purpose. However, after the auxiliary winding disconnects, the motor continues to operate, creating its own rotating flux as it runs. The auxiliary-start windings for the resistance-start and capacitor-start motors are designed for starting duty only. If the centrifugal switch or relay fails to open, the starting winding will burn out in a very short time.

The single-phase motors shown in Fig. 5-9 may be reversed for clockwise (CW) rotation by interchanging leads T_1 and T_4 of the main winding or T_5 and T_8 of the auxiliary winding.

The thermal protector shown in Fig. 5-9b, d, and f is a built-in thermal overload relay that protects the machine against overloads by automatically opening when the temperature or current exceeds predetermined safe values. After tripping, the protector continues to sense the heat from the motor windings and does not allow the motor to operate until it has cooled. If the motor has a manual reset protector, pushing the red flag to reset the protector will restart the motor. However, if the motor has not cooled sufficiently, the reset will not lock in. If an automatic protector is used, it will reset automatically when the motor cools to a safe operating temperature. Never short-circuit the protector to keep a tripped motor running, or to prevent tripping; such action will result in a burned-out motor. If the protector trips, search for the fault and correct it.

The standard connections for high and low nameplate voltages of double-voltage machines are represented by the capacitor-start motor in Fig. 5-9g and h.

Most split-phase motors are fractional-horsepower units used for oil burners, blowers, refrigerators, etc. Such motors require very little main-

[1] Terminal Markings For Electrical Apparatus, USAS C6.1-1956.

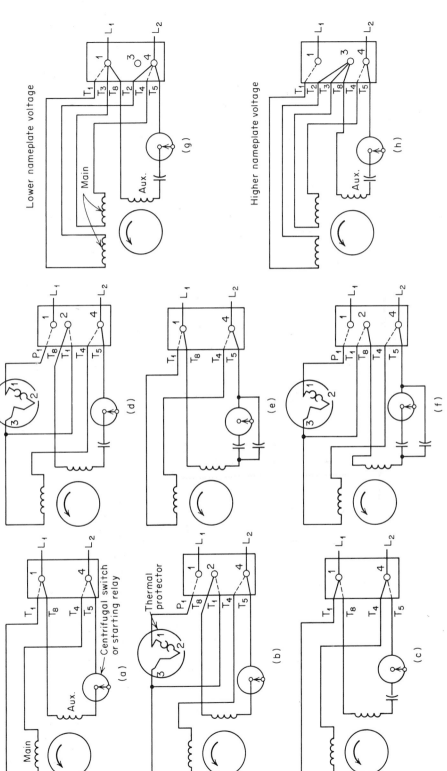

FIG. 5-9 Standard connection diagrams for single-phase induction motors (counterclockwise rotation as observed when facing the end opposite the drive shaft).

tenance. Once a year the outside frame and ventilating ducts of the machine should be cleaned, all bolts tightened, and sleeve bearings lubricated with SAE-10 electric-motor oil. This annual oiling should start after the 3- or 5-year period specified by the manufacturer. Ball bearings should be repacked with electric-motor grease once every 10 years (see Sec. 12-5).

5-9 Synchronous Motors

The construction details of a three-phase synchronous motor are shown in Fig. 5-10, and are similar to those of a three-phase generator. In fact, when two three-phase generators are in parallel and one of them is motorized, it becomes a synchronous motor. The rotor consists of a set of magnets, and a squirrel-cage winding of copper bars (called an *amortisseur* or *starting winding*) embedded in the pole faces. The stator is energized from the three-phase system, and the field magnets are energized by a dc exciter to form alternate north and south poles.

Figure 5-11 is an elementary circuit diagram showing the minimum rotor and stator connections required for operation. To start the motor,

60 pole – 120 rpm

(a)

FIG. 5-10 Synchronous motor. (a) ~~Magnet member~~; (b) stator. (*Westinghouse Electric Corp.*)

Field assembly

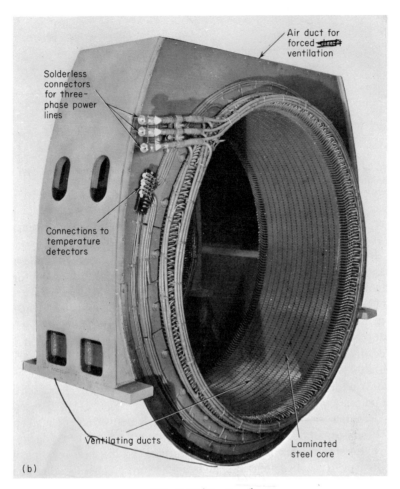

Air duct for forced ~~draft~~ ventilation

Solderless connectors for three-phase power lines

Connections to temperature detectors

Ventilating ducts

Laminated steel core

(b)

FIG. 5-10 (Continued)

first make sure that the dc circuit breaker that supplies the magnets is open, then close the three-phase breaker. Energizing the armature winding causes a rotating magnetic field in the stator that sweeps the stationary squirrel-cage rotor. The resultant induction-motor torque accelerates the rotor to a speed somewhat less than the synchronous speed of the rotating flux. As soon as acceleration by induction-motor action is completed, the dc circuit breaker should be closed, energizing the magnets in the rotor. The strong magnetic attraction between the north poles of the stator and the south poles of the rotor, and vice versa, causes the rotor magnets to lock in with the rotating flux of the stator. The rotor poles are then dragged around by the rotating poles produced by the stator flux, at the synchronous speed of the stator. To change the speed of a synchronous

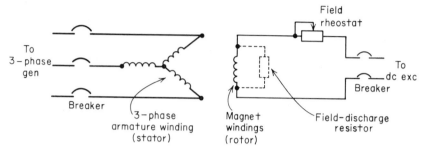

FIG. 5-11 Elementary circuit diagram showing the rotor and stator connections for a synchronous motor.

motor, it is necessary to change the frequency of the three-phase generator that supplies it.

If a synchronous motor fails to pull into synchronism when the direct current is applied, the rheostat resistance may be set too high for the particular load on the motor shaft.

To stop a synchronous motor, first deenergize the ~~magnets~~ *field* by opening the dc breaker, and then open the three-phase breaker. The field-discharge resistor is used to dissipate the energy stored in the magnetic field when the dc breaker is opened; it is also used to reduce the high voltage (5,000 to 30,000 volts) that would otherwise be produced in the field circuit by the rotating magnetic field of the stator when starting. For additional information concerning this inductive effect, see Sec. 1-19. The energy stored in the ~~magnetic~~ field ~~of the magnets~~ *field* is converted to heat energy in the resistor and dissipated to the atmosphere. A broken (open) discharge resistor results in arcing and burning at the breaker tips when the breaker is opened, and the high voltage may cause serious damage to the insulation of the ~~magnet~~ *field* windings.

The direction of rotation of a synchronous motor is reversed by first stopping it, interchanging any two of the three ac line leads, and then restarting. Interchanging two of the ac line leads reverses the phase rotation, causing the stator flux to rotate in the opposite direction. A motor-rotation indicator, discussed in Sec. 5-2 for induction motors, may also be used to predict the direction of rotation of synchronous motors.

In addition to adjusting the hold-in strength of the magnets, changing the amount of direct current in the ~~magnet~~ *field* windings causes a change in the power factor of the stator. Low values of ~~magnet~~ *field* current result in less hold-in strength and a lower power factor; high values of ~~magnet~~ *field* current result in greater hold-in strength and a higher power factor. This characteristic of a synchronous motor is a useful fringe benefit when power-factor adjustment of a system is required (see Sec. 15-4).

The value of torque load that causes the rotor to pull out of synchronism is called the *pullout torque*. If the rotor pulls out of synchronism, the relative motion between the slower-moving rotor poles and the faster-moving stator poles will cause serious and possibly damaging vibrations. If this occurs, the dc breaker should be opened, the field rheostat adjusted to allow a higher field current, and the rotor resynchronized by closing the dc breaker. Increasing the field current increases the hold-in strength of the magnetic poles, permitting higher torque loads to be carried without pulling out of synchronism. The squirrel-cage winding is designed only for starting duty. Hence, if operated as an induction motor for more than a short period, severe overheating and failure of the squirrel-cage winding will occur. Furthermore, without a field the stator current is considerably greater than normal, causing excessive stator temperature.

5-10 Checking Motor Performance

Operators of electrical machinery should keep a constant check on the load that each machine is carrying. Exceeding nameplate values shortens the useful life of a machine. A hook-on volt-ammeter, shown in Fig. 5-12, is a very convenient instrument for periodic checks of motor current; shorted or opened coils in three-phase motors are indicated by unbalanced (unequal) currents in the three motor lines, assuming that the three line-to-line voltages are equal. The instrument shown is for alternating current and has an easily adjusted pointer stop for reading motor starting current. The jaws of the instrument are simply hooked around the insulated conductor; there is no interruption of service. The hook is not used when the instrument leads are connected for voltage measurements. The principle of operation of a hook-on ammeter is explained with the aid of Fig. 5-12*b*. The alternating current in the primary conductor causes an alternating magnetic field in the iron core. This changing field, passing through the window of the secondary coil, generates a voltage that causes a current in the meter. The meter is calibrated to indicate the rms current in the primary conductor. Test leads may be connected to appropriate terminals to convert the instrument to a voltmeter. A selector switch provides a wide range of measurements for both current and voltage. To measure very low currents, thread the conductor through the hook two or more times, and then divide the current reading by the number of conductors through the hook.

Single- or three-phase power (kw) can be measured directly with a polyphase wattmeter or a hook-on wattmeter, as shown in Fig. 5-13. When using a hook-on wattmeter to measure balanced three-phase power, two conductors are hooked (one straight through and the other reversed),

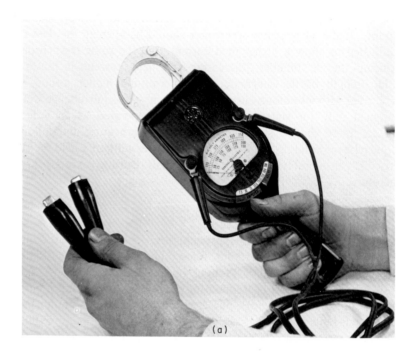

(a)

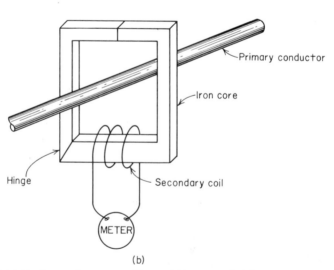

Primary conductor

Iron core

Hinge

Secondary coil

METER

(b)

FIG. 5-12 (a) Hook-on volt-ammeter (*General Electric* Co.); (b) construction details of a hook-on ammeter.

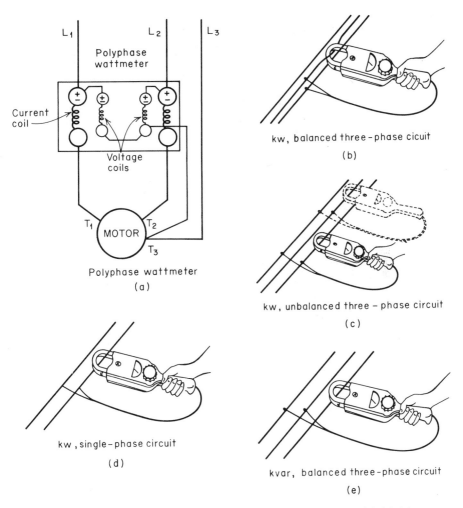

kw, balanced three-phase cicuit

(b)

kw, unbalanced three – phase circuit

(c)

kw, single-phase circuit

(d)

kvar, balanced three-phase circuit

(e)

Polyphase wattmeter

(a)

FIG. 5-13 Ac power measurement. (a) Polyphase wattmeter connections; (b), (c), (d) hook-on wattmeter connections, (e) kvar connections. (*General Electric* Co.)

and the voltage leads connected to the corresponding lines, as shown in Fig. 5-13*b*. A *balanced* circuit is one that has the same amount of current in each of the three lines. Motors that are not shorted or opened represent balanced three-phase loads.

An *unbalanced* three-phase circuit requires two readings, as shown in Fig. 5-13*c*; if it is necessary to reverse the potential connections to obtain an upscale reading, this reading should be subtracted from the other one.

To measure the power in a single-phase circuit, such as that drawn by a single-phase motor, connect the instrument as shown in Fig. 5-13*d*.

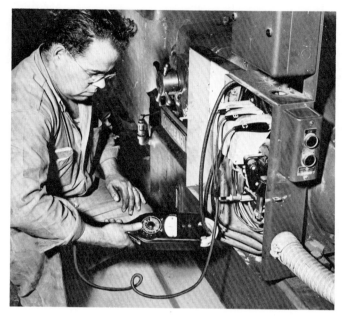

FIG. 5-14 Hook-on power-factor meter for direct power-factor readings. (*General Electric Co.*)

The kvars drawn by a three-phase motor or other balanced three-phase loads can be measured with a hook-on wattmeter by using the circuit shown in Fig. 5-13*e*. One conductor is hooked, and the potential leads are connected to the other two conductors. The meter indication must be multiplied by 1.73.

Low power factor causes relatively high line currents, resulting in poor voltage regulation, wasted power in needless heating of power cables, and greater demand requirements (see Sec. 15-4). A hook-on instrument that measures power factor directly, without interruption of service, is shown in Fig. 5-14; three voltage leads are connected to the supply lines, and one load line is placed on the hook. Figure 5-15 illustrates a type of industrial analyzer that includes an ammeter, a voltmeter, a wattmeter, and a power-factor meter.

Thermometer readings are also very helpful in detecting impending trouble. If the temperature rise of the machine increases for the same load condition, it may mean that the air passages and ventilating screens are clogged with dirt, the flow of water to the coolers is cut off, the strip heaters are left on, etc. If embedded temperature detectors are not installed, thermometers may be secured to the stationary parts of the windings for a temperature check, or they may be fastened to the

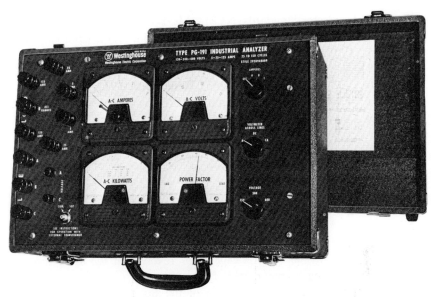

FIG. 5-15 Industrial analyzer for testing ac apparatus. (*Westinghouse Electric Corp.*)

FIG. 5-16 Temperature check of electrical machines using a bulb thermometer.

hottest part of the frame if the machine is totally enclosed. Figure 5-16 illustrates a temperature check with a bulb thermometer attached to the motor frame with electrician's putty.

Bibliography

Hubert, C. I.: "Operational Electricity," John Wiley & Sons, Inc., New York, 1961.

Motor and Generator Standards, NEMA MG1-1963.

Siskind, C. S.: "Induction Motors," McGraw-Hill Book Company, New
York, 1958.

Shoults, D. R., C. J. Rife, and T. C. Johnson: "Electric Motors in Industry," John Wiley & Sons, Inc., New York, 1942.

Veinott, C. G.: "Fractional Horsepower Electric Motors," McGraw-Hill
Book Company, New York, 1948.

Questions

5-1. Explain how a rotating magnetic field is produced in the stator of a squirrel-cage motor.

5-2. Because there are no electrical connections from the generator to the squirrel-cage rotor, what causes the rotor current? Explain.

5-3. Explain why the squirrel-cage rotor revolves in the same direction as the rotating magnetic field of the stator.

5-4. Explain why interchanging two of the three line leads to a three-phase squirrel-cage motor causes it to run in the opposite direction.

5-5. Explain why a two-pole machine runs at a higher speed than a four-pole machine when both are connected to the same supply line.

5-6. What two methods are used to change the speed of a three-phase squirrel-cage induction motor?

5-7. What is the purpose of a consequent-pole connection?

5-8. What factors determine whether or not a squirrel-cage induction motor may be started at full line voltage?

5-9. Define rated breakdown torque and rated locked-rotor torque as they pertain to squirrel-cage induction motors. Which NEMA-design motor develops the greatest locked-rotor torque?

5-10. Referring to Table 5-1, what effect does the loading of a motor have on (a) the motor efficiency and (b) the power factor?

5-11. State the difference in construction details between the squirrel-cage rotor and the wound rotor as used in induction motors.

5-12. What are the advantages of the wound-rotor motor over the squir-rel-cage motor? State an application for each of the two types of motors.

5-13. State three methods that can be used for adjusting the speed of a wound-rotor motor.

5-14. State the correct procedure for (a) starting a wound-rotor induction motor and (b) reversing a wound-rotor motor.

5-15. (a) What effect does a decrease in supply voltage have on the starting and running torque of an induction motor? (*b*) What effect does low voltage have on the full-load current and the operating temperature? (*c*) What effect does low voltage have on the operating speed?

5-16. Assuming that the three line-to-line voltages that supply a three-phase induction motor are all different, what adverse effect does this have on the machine?

5-17. If one supply line to a three-phase motor accidentally opens, will the motor continue to run? Is continued operation with one line open harmful?

5-18. What is the difference in construction details between a split-phase motor and a three-phase squirrel-cage motor? What is the purpose of the rotor-mounted centrifugal mechanism?

5-19. Explain how a rotating magnetic field is produced in the stator of a split-phase motor.

5-20. What is a motor thermal protector? How does it work?

5-21. What is the difference in construction details between a three-phase squirrel-cage motor and a three-phase synchronous motor?

5-22. State the correct procedure for (*a*) starting a three-phase synchronous motor, (*b*) stopping the motor, and (*c*) reversing the motor.

5-23. Explain the purpose of a field-discharge resistor that is connected across the field circuit of a synchronous motor.

5-24. What may cause a synchronous motor to fail to pull into synchronism?

5-25. Assume that a synchronous motor is operating at rated load and 1.0 *pf*. What effect does increasing the field current to the magnets have on the pullout torque and the power factor?

5-26. What damaging effects may be produced when a synchronous motor pulls out of synchronism? Assume that the armature and field circuits are still energized.

5-27. What remedial action should be taken when a synchronous motor pulls out of synchronism?

5-28. Explain the principle of operation of a clip-on ammeter.

5-29. What instruments are useful for checking motor performance?

Problems

5-1. Determine the synchronous speed of a 10-hp 450-volt four-pole induction motor when operating from (*a*) a 60-hertz line and (*b*) a 50-hertz line.

5-2. A 25-hp induction motor runs at 3,575 rpm when operating at rated load from a 2,300-volt 60-hertz supply. (*a*) Determine the number of poles in its stator winding. (*b*) Determine the slip speed. (*Hint:* The operating speed of an induction motor is always slightly lower than synchronous speed.)

5-3. A partial blackout, or dimout, in an electrical utility system results in 400 volts instead of 460 volts at the terminals of a three-phase motor. If the locked-rotor torque at 460 volts is 150 percent of rated torque, what is the locked-rotor torque at 400 volts?

5-4. An air compressor driven by a 550-volt three-phase design C induction motor requires 210 percent of rated torque when starting. As shown in Fig. 5-6, the maximum starting torque that the motor can develop when connected across 550 volts is 250 percent of rated load. (*a*) If the available line voltage drops to 490 volts, how much torque can the motor develop? (*b*) Will the motor start?

six

trouble shooting and emergency repairs of ac machinery

This chapter is devoted to logical methods for determining some of the common troubles that occur in ac machinery, and to suggestions for emergency repairs that will keep the equipment in operation until time is available for a major overhaul. Difficulties of a mechanical nature, such as brush tension, slip-ring wear, bearing wear, and vibration, are covered in other chapters.

Electrical apparatus itself does not reverse the connections to its field, armature, and control system. Hence, unless there is reasonable suspicion that the connections were changed, it may be assumed that they are correct. The nameplate data of the machine should always be checked against the actual operating values of voltage, current, frequency, temperature, speed, etc. Poor performance may often be traced to operating above or below the rated nameplate values.

6-1 Trouble-shooting an AC Motor

When a motor fails to start, the branch circuit fuses (or molded case breaker), the thermal overload relay, and the small fuses on the motor control panel should be the first things checked (see Secs. 14-3, 14-9, and 14-12). If the fuses are good, and the breakers and relays not tripped, the control panel should be checked (see Sec. 13-5).

To check the motor, disconnect the motor leads from the controller, and make continuity and ground tests. The ground test is made by connecting one terminal of the megohmmeter to the shaft or an unpainted part of the motor frame and connecting the other terminal to the stator leads. A zero reading indicates a grounded circuit.

The continuity test is made by testing with a megohmmeter between the phases of each motor circuit. For example, referring to the three-speed circuit in Fig. 5-5d, test between T_1 and T_2, T_3; test between T_{11} and T_{12}, T_{13}, T_{17}; test between T_{21} and T_{22}, T_{23}. A zero reading indicates a complete circuit. A high reading on the megohmmeter indicates an open circuit.

A listing of some of the common troubles experienced with ac motors and their possible causes is given in Sec. 6-14.

6-2 Locating Shorted Stator Coils

Shorted coils in the stator of an ac machine are generally indicated by higher than normal current. A shorted coil that is not identified by burning, overheating, or discoloration of its insulation may be located with a growler. A growler consists of a coil of wire wound around an iron core and connected to a source of alternating current. When placed on the armature core, the growler coil acts as the primary of a transformer, and the armature coils act as the secondary. Figure 6-1 illustrates the location of a shorted coil, using a portable growler with a built-in feeler. The growler is moved from slot to slot and vibrates with a very loud growling noise when directly over a slot that contains a shorted coil.

FIG. 6-1 Testing stator coils for short circuits, using a portable growler. (*United States Merchant Marine Academy.*)

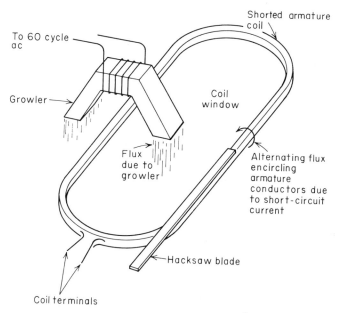

FIG. 6-2 Principle of growler action.

The built-in feeler is particularly handy for very small stators, where there is insufficient room for a separate feeler. A separate feeler fashioned from a hacksaw blade provides a high degree of sensitivity when larger stators are tested. If a growler is not available, the application of reduced voltage to the stator, approximately 25 percent of rated voltage, will cause the shorted coil to become noticeably warmer than the others. The rotor must be removed for this test.

The principle of growler action may be explained by reference to Fig. 6-2, where the alternating magnetic flux set up by the growler is shown passing through the window of an armature coil. The changing flux through the coil window generates an alternating voltage in the armature coil. If the coil is shorted, the circuit is complete and an alternating current will appear; the resultant magnetic field encircling the armature conductors will alternately attract and release the hacksaw blade, causing it to vibrate in synchronism with the alternating current. A vibrating noise (growling) indicates that the coil is shorted.

6-3 Locating Grounded Stator Coils

A grounded stator coil may be located by applying reduced voltage between the motor terminals and ground, as illustrated in Fig. 6-3. The rheostat should be adjusted to permit sufficient current to flow through the defective coil to cause overheating and smoking at the point of grounding. Before making this test, the stator should be insulated from ground

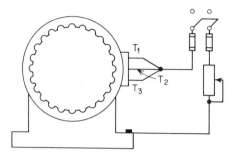

FIG. 6-3 Circuit connections for smoking out a ground.

with dry wood blocks or other heavy insulating material. Large machines that are supplied by a three-phase system with a grounded neutral generally have ground and phase-balance relays that trip the machine off the line when such faults occur. If the fault cannot be detected by burned or discolored insulation, the protective relays may be removed from the circuit, and the machine may be operated at low voltage and low speed until the faulty coil is identified by smoking or overheating. The coils should not be touched while the power is on.

6-4 Emergency Repair of a Shorted or Grounded Stator

An emergency repair of a shorted or grounded stator may be made by cutting the defective coil in half. This can be done with a diagonal type of plier, bolt cutter, chisel, or hacksaw, and depends on the size of the coils. Figure 6-4 illustrates how this cut may be made. The open ends should be separated from each other, properly insulated, and tied down to prevent movement. The coil leads should be disconnected, and a jumper should be used to complete the circuit, as shown in Fig. 6-5. The success of such emergency repairs depends to a good extent on the number of coils in the stator. Cutting one coil out of the circuit of a machine that has only a few coils has a serious effect on its operation; whereas a large machine, such as a 10,000-hp synchronous propulsion motor of an electric-drive ship, operates satisfactorily at reduced load, even though several coils may be cut out of the circuit. However, whenever possible, the manufacturer's recommendations for emergency repairs should be followed.

6-5 Locating Open Stator Coils

Single-circuit stators. If the open occurs in a single-circuit stator, as shown in Fig. 6-6, the machine will not start. It will hum and have symptoms identical to those produced by a blown fuse in one of the three line leads. An open that occurs while the machine is carrying load causes arcing and generally burns the insulation in the vicinity of the fault. However, if the open is not apparent by visual inspection, a series of tests must be made to determine the exact location of the fault.

FIG. 6-4 Cutting a defective coil as part of an emergency repair of a shorted stator. (*United States Merchant Marine Academy.*)

A preliminary test should be made to determine the type of winding, wye or delta. To do this, megohmmeter readings should be taken between T_1 and T_2, T_2 and T_3, T_3 and T_1. A delta connection will indicate zero for all combinations, whereas a wye connection will indicate zero for one combination only, T_1 and T_3 of Fig. 6-6a. If the stator is delta-con-

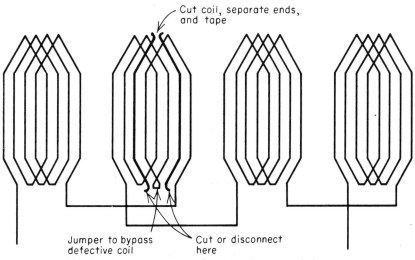

FIG. 6-5 Emergency repair of a shorted or grounded stator.

nected, the connections at *A*, *B*, and *C* must be opened, and the nine wires (including the three line leads) must be carefully marked so that they may be reconnected in the same manner after the repair is made. The six stator wires should then be tested with a megohmmeter for continuity; the two wires that indicate infinity, when tested with each of the others, connect to the phase that contains the open. The actual location of the open may be determined by testing the faulty phase with a megohmmeter and needle-point probes, as illustrated in Fig. 6-7. The needle points easily penetrate the insulation. One probe should be connected to one end of the phase, and the other probe should be shifted from junction to junction until the megohmmeter indicates a high reading. As indicated in Fig. 6-7, the fault is between adjacent junctions that have zero and high readings, respectively, on the megohmmeter.

If the stator is wye-connected, the open is in the phase that indicates infinity when megohmmeter-tested with the other two phases. The actual location of the open may be determined by testing the faulty phase with a megohmmeter and needle-point probes as illustrated in Fig. 6-7.

Multicircuit stators. Large machines often have the coils of each phase connected in series-parallel arrangements, as illustrated in Fig. 6-8. Such machines will operate with an open in one coil but will not deliver rated power. The fault may be determined by opening the connections at *A*, *B*, and *C* and following the procedure outlined for single-circuit machines. The wires should be marked so that proper reconnection is possible after the repair is made.

6-6 Emergency Repair of an Open Stator

An emergency repair of an open stator may be made by following the repair procedure outlined for a shorted coil.

6-7 Squirrel-cage-rotor Troubles

Broken rotor bars cause a squirrel-cage motor to operate unsatisfactorily. The machine is noisy, has a low starting torque, and does not come up to speed, even though rated voltage, rated frequency, and rated load are applied. Open rotor bars that are not apparent by visual inspection may be detected by applying 25 percent of rated current to only two phases, and then, the line current measuring as the rotor is turned slowly by hand. This is illustrated in Fig. 6-9. A broken rotor bar causes variations in the ammeter readings as the rotor is turned. The number of variations in one revolution is equal to the number of poles in the stator. See Sec. 5-3 for the determination of poles for a given speed and frequency. An

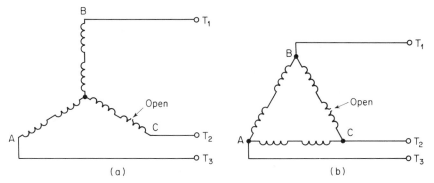

FIG. 6-6 Single-circuit three-phase stator connections, showing open coils. (a) Wye connection; (b) delta connection.

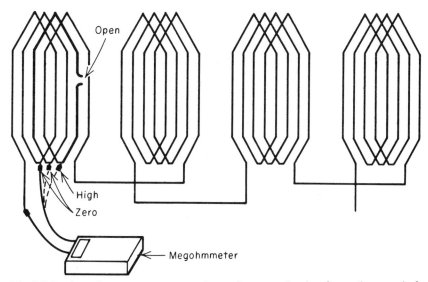

FIG. 6-7 Tracking down an open in one phase of a stator that has four coils per pole for each phase.

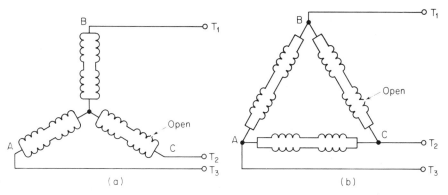

FIG. 6-8 Two-circuit three-phase stator connections, showing open coils. (a) Wye connection; (b) delta connection.

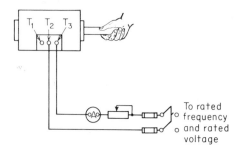

To rated
frequency
and rated
voltage

FIG. 6-9 Ammeter test for broken rotor bars.

emergency repair of bar-type rotors may be made by rebrazing the defective bars to the end ring. Cast-aluminum rotors can seldom be repaired.

6-8 Wound-rotor Troubles

When a wound-rotor motor fails to start or operates at reduced speed, its rheostat should be checked for a possible open, and the brushes should be checked for freedom of movement and good contact with the slip rings. If the rheostat is at fault, short-circuiting the three slip rings will cause the motor to run. If shorting the slip rings does not start the machine, there is an open in either the stator or the rotor circuit. Note that the starting torque with all resistance out is relatively low. Hence, the motor should have no load or be only lightly loaded when making this test.

6-9 Single-phase Induction-motor Troubles

When a single-phase induction motor fails to start, the trouble may be in the starting winding, running winding, centrifugal mechanism, centrifugal switch, or capacitor if one is used.

To diagnose the trouble, the rotor should be spun rapidly by hand, and then the line switch should be closed. If the motor accelerates to rated speed and then continues to run, the trouble is in the starting circuit. However, if the motor accelerates to about one-half or three-fourths speed and then slows down to a very slow speed, whence it starts to accelerate again, etc., the trouble is in the running winding.

If the trouble is in the starting circuit, the machine should be disassembled, and the starting winding tested for continuity. The centrifugal switch should be checked for burned or worn contacts, and the capacitor should be tested if one is used (see Sec. 1-18). Substituting a replacement capacitor that is known to be good will resolve any doubts as to whether or not the capacitor in question is at fault.

In some installations, such as small refrigerator units or water coolers, the switch that cuts the starting winding in or out may be external to the motor and operated by a coil connected in series with the machine. The high starting current drawn by the running winding actuates the relay, causing it to close and excite the starting winding. When the machine attains its rated speed, the countervoltage produced by motor action reduces the current in the windings. The relay coil releases the starting-circuit contacts, and the starting circuit opens.

6-10 Locating Shorted Field Coils

To check for short circuits in the field coils of a salient-pole alternator or synchronous motor, a 120- or 240-volt 60-hertz voltage should be applied to the collector rings, and the voltage drop measured across each field coil, as illustrated in Fig. 6-10. Because each field coil has the same number of turns, the voltage across each should be the same.

• *Example 6-1.* For the 10-pole field shown in Fig. 6-10, if the voltage applied to the collector rings is 240 volts, the average voltage across each coil should be

$$\frac{240}{10} = 24 \text{ volts}$$

all

If all coil voltages are equal, it can be assumed that (la) 10 coils are free of shorted turns. A shorted turn in one coil causes its voltage and that of the adjacent coils to be noticeably lower than the average value. If the voltage drop across any coil is 10 percent less than the average, one or more turns may be shorted.

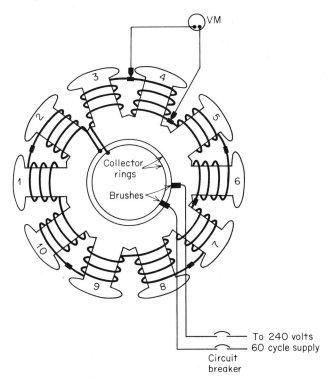

FIG. 6-10 Locating a shorted field coil.

In Example 6-1, the average voltage, determined by dividing the impressed voltage by the number of field poles, is 24 volts. Hence a voltage of less than 21.6 volts (24 − 2.4) across a field coil is indicative of shorted turns. A tabulation of representative voltage drops in Table 6-1, for a machine with a shorted field coil, indicates that coil 4 has shorted turns.

Table 6-1
Voltage Drops

Coil	Voltage
1	25
2	25
3	24
4	17*
5	24
6	25
7	25
8	25
9	25
10	25

* Shorted turn or turns.

Faulty field coils should be removed, and the coils unwound to note the general condition of the insulation and to locate the fault. If the insulation is very brittle and cracked, the other coils may be similarly deteriorated and, though not shorted, should be replaced. However, if it is an old machine that is not a critical component in plant operation, it may be economically justifiable to let it run as is, provided that overheating and vibration do not occur.

Making this test with direct current does not produce any significant results; one shorted turn has very little effect on the voltage drop of a coil. If each coil in Example 6-1 has 200 turns, one shorted turn will result in a resistance change of only one two-hundredth of the coil resistance. This very small change in resistance will have no noticeable effect on the voltage measured across the coil. However, with alternating current applied, the transformer action of the shorted turn appreciably lowers the impedance of the affected coil, and the adjacent coils, resulting in noticeably lower voltages (see Sec. 7-2).

6-11 Locating an Open Field Coil

An open coil may be located by connecting the field circuit to its rated voltage or less and testing with a voltmeter from one line terminal to successive coil leads, as shown in Fig. 6-11. The voltmeter will indicate zero until it passes the coil with the open; then full line voltage will be indicated. The test voltage may be either ac or dc. An open coil may also be located with a megohmmeter test across each one; a closed coil

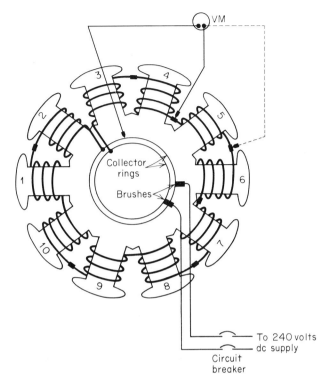

FIG. 6-11 Locating an open field coil.

will indicate zero, and an open coil will indicate infinity. An emergency repair of an open field coil can be made by unwinding the coil to expose the break and then soldering the broken ends together.

6-12 Locating a Grounded Field Coil

A grounded coil may be located by connecting the field circuit to its rated voltage or less and measuring the voltage drop between each coil and ground, as shown in Fig. 6-12. A grounded coil will indicate very low or no voltage when tested from its terminals to ground.

A grounded coil may be located with a megohmmeter, but each coil must be disconnected from the others and tested to ground individually. A grounded coil will indicate zero. The megohmmeter test is recommended for shipboard use, because the voltage test requires the field frame to be insulated from the ship's hull or floor plates, and this is often difficult to do.

A ground caused by dirt and carbon dust can generally be cleared by cleaning and revarnishing. However, a coil that is grounded by chafing of insulation against the pole iron should be replaced if the copper shows any signs of having been worn away.

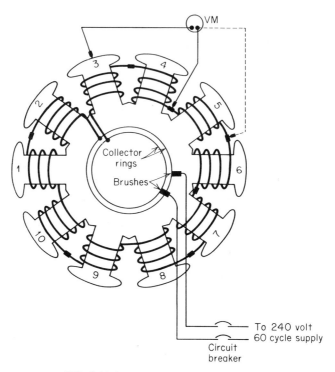

FIG. 6-12 Locating a grounded field coil.

6-13 Testing for a Reversed-field Pole

A reversed-field pole in a generator or motor may be detected by testing with a magnetic compass. To make the test, the field should be connected to a source of direct current no higher than the rated voltage, and the polarity of each pole should be tested with a compass. The poles should test alternately north and south around the frame. The compass should not be held too close to the field, or its own polarity may be reversed. An alternate method for testing magnetic polarity makes use of two identical short steel bolts bridged across the gap between adjacent field poles; correct polarity will cause a strong attraction between the bolts, but a reversed coil will cause the ends to repel. Because of the very strong magnetic field at the pole faces, testing with a single bolt may give misleading results. Two bolts should be used. See Fig. 11-6.

6-14 Trouble-shooting Chart

The trouble-shooting chart in Table 6-2 lists the most common troubles, their possible causes, and the pertinent book section.

Table 6-2   Trouble-shooting Chart

Trouble	Possible cause	Section
Synchronous motors		
Fails to start	1. Blown fuses	14-3
	2. Open in one phase	6-5
	3. Overload	
	4. Low voltage	
Runs hot	1. Overload	
	2. Clogged ventilating ducts	2-18
	3. Shorted stator coils	6-2, 6-4
	4. Open stator coils	6-5
	5. High voltage	
	6. Grounded stator	6-3, 6-4
	7. Field current set too low	5-9
	8. Field current set too high	5-9
	9. Uneven air gap	12-4
	10. Rotor rubbing on stator	
Runs fast	1. High frequency	4-10
Runs slow	1. Low frequency	4-10
Pulls out of synchronism	1. Overload	5-10
	2. Open in field coils	6-11
	3. No exciter voltage	8-15
	4. Open in field rheostat	
	5. Rheostat resistance set too high	5-9
Will not synchronize	1. Field current set too low	5-9
	2. Open in field coils	6-11
	3. No exciter voltage	8-15
	4. Open in field rheostat	
Vibrates severely	1. Out of synchronism	5-9
	2. Open armature coil	6-5
	3. Open phase	6-5
	4. Misaligned	12-6
Three-phase squirrel-cage induction motors		
Fails to start	1. Blown fuses	14-3
	2. Open in one phase	6-5
	3. Overload	
Runs hot	1. Overload	
	2. Clogged ventilating ducts	2-18
	3. Shorted stator coils	6-2, 6-4

Table 6-2   Trouble-shooting Chart (continued)

Trouble	Possible cause	Section
Three-phase squirrel-cage induction motors (continued)		
Runs hot	4. Low voltage	4-10, 5-6
	5. High voltage	4-10, 5-6
	6. Low frequency	4-10, 5-6
	7. Open stator coils	6-5
	8. One phase open	6-5
	9. Grounded stator	6-3, 6-4
	10. Uneven air gap	12-4
	11. Rotor rubbing on stator	
Runs slow	1. Overload	
	2. Low voltage	4-10, 5-6
	3. Low frequency	4-10, 5-6
	4. Broken rotor bars	6-7
	5. Shorted stator coils	6-2, 6-4
	6. Open stator coils	6-5
	7. One phase open	6-5
Wound-rotor induction motors		
Fails to start	1. Blown fuses	14-3
	2. Open in one phase of stator	6-5
	3. Overload	
	4. Open in rheostat	
	5. Inadequate brush tension	10-13
	6. Brushes do not touch collector rings	10-6
	7. Open in rotor circuit	6-8
Runs hot	1 Overload	
	2. Clogged ventilating ducts	2-18
	3. Low voltage	4-10, 5-6
	4. High voltage	4-10, 5-6
	5. Uneven air gap	12-4
	6. Shorted stator coils	6-2, 6-4
	7. Open stator coils	6-5
	8. One phase open	6-5
	9. Low frequency	4-10, 5-6
	10. Grounded stator	6-3, 6-4
	11. Rotor rubbing on stator	
Runs slow	1. Overload	
	2. Low voltage	4-10, 5-6
	3. Low frequency	4-10, 5-6
	4. Too much resistance in rheostat	5-5
	5. Shorted stator coils	6-2, 6-4

Table 6-2   Trouble-shooting Chart (continued)

Trouble	Possible cause	Section
Wound-rotor induction motors (*continued*)		
Runs Slow	6. Open stator coils	6-5
	7. One phase open	6-5
	8. Open in rotor circuit	6-8
Single-phase induction motors		
Fails to start	1. Blown fuses	14-3
	2. Defective starting mechanism	5-8, 6-9
	3. Open in auxiliary winding	5-8, 6-9
	4. Open in main winding	6-9
	5. Shorted capacitor	1-18, 6-9
	6. Open capacitor	1-18, 6-9
	7. Overload	
Runs hot	1. Overload	
	2. Starting mechanism does not open	6-9
	3. Low voltage	5-6
	4. High voltage	5-6
	5. Clogged ventilating ducts	2-18
	6. Shorted stator coils	6-2, 6-4
	7. Worn bearings	12-4
	8. Low frequency	5-6
	9. Rotor rubbing on stator	
Runs slow	1. Overload	
	2. Low voltage	5-6
	3. Low frequency	5-6
	4. Broken rotor bars	6-7
	5. Shorted stator coils	6-2, 6-3

Bibliography

Dudley, A. M., and S. F. Henderson: "Connecting Induction Motors," McGraw-Hill Book Company, New York, 1961.

Graham, K. C.: "Understanding and Servicing Fractional Horsepower Motors," American Technical Society, Chicago, Ill., 1961.

Rosenberg, R.: "Electric Motor Repair," Holt, Rinehart and Winston, Inc., New York, 1960.

Stafford, H. E.: "Troubles of Electrical Equipment," McGraw-Hill Book Company, New York, 1947.

Questions

6-1. What is the first thing that should be checked when a motor fails to start?

6-2. Describe a continuity test applicable to a squirrel-cage induction motor.

6-3. What type of fault (ground, short, or open) or combination of faults is identified by burning, overheating, or discoloration? Explain.

6-4. Where and for what purpose is a phase-balance relay used?

6-5. Explain the principle of growler operation.

6-6. Describe a method for locating a grounded stator coil. Using suitable sketches, state how an emergency repair can be made.

6-7. Describe a test for locating an open stator coil in a single-circuit stator. Using suitable sketches state how an emergency repair can be made.

6-8. How would you locate an open coil in a multicircuit stator?

6-9. What effect do broken rotor bars have on the performance of a squirrel-cage induction motor?

6-10. Describe a test for detecting the presence of broken rotor bars in a squirrel-cage rotor. Can broken rotor bars be repaired? Explain.

6-11. What faults other than blown fuses or voltage failure can prevent a wound-rotor motor from starting?

6-12. Describe a simple test to determine whether or not the starting rheostat for a wound-rotor motor is defective.

6-13. State the possible faults that would prevent a single-phase induction motor from starting.

6-14. When a single-phase induction motor fails to start, what test procedure may be used to diagnose the trouble?

6-15. Explain the operation of a starting relay used with small refrigerator units.

6-16. Describe a method for testing a motor-starting capacitor.

6-17. Describe a test for locating shorted field coils in the rotating-field member of a salient-pole alternator.

6-18. Explain why alternating current is better than direct current when testing for a shorted field coil.

6-19. Describe a test for locating an open field coil; state the procedure for making an emergency repair.

6-20. Describe a test for locating a grounded field coil; state the procedure for making an emergency repair.

6-21. Describe two different test procedures for checking the polarity of the field poles of a synchronous motor.

seven

transformers, regulators, and current- limiting reactors

A transformer is an electromagnetic device, without continuously moving parts, which by electromagnetic induction transfers electric energy from one or more circuits to one or more other circuits at the same frequency, usually with changed values of voltage and current.

7-1 Principle of Transformer Operation

The principle of transformer operation may be explained by reference to Fig. 7-1a, b, and c for different coil and core arrangements; the alternating flux set up by the primary current is shown passing through the window of the secondary coil. The changing flux through the window generates an alternating voltage in the secondary coil. This behavior is similar to that of the ac generator (see Sec. 4-1) except that in the latter case the changing flux through the window of the armature coil is caused by the physical rotation of the magnets. The construction details of a single-phase distribution transformer equipped with a tap changer is shown in Fig. 7-1d. The tap changer, which is manually operated, is used to obtain slightly different voltage ratios, so that adjustments can be made to compensate for differences in system voltages. Because there is a variety of different tap changers in use, each requiring different motions of the operating handle, be sure to read the operating instructions before changing taps.

> **Warning.** Unless specified otherwise by the manufacturer, tap changers must not be operated while the transformer is energized;

operating a tap changer while energized causes a high short-circuit current in the turns connected to the taps being switched. This results in severe arcing and burning, with the possibility of an explosion damaging the transformer and seriously injuring the operator.

The iron core is laminated to reduce eddy currents (circulating currents within the iron core). Low-loss silicon steel is used to minimize hysteresis loss (heat generated by molecular motion as the magnetic domains shift in response to the alternating current).

The magnitude of this magnetically induced voltage is proportional to the number of turns of wire in the coil and the rate of change of flux through the window (Faraday's law). Thus, for the secondary winding,

$$e_S = N_S \left(\frac{d\phi}{dt}\right)_S \qquad \text{volts} \tag{7-1}$$

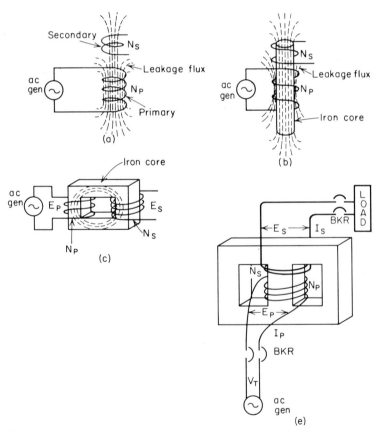

FIG. 7-1 Transformer arrangements and circuits. (a) With no iron core; (b) with iron core; (c) iron core forms a closed loop through the coils; (d) construction details; (e) loaded transformer.

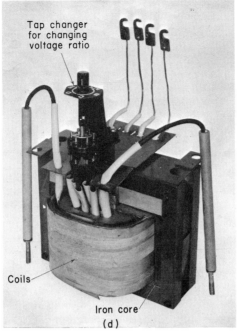

Tap changer
for changing
voltage ratio

Coils

Iron core
(d)

FIG. 7-1 (Continued)

where e_S = generated voltage in secondary coil

N_S = number of turns of wire in secondary coil

$\left(\dfrac{d\phi}{dt}\right)_S$ = rate of change of flux through window of secondary coil

The flux that passes through the primary window but not through the secondary window is called *leakage flux* and does not contribute to the development of secondary voltage.

In a practical transformer (see Fig. 7-1c, d, and e) the iron core forms a complete loop through both coils. This "close coupling" of both coils eliminates almost all leakage, and the flux through the windows of both windings may be assumed to be equal. Under such conditions, the rate of change of flux $(d\phi/dt)$ through both windows is the same, causing an equal voltage to be generated in every turn of both coils. Thus, regardless of the number of turns, the ratio of volts per turn is the same for both coils:

$$\frac{E_S}{N_S} = \frac{E_P}{N_P}$$

or

$$\frac{E_S}{E_P} = \frac{N_S}{N_P} \tag{7-2}$$

where E_S = rms voltage generated in secondary coil
  E_P = rms voltage generated in primary coil
  N_S = number of turns of wire in secondary coil
  N_P = number of turns of wire in primary coil

The voltage E_P' is in effect a counter emf that opposes the driving voltage V_T, limiting the magnitude of the primary current. The voltage E_s is the driving voltage for the load.

7-2 Transformer Behavior under Load

When a load is connected to the secondary of a transformer, the resultant current in the secondary sets up a flux of its own in opposition to the primary flux. This reduces the overall flux through both coils, causing the voltages E_P and E_s to drop momentarily. The decrease in E_P enables the driving voltage V_T to force more current through the primary coil. Thus, connecting a load to the secondary has the effect of reducing the impedance of the primary. The increased primary current reestablishes the flux and provides the energy for the load.

Neglecting the relatively small losses in transformers used for most power and distribution applications, the power input is equal to the power output:

$$E_P I_P pf = E_s I_s pf \tag{7-3}$$

Furthermore, because the power factor pf of the primary may be assumed to equal that of the secondary, Eq. (7-3) reduces to

$$E_P I_P = E_s I_s \tag{7-4}$$

7-3 Autotransformer

A transformer that accomplishes voltage transformation with a single winding is called an autotransformer. This is shown in Fig. 7-2, where part of the total winding is common to both the primary and secondary circuits.

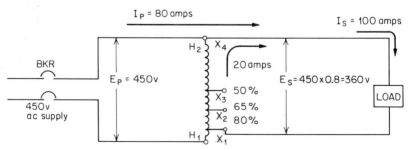

FIG. 7-2 Circuit diagram for an autotransformer, using step-down operation with the load connected to the 80 percent tap.

The voltage and power relationships for the two-winding transformer, as expressed in Eqs. (7-1), (7-2), and (7-4), are directly applicable to the autotransformer. Autotransformers require less copper and are more efficient than two-winding transformers. However, they are not safe for supplying a low voltage from a high-voltage source. If the winding that is common to both primary and secondary is accidentally opened, the full primary voltage will appear across the secondary terminals. However, autotransformers are very useful as booster transformers on distribution systems to compensate for voltage drops along the line. They are also used for reduced-voltage starting of ac motors (see Sec. 13-2). Standard autotransformers, used for starting ac motors of 50 hp and below, are equipped with 65 and 80 percent voltage taps. For motors above 50 hp, the standard taps are 50, 65, and 80 percent.

Power supplied to a load through the medium of an autotransformer is partly conducted and partly transformed. Referring to the circuit in Fig. 7-2, the power supplied to the load (assuming a resistor load) is

$$P = E_S I_S \quad \text{watts}$$
$$P = 360 \times 100 = 3{,}600 \text{ watts}$$

Because the power input must equal the power output (neglecting losses),

$$E_P I_P = E_S I_S$$
$$450 I_P = 3{,}600$$
$$I_P = \frac{3{,}600}{450} = 80 \text{ amp}$$

Although 100 amp is drawn by the load, only 80 amp is conducted from the supply line. The remaining 20 amp is supplied by transformer action. Thus:

Power conducted to load $= 80 \times 360 = 28{,}800$ watts
Power transformed to load $= 20 \times 360 = 7{,}200$ watts
Total power supplied $= 100 \times 360 = 36{,}000$ watts

7-4 Terminal Markings and Transformer Connections

In order to parallel transformers successfully or connect them properly in polyphase arrangements, the polarity of the respective primary and secondary terminals must be known. High-voltage terminals are designated by the letter H, and low-voltage terminals by the letter X, with corresponding subscripts denoting the same instantaneous polarity. The standard terminal markings for a transformer are shown in Fig. 7-3.[1] Terminals H_1 and X_1 have the same instantaneous polarity; likewise H_2 and X_2. Thus

[1] Transformers, Regulators, and Reactors, NEMA TR1-1968.

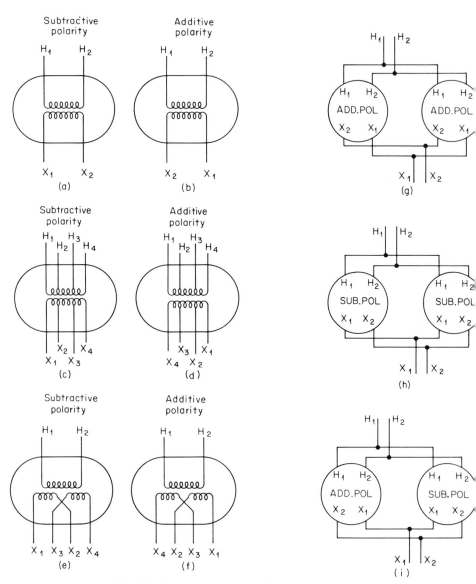

FIG. 7-3 Standard terminal markings for transformers.

at any given instant during most of each half-cycle, if the current is entering H_1, it is leaving X_1 as though the two leads formed a continuous
line.

The lowest and highest subscript numbers mark the respective ends
of the full winding, and the intermediate numbers mark the fractions
of windings or taps. The polarity of any lead with a lower number, with

respect to any lead with a higher number, has the same sign ($+$ or $-$) at any given instant. Referring to Fig. 7-3e and f, terminals X_1, X_2 and X_3 have the same instantaneous polarity with respect to X_4.

The position of the terminals with respect to the high- and low-voltage sides of the transformer affects the voltage stress on the external leads, especially in high-voltage transfomers. In Fig. 7-3a, c, and e, the terminals with the same instantaneous polarity are opposite each other, and if accidental contact of two adjacent terminals occurs, one from each winding, the voltage across the other ends will be the difference between the high and low voltages. Hence this arrangement of terminals is called *subtractive polarity*.

However, if the terminals are arranged as shown in Fig. 7-3b, d, and f, accidental contact between adjacent terminals of opposite windings will result in a voltage across the other ends equal to the sum of the high and low voltages. This arrangement of terminals is called *additive polarity* and has the disadvantage of producing high-voltage stresses between windings. For this reason, and with few exceptions, transformer terminals are arranged for subtractive polarity.

However, regardless of additive or subtractive polarity, when transformers are connected in parallel, all terminals with the same letter and subscript marking must tie together. This is shown in Fig. 7-3g, h, and i.

7-5 Three-phase Connections of Single-phase Transformers

Single-phase transformers can be used to transform three-phase power by connecting them in delta-delta, wye-wye, delta-wye, or wye-delta arrangements, as shown in Fig. 7-4. The line-to-line voltage, line current, and phase-angle displacement between the high voltage HV and low voltage LV lines in each case are:

Delta-delta HV LV	$E_{HV} = E_{LV} \times$ turn ratio $I_{HV} = I_{LV} \div$ turn ratio Phase-angle displacement is 0°
Wye-wye HV LV	$E_{HV} = E_{LV} \times$ turn ratio $I_{HV} = I_{LV} \div$ turn ratio Phase-angle displacement is 0°
Delta-wye HV LV	$E_{HV} = E_{LV} \times$ (turn ratio $\div$ 1.73) $I_{HV} = I_{LV} \times$ (1.73 $\div$ turn ratio) Phase-angle displacement is 30°
Wye-delta HV LV	$E_{HV} = E_{LV} \times$ (turn ratio $\times$ 1.73) $I_{HV} = I_{LV} \div$ (turn ratio $\times$ 1.73) Phase-angle displacement is 30°

The delta-delta bank has the advantage of being able to operate

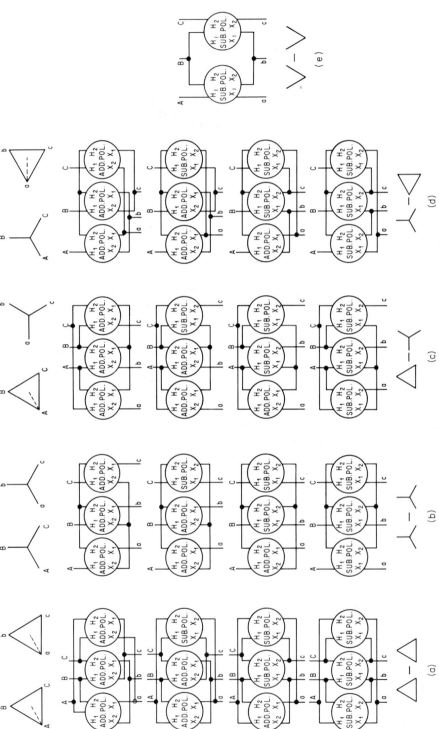

FIG. 7-4 Three-phase connections of single-phase transformers. (a) Delta-delta; (b) wye-wye; (c) delta-wye; (d) wye-delta; (e) open-delta (V connection).

continuously with one of the three transformers disconnected from the circuit. This open-delta connection, also called a V connection, provides a convenient means for inspection, maintenance, testing, and replacing of transformers, one at a time, with only a momentary power interruption. Disconnecting one transformer converts the circuits shown in Fig. 7-4*a* to the circuit shown in Fig. 7-4*e*. Although the two transformers supply the same balanced voltages as does the delta-delta bank, the kva rating of the bank must be de-rated to 58 percent of the delta-delta rating. Thus, if the delta-delta bank is loaded to its rated capacity, the load must be reduced to 58 percent of rated capacity before the bank can be safely reconnected for open-delta operation. Be sure to open the circuit breaker in the primary circuit before changing connections.

The open-delta connection is also used to provide three-phase service where a considerable future increase in load is expected. The increase may be accommodated by adding the third transformer to the bank at a later date.

7-6 Three-phase Connections of Transformers When Terminals Are Unmarked

If the terminals of the transformers are unmarked, it is necessary to test each secondary connection before moving on to the next one. First connect the three high-voltage windings in wye or delta, as the case may be; then connect one secondary terminal of each of any two secondaries together. Excite the high-voltage windings with rated or below-rated voltage. Measure, with a voltmeter, the voltage of one secondary and then the voltage across the open ends of the two transformer secondaries just connected. The voltage will be equal to that of one transformer, or 1.73 times the voltage of one transformer. If a wye-connected secondary is desired, the voltage across the open ends should be equal to 1.73 times the voltage across one transformer. If it is not equal to this value, reverse one of the secondaries just connected. Connect one terminal of the third transformer secondary to the junction point of the other two. Measure the line-to-line voltage across the three remaining open ends. All three line-to-line voltages should equal the value of one transformer secondary times 1.73. If the voltage is not correct, reverse the secondary connections of the third transformer. The transformer secondary is now wye-connected.

If a delta-connected secondary is desired, connect two secondaries in series, so that the voltage across the open ends will be equal to the voltage across one transformer. Connect the third secondary in series with the other two. The voltage across the remaining open ends will be

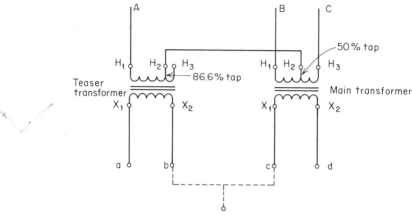

FIG. 7-5 Scott-connected transformers for three-phase to two-phase transformation, and vice versa.

equal to zero,[1] or twice the voltage of one transformer secondary. If the voltage is zero, the remaining ends may be closed; if it is equal to double the voltage of one secondary, reverse the third transformer secondary. With zero volts across the open ends, the delta may be closed.

When making any of these tests, use a voltmeter that has a range equal to double the expected voltage of the high-voltage winding.

7-7 Scott Connection

A Scott connection of transformers is the most common method used to transform three-phase to two-phase, and vice versa. Two transformers are required, as shown in Fig. 7-5. The *main transformer* has a 50 percent tap, and the *teaser transformer* an 86.6 percent tap. One terminal of each transformer on the two-phase side may be joined together, as shown with a broken line, to provide a two-phase three-wire system.

7-8 Transformer Types of Voltage Regulators

Transformer types of voltage regulators are used on feeders between the substation and the load to maintain the voltage at the load within acceptable limits. The adjustable range of the voltage regulator is generally from 10 percent above to 10 percent below the normal system voltage. The voltage regulators are basically autotransformers that use either tap changing or adjustment of the relative positions of the windings to control

---

[1] If the primary is wye-connected and the neutral is not connected to the generator, a third harmonic voltage will prevent a zero indication. However, it will be less than that of a single transformer secondary.

the voltage of the regulated circuit. The voltage regulator raises or lowers the voltage on distribution systems to compensate for changes in the voltage drops along the line as load on the system is increased or decreased.

Step voltage regulators. Transformers that are designed to provide small changes in the turns ratio of a transformer automatically, to compensate for changes in system voltage, are called step voltage regulators. The motor-operated tap changer, also called the ratio adjuster, is controlled

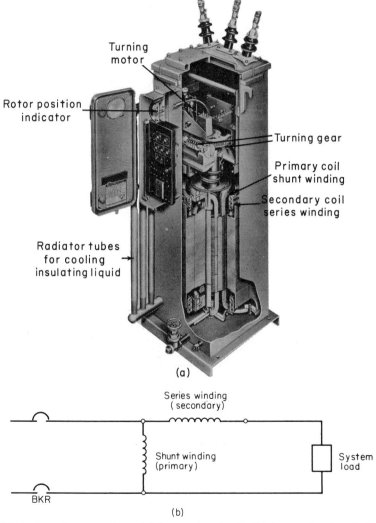

(a)

(b)

FIG. 7-6 Induction voltage regulator. (a) Construction details; (b) shunt and series winding connections. (*Westinghouse Electric Corp.*)

by a contact-making voltmeter and reversing relays. Because high-current arcing always occurs when the ratio adjuster shifts from tap to tap, the tap changing contacts should be examined whenever the transformer is deenergized and opened for inspection; the tap changer should be rotated through all positions to make sure that the contacts are in good condition and that the spring tension is adequate.

Induction voltage regulator. An induction voltage regulator is shown in Fig. 7-6. The primary coil (rotating member) is connected to the incoming line, and the series coil (stationary member) is connected in series with the system load. The series coil adds its voltage to, or subtracts it from, the line voltage. A small motor is used to change the angular position of the shunt coil with respect to the series coil. The motor is controlled by a contact-making voltmeter and reversing relays. Changing the angular position of the rotor changes the coupling between it and the stator, resulting in a different voltage ratio.

7-9 Current-limiting Reactor

A current-limiting reactor, shown in Fig. 7-7, consists of a coil of insulated cable connected in series with the equipment or circuit it is designed to protect. Iron cores are not used, because saturation of the iron at high currents would cause it to lose its effectiveness. The purpose of the reactor is to place an upper limit on the available short-circuit current that can occur under fault conditions.[1] It accomplishes this by contributing

[1] If the available short-circuit current is greater than the interrupting capacity of the breaker, the breaker can explode (see Sec. 14-5).

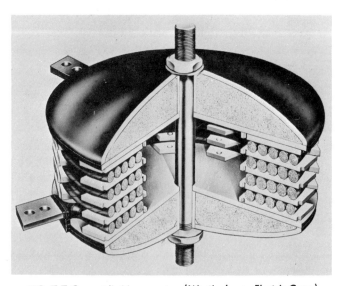

FIG. 7-7 Current-limiting reactor. (*Westinghouse Electric Corp.*)

additional inductive reactance to the circuit. The layers of cable are rigidly clamped between reinforced concrete disks to prevent the adjacent turns from moving when carrying short-circuit current. The severe mechanical forces produced by the intense magnetic field can damage the coil if it is not adequately braced.

7-10 Instrument Transformers

Instrument transformers are used to transform high currents and voltages to low values for instrumentation and control. Instrument potential transformers are used in voltage measurements, and instrument current transformers are used in current measurements. Both types also serve to insulate the low-voltage instruments from the high-voltage system. The two types of instrument transformers and their methods of connection to a circuit are shown in Fig. 7-8. The instrument or relay connected to the secondary is called the *burden*.

In those applications involving currents in excess of 2,000 amp, the return cable L_1 (see Fig. 7-8e) should be kept some distance away from the current transformer; this will prevent the relatively large magnetic flux about L_1 from entering the iron core, causing an error in the instrument reading.

The relative instantaneous polarities of the transformer terminals are indicated with $\pm$ markings or H_1 and X_1 markings, and are generally supplemented with a bright paint marker. For instrument transformers the letter H_1 always denotes the primary polarity lead or terminal and X_1 the secondary polarity lead or terminal. At the instant that current enters the polarity terminal of the primary coil, current leaves the polarity terminal of the secondary coil, as though the two leads formed a continuous line.

A polarity test should always be made on rebuilt or repaired instrument transformers to make sure that the polarity marked terminals are indeed correct. Extensive damage can be caused to ac generators and distribution systems that utilize protective relaying if the instrument transformers have incorrect polarity.

To make a polarity check of a potential transformer, deenergize the system, and disconnect the primary and secondary terminals; then connect a dc voltmeter to one winding of the transformer and a battery and pushbutton to the other winding, as shown in Fig. 7-8c. Make a momentary contact by depressing and releasing the pushbutton quickly. If the polarity is correct, the deflection upon pushing the button will be upscale and upon releasing the button will be downscale. To prevent permanently magnetizing the iron, no more than one or two No. 6 dry cells (for a total of 3 volts) should be used. A polarity check of a current transformer may be made in a similar manner; the test circuit is shown in Fig. 7-8f.

The secondary winding of a current transformer must always be connected to a burden, or it must be short-circuited at the terminals, as shown in Fig. 7-8e. Opening the secondary circuit while current is in the primary not only causes dangerously high voltages at the secondary terminals but may also permanently magnetize the transformer iron, introducing errors in the transformer ratio. Before servicing instruments and relays connected to the secondary of a current transformer, the secondary must be shorted.

> **Warning.** Although short-circuiting a current transformer does not harm it, do not short-circuit a potential transformer, because it will burn out.

7-11 Maintenance of Transformers

Although transformers require less attention than most other electrical apparatus, some routine maintenance is required. Figure 7-9 shows damage due to neglected maintenance; deterioration of the insulation resulted in a severe short circuit, rendering the transformer a total loss.

To help reduce accidental outages, monthly or bimonthly inspections

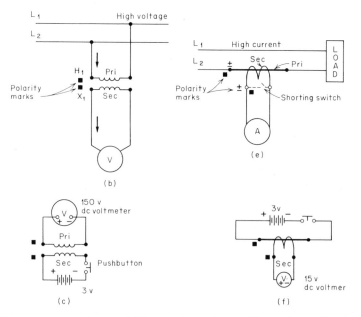

FIG. 7-8 Instrument transformers. (a) Potential transformer (*Westinghouse Electric Corp.*); (b) potential transformer connections; (c) potential transformer polarity test; (d) current transformer (*General Electric Co.*); (e) current transformer connections; (f) current transformer polarity check.

(a)

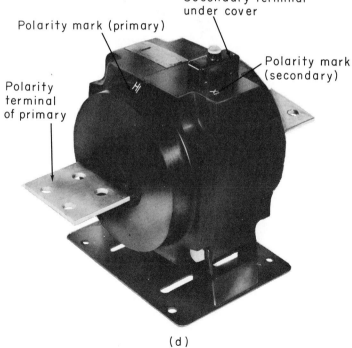

(d)

FIG. 7-8 (Continued)

FIG. 7-9 Damage to transformer caused by lack of maintenance. (*Mutual Boiler and Machinery Insurance Co.*)

should be made. These inspections should include checks of temperature, liquid level, and leaks in liquid-filled transformers. The inspector should check for dirt accumulations on high-voltage bushings (see Sec. 15-7), rusting, discolored connections, which indicate excessive heat, and the accumulation of refuse on transformer lids by birds, squirrels, etc. Dirty bushings or the accumulation of conducting refuse on transformer lids can cause a flashover at the high-voltage terminals.

Maintenance of dry-type transformers. Dry-type transformers are designed for installation in dry locations.[1] Hence care should be exercised to prevent the entrance of water by splashing from open windows or from leaky or broken steam lines. Dry-type transformers should be located at least 12 in. away from walls, so that the free circulation of air around and through the transformer is not impaired. They should be sheltered from dust and chemical fumes. Dust settling on the windings, core, and enclosing

[1] This section is based on the Proposed Guide for Operation and Maintenance of Dry Type Transformers with Class B Insulation, USAS C57.94-1958.

case reduces heat dissipation and results in overheating. Periodic inspections of the windings and core should be made at least once a year to detect incipient failures. To do this, the transformer should be disconnected from the line (and load if in parallel with other transformers), the covers removed, and inspection made for accumulations of dirt, corrosion, loose connections, discoloration caused by excessive heat, and carbonized paths caused by electron creepage over the insulation surfaces. The windings may be cleaned with a vacuum cleaner or dry compressed air not exceeding 25 psi. This will clear the air passages and prevent overheating from such causes. All insulating surfaces should be wiped clean with a dry cloth. The use of solvents should be avoided, because they may have a deteriorating effect on the insulation. Strip heaters should be used during periods of prolonged shutdown to avoid condensation of moisture on the windings.

If insulation-resistance tests indicate that the windings have absorbed moisture or if the windings have been subject to unusually damp conditions, they should be dried by external heat, internal heat, or a combination of both. However, drying with internal heat is not advisable if the insulation resistance is less than 50,000 ohms when cold (20°C or 68°F). The preferred method, with external heat, may be accomplished by directing heated air into the bottom ventilating ducts of the transformer housing or by placing the transformer in a baking oven. The temperature of the air used for drying must not exceed 110°C (see Sec. 2-20).

Drying with internal heat is a slower process. It is accomplished by short-circuiting one winding and applying reduced voltage to the other to obtain approximately rated transformer current. Due care must be taken to avoid contact with dangerously high voltages. The winding temperature must not be allowed to exceed 100°C as measured by spirit thermometers placed in the ducts between the windings. Mercury thermometers must not be used, because induced currents in the mercury will result in an erroneous indication. A combination of external and internal heat provides the quickest method for drying and also reduces the short-circuit current. A transformer must be under constant surveillance during the drying process, so that emergency action may be taken in the event of fire. A CO_2 extinguisher should be available for immediate use.

Maintenance of liquid-filled transformers. Liquid-filled transformers[1] require more attention than the dry type. The liquid used for the insulating medium is either mineral oil or an inert synthetic liquid manufactured under trade names such as Inerteen, Pyranol, and Chlorextol. These liquids

[1] Guide for Installation and Maintenance of Oil-immersed Transformers, USAS C57.93-1958.

should be checked at least once a year for the presence of moisture and sludge. The accumulation of sludge on transformer coils and in cooling ducts reduces the heat-transfer capability, causing higher operating temperatures. Moisture may be detected by testing samples of liquid at high voltage to determine its dielectric strength. The test requires special equipment and is described in Sec. 2-16. If moisture is indicated, all gaskets and bushings should be checked to determine the source of entry. Sludging of the liquid is caused by oxidation through exposure to air. If badly sludged, the liquid should be drained, and the inside of the tank and the winding should be hosed down with clean liquid from a filter press. The old liquid should then be returned to the tank through the filter (see Sec. 2-17 for instructions on filtering and drying insulating liquids).

Liquid-filled transformers should be inspected under-the-cover at least once every 10 years; the transformer should be deenergized, the lid removed, and enough of the insulating liquid should be drained to expose the top of the core and coils. To prevent the absorption of moisture, do not leave the coils exposed to the air any more than is necessary to make the inspection.

Liquid-filled transformers are either self-cooled or water-cooled. The self-cooled type depends on the free circulation of air about the tank and cooling fins. The water-cooled type uses cooling coils through which water circulates. Accumulations of scale in the cooling coils reduce the flow of cooling water and result in overheating. To remove the scale, drain the water by siphoning or blowing with compressed air, and then fill the tubes with a muriatic acid solution. Recommended strength, 1.10 specific gravity, may be obtained by mixing equal parts of commercially pure muriatic acid and water. The solution should be allowed to stand for about 1 hr and should then be flushed out with clean water. While this operation is performed, the inlet and outlet pipes of the cooling coil should be disconnected from the water system and temporarily piped away from the transformer. This reduces the possibility that any acid or water will enter the transformer. The pipes should not be capped when filled with the acid solution, because the chemical reaction will build up excessive pressure.

Leaks at welded joints may often be stopped by peening over with a drift and a ball-peen hammer or by soldering with a hard solder such as Iron Fix. Large leaks generally require welding, which should be done without removing the liquid. If the transformer is oil-filled, draining the oil will leave an explosive mixture of oil vapor and air, which may be ignited by welding. Leaks in cast iron may be repaired by drilling, tapping, and inserting a pipe plug. Leaks at gaskets may be stopped by pulling up on the bolts or by installing new cork gaskets. Liquid-filled transformers that have been flooded or have otherwise taken in water should be dried by following the manufacturer's instructions. However, if circumstances require that the drying procedure be started immediately, drain the liquid into another container, and follow the procedure outlined for the dry-type transformer. The liquid should be dried with a filter press and tested for dielectric strength before it is pumped back into the transformer tank.

7-12 Testing Transformer Windings

A Megger test of the transformer windings to ground will determine the relative conditions of the insulation. For testing and evaluating insulation, see Secs. 2-6 to 2-13. However, the magnitude of the exciting current will indicate whether or not the winding is shorted. To determine this, disconnect the secondary from the load and measure the primary current at rated voltage and frequency. If the reading is higher than it should be for the no-load condition, the transformer is shorted. If a dead short is suspected, use a heavy-duty rheostat in series with the ammeter, and gradually cut out the resistance; without a current-limiting device, the ammeter may be damaged. The indication will be the same whether the short is in the primary or secondary windings. Hence the actual determination of the defective winding must be done with a low-resistance ohmmeter, such as a Kelvin double bridge, or by the voltmeter-ammeter method using direct current. The resistance of each winding should be measured and compared with the manufacturer's values for the dc resistance. A shorted transformer should be reinsulated or rewound.

Bibliography

Gibbs, G.: "Transformer Principles and Practice," McGraw-Hill Book Company, New York, 1950.

Hubert, C. I.: "Operational Electricity," John Wiley & Sons, Inc., New York, 1961.

Questions

7-1. What is the function of a transformer? How is it constructed? Why is the core laminated?

7-2. Using suitable sketches, explain the principle of transformer operation. What is leakage flux?

7-3. Why are some transformers provided with tap changers? What precautions should be observed when operating a tap changer?

7-4. Explain why increasing the load on the secondary of a transformer causes the primary current to increase.

7-5. What is an autotransformer? State several applications of autotransformers.

7-6. Differentiate between subtractive and additive polarity as it applies to transformers.

7-7. What maintenance advantage does a delta-delta connection of single-phase transformers have over other three-phase arrangements? Explain.

7-8. Sketch the circuits for wye-wye, wye-delta, delta-wye, delta-delta, and open-delta connections of single-phase transformers.

7-9. State the procedure for determining the correct connections for three single-phase transformers with unmarked terminals that are to be connected in wye-delta.

7-10. Will a delta-delta bank of single-phase transformers continue to supply three-phase power when an open occurs in one transformer secondary?

7-11. State the procedure for determining the correct connections for three single-phase transformers with unmarked terminals that are to be connected in delta-delta.

7-12. Sketch the circuit for a Scott connection of transformers, and indicate the taps used. What is the purpose of this type of transformer arrangement?

7-13. Differentiate between a step-voltage regulator and an induction-voltage regulator. Where are they used?

7-14. What type of apparatus is connected in series with a power line to limit the available short-circuit current? How is it constructed?

7-15. Sketch a circuit showing the correct connections for current and potential transformers.

7-16. Describe a test that may be used to check the polarity of a current transformer and a potential transformer.

7-17. Why should the secondary terminals of a current transformer always be closed, either by a burden or a short circuit? What will happen if the secondary terminals of a potential transformer are short-circuited?

7-18. What should an electrical inspector look for when making an external check of transformers?

7-19. What care, if any, is required by the dry-type transformer?

7-20. State the correct procedure for an under-the-cover inspection of a liquid-filled transformer.

7-21. A transformer that has been idle for 12 months is to be placed into service. The insulation is damp. Describe a method for removing the accumulated moisture using external heat. What is the maximum allowable temperature?

7-22. If equipment for the external heating of transformers is not avail-

able, state the procedure and precautions to be followed when drying
with internal heat.

7-23. Why are spirit thermometers preferred over mercury thermometers when measuring the winding temperature of an energized
transformer?

7-24. What types of insulating liquids are used in liquid-filled transformers?

7-25. What causes sludging of insulating liquids, and what can be done
about it?

7-26. What method should be used to remove the scale from the inside
surface of the cooling coils of a water-cooled transformer?

7-27. How may small leaks at the welded joints of a liquid-filled transformer be repaired?

7-28. Describe the procedure used to determine a short in the primary
or secondary windings of a transformer.

Problems

7-1. A 440-volt supply is connected to the 880-turn primary winding
of a transformer. (a) What is the voltage across the 220-turn secondary? (b) What are the volts per turn of each winding?

7-2. A 120/12-volt bell-ringing transformer has 400 turns of wire in the
primary winding. How many turns are in secondary?

7-3. A 4,160/460-volt transformer supplies electric energy from a 4,160-
volt line to a load that draws 200 amp at 460 volts. (a) Sketch the
circuit; (b) determine the primary current.

7-4. A 450/110-volt transformer is operating at rated voltage. A clip-on
ammeter indicates a primary current of 25 amp. (a) Sketch the
circuit, and determine the secondary current. (b) What is the ratio
of transformation?

7-5. An autotransformer with 80 and 65 percent taps is connected to
a 230-volt 60-hertz supply voltage. Sketch the circuit, and determine
the two tap voltages.

7-6. An autotransformer with a 20 percent tap supplies 50 amp to a
resistor load from a 450-volt supply line. Sketch the circuit, and
determine (a) the secondary voltage (tap voltage), (b) the primary
current, (c) the power supplied to the transformer (neglect losses),
(d) the power conducted, and (e) the power transformed.

7-7. A wye-delta transformer bank reduces a 2,300-volt supply to 450
volts for use at a load center. Sketch the circuit, and determine (a)
the turns ratio of the transformers and (b) the line current on the
high-voltage side if the line current to the load is 300 amp.

7-8. A delta-delta transformer bank reduces a 460-volt supply to 115 volts for lighting circuits. Sketch the circuit, and determine (*a*) the turns ratio of the transformers, (*b*) the line current on the high-voltage side if the line current to the load is 100 amp, and (*c*) the kva delivered to the load.

7-9. Refer to the delta-delta bank in Prob. 7-8. (*a*) If one transformer (primary and secondary) is disconnected, what effect will it have on the line currents and line voltages? (*b*) If the kva delivered to the load is the rated kva of the bank, what is the maximum safe loading of the bank when connected open-delta? Sketch the circuit.

eight

dc generator

The conventional dc generator is basically an ac machine that comes equipped with a commutator to convert the alternating voltage generated within the armature to a direct voltage for use in the load circuit.

8-1 Construction

The construction details of a dc generator are shown in Fig. 8-1. Except for the commutator, the armature of a dc generator is similar in construction to that of an ac generator with a stationary field. The armature coils are soldered to the commutator, and carbon brushes riding on the commutator pick off the current to supply the connected loads. Each armature coil spans a circumferential distance equal to the circumference divided by the number of main field poles.

The commutator is composed of alternate copper bars and mica insulation. The armature coils are connected to the commutator bars at the risers. A mica shell and mica end cones insulate the copper bars from the steel shell and clamping ring. A string band keeps the mica-cone insulation in place. The side mica is undercut below the surface of the bars to prevent interference with the carbon brushes that ride the brush surface.

The field coils supply the magnetic flux for generator action.

8-2 Generated Voltage

The arrangement of the armature coils and their connection to the commutator bars for a four-pole dc generator are shown in Fig. 8-2. Figure 8-2a shows a typical armature coil and its connections to the commutator.

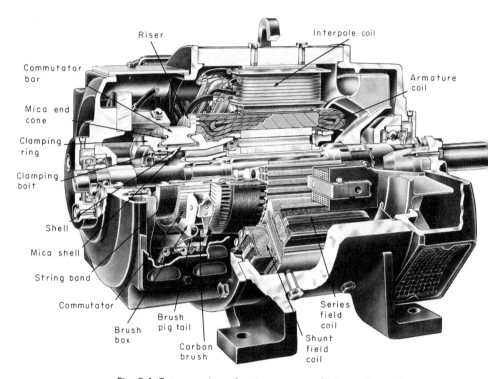

Fig. 8-1 Cutaway view of a dc generator. (*Reliance Electric.*)

Figure 8-2*b* is the commutator-end view of the armature and the field structure, showing the overlapping of coils and their connections to the commutator. As the armature rotates, the net flux passing through the coil windows increases, decreases, and reverses as each coil approaches and then sweeps past the alternate north and south poles. The changing flux through the window of the coil generates a voltage that is proportional to the number of turns of wire in the armature coil and the rate of change of flux through the window (Faraday's law):

$$v = N \left(\frac{d\phi}{dt} \right)$$

where v = generated voltage
N = number of turns of wire in coil
$\left(\frac{d\phi}{dt} \right)$ = rate of change of flux through window

A laid-out view (development) of the field poles and armature coils in Fig. 8-2c, illustrates how the changing flux through the windows can be used to determine the instantaneous direction of the voltage and current in each coil. As the coils shift position with respect to the field poles, moving from positions 1 to 2 to 3 to 4, etc., in turn, the net flux through the window is respectively five lines downward in 1, two lines downward in 2, two lines upward in 3, etc. Knowing how the flux is changing, the direction of the voltage generated in each armature coil and its associated current can be determined from Lenz's law;[1] although the coil current and coil flux will be zero if there is no connected load, for purposes of determining the direction of the generated voltage, a current may be assumed. If the net flux through the window is decreasing, the current in the coil will set up a flux of its own that will delay the decrease, and if increasing will delay the increase.

As the coil moves from position 1 to position 2, the net downward flux through its window is reduced. This causes a voltage to be generated in the coil in a direction that tends to provide additional downward flux, thus delaying the change. The directions of generated voltage, current (electron flow), and flux contribution are indicated by arrows. As the coil moves from position 2 to position 3, the flux through the window changes from two lines downward to two lines upward. To delay this change, the generated voltage has to be in the same direction as before, providing additional downward flux. Similarly, when the coil moves from position 4 to position 5, the net flux through the window changes from five lines upward to two lines upward (three up minus one down); this causes a voltage to be generated in the coil in a direction that tends to provide additional upward flux, thus delaying the change. Note that the directions of generated voltage, current, and flux contribution for the coil in position 5 are opposite that for position 3.

The voltage generated in an armature coil drops to zero and reverses direction each time the center of the coil window passes the center of a main field pole. The alternating current and voltage within the armature coils are made unidirectional in the output or load circuit by means of the commutator and carbon brushes.

Increasing the speed of rotation of the armature increases the rate of change of flux through the window, and a higher voltage is induced. The same increase in voltage may be accomplished without changing the speed by using stronger field magnets; because the armature is rotating at the same speed, a greater flux causes a greater rate of change through the window and hence a higher voltage. Still another means for obtaining a higher voltage without changing the strength of the magnet or changing

[1] The direction of a magnetically induced voltage is always generated in a direction to delay the change in flux that caused it.

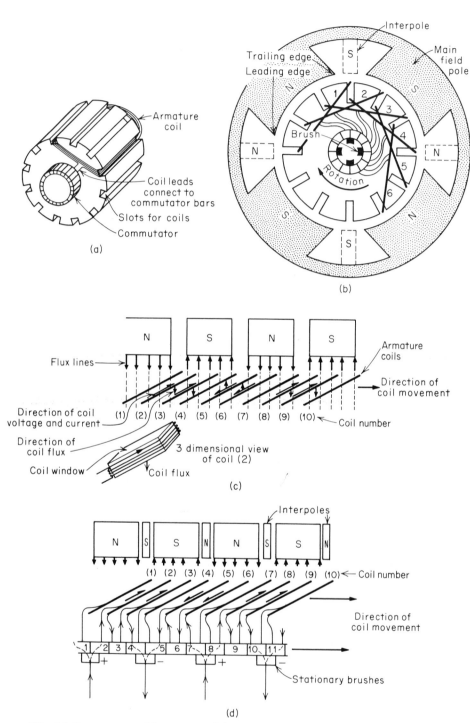

FIG. 8-2 Dc generator. (a) Armature; (b) commutator-end view of armature and field structure; (c) laid-out view of field poles and armature coils; (d) rectifying action of the commutator and brushes.

the speed is to increase the number of turns of wire in the armature coil. Practical generators use dc electromagnets whose strength is adjusted by varying the current to the magnet coils. The small generator frequently used to supply the current to the magnets is called an *exciter generator* or *exciter*. The exciter builds up its own voltage from residual magnetism.

The voltage generated in a dc machine can be conveniently expressed in terms of speed, flux, and a machine constant:

$$E = n\Phi K_G \qquad \text{volts}$$

where n = rpm
Φ = flux per pole (webers)[1]
K_G = generator constant

The generator constant includes such design factors as the number of conductors, arrangement of armature windings, and the number of poles. If the generated voltage for a given speed and flux is known, the voltage for some other speed and flux can be determined by substituting in the following equation:

$$\frac{E_2}{E_1} = \frac{n_2\Phi_2}{n_1\Phi_1}$$

The subscript 1 denotes the original combination of voltage, speed, and flux; the subscript 2 denotes the new combination.

• *Example 8-1.* A dc generator running at 1,800 rpm generates 240 volts. If the speed is reduced to 1,700 rpm and the field current is not changed, determine the new voltage.

• *Solution*

$$\frac{E_2}{240} = \frac{(1,700)\Phi_2}{(1,800)\Phi_1}$$

Because Φ_1 and Φ_2 are equal,

$$E_2 = 240 \frac{(1,700)}{(1,800)} = 227 \text{ volts}$$

The rotating commutator and associated stationary brushes constitute a rotary switch that, when properly installed and adjusted, provides an effective rectifying action. This switching action, also called the *commutation process,* switches the internal alternating current to direct current for the external circuit.

For optimum performance as a reversing switch, these stationary brushes must be positioned with respect to the field poles to make contact

[1] 10^8 maxwells (lines of force) equals one weber.

with the armature coils only during the brief instant that the center of a coil window is lined up with the center of a main field pole. This brush position, shown in Fig. 8-2b, is called the *brush neutral position*. For instructions on the adjustment of the brush neutral position, see Secs. 10-14 to 10-17.

Although the coil is momentarily shorted by the brush during this brief instant, the voltage generated in the coil is zero, and no short-circuit current results. If the brushes are not in the neutral position, they will short-circuit coils that are generating a voltage, and a large short-circuit current will occur, resulting in severe sparking and burning of the brushes and commutator.

Figure 8-2d illustrates the rectifying action of the commutator and brushes. Note that when coil 3 moves to the position occupied by coil 5, the current in coil 3 is reversed. However, the current supplied to the external circuit by coil 3 is in the same direction as before. The process of current reversal, or switching, that takes place when the brush is in contact with a coil via its respective commutator bars is called *commutation*.

8-3 Commutating Poles

When the generator is connected to a load, the direct current that appears in the load circuit is an alternating current within the armature coils. Hence, at the instant that a coil is shorted by a brush, the inductive property of the armature coil delays the reversal of the coil current. For a discussion of the inductive property of a circuit, see Sec. 1-19. Unless this inductive effect is neutralized, severe sparking and burning of brushes and commutator will occur.

Small poles, called commutating poles or interpoles, are placed in the region between the main poles (see Figs. 8-1 and 8-2b and d) to force the current to go to zero and reverse during the brief instant of time it is shorted by the brush. Because the commutating pole occupies a very small space, compared with the main field poles, its flux through the window of the armature coil changes from all-in to all-out during the very short interval of time that the coil is shorted by the brush. Hence the commutating poles generate a neutralizing voltage in the armature coil only while it is being switched by the brushes.

As indicated in Fig. 8-2d, the commutating poles have the same polarity as the respective main field poles that follow it in the direction of rotation. In order to prevent sparking at all loads, from no load to full load, the interpole coils are connected in series with the armature, so that its flux is always in proportion to the armature current.

8-4 Self-excited and Separately Excited Generators

An elementary circuit diagram for a self-excited dc generator is shown in Fig. 8-3a. The terminal markings and polarity correspond to a clockwise shaft rotation as viewed from the end opposite the drive shaft.[1] During normal operation, the main field winding, also called shunt field winding, is self-excited, obtaining its current from the armature.

If the prime mover is not turning, no voltage is generated, and the only magnetic field present is a small amount of residual magnetism in the iron poles of the field structure. The buildup of voltage starts with rotation of the prime mover. Assuming that the prime mover is operating at rated speed, the rotating coils of the armature sweep past the field poles, causing the changing flux (residual) through the coil windows to generate a low voltage.

However, this low voltage forces a small current through the field windings in a direction that causes a small amount of flux to be added to that already present in the form of residual magnetism. This slightly greater flux produces a somewhat higher voltage, which in turn causes a still higher field current and a further increase in flux. The process

[1] Terminal Markings for Electrical Apparatus, USAS C6.1-1956.

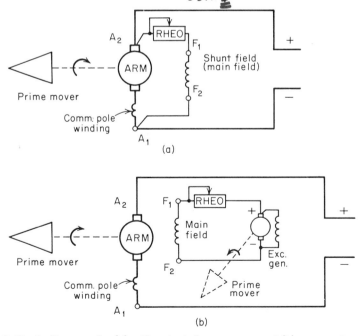

FIG. 8-3 Circuit diagrams for (a) self-excited dc generator and (b) separately excited dc generator.

of voltage buildup continues until there is no further flux increase in the iron field poles. The rheostat is used to adjust the field current, thereby adjusting the amount of flux and changing the output voltage.

The circuit for a separately excited generator is shown in Fig. 8-3*b*. A small self-excited generator, called an *exciter,* is used to supply the current for the main field winding.

8-5 Voltage Regulation

To prevent the dimming of lights and the malfunctioning of motors and other electrical equipment, the output voltage of a dc generator must be held relatively constant as electrical loads are switched on and off. Unfortunately, when load is connected to a generator, two undesirable reactions take place. First, the prime mover slows down, causing a lowering of the generated voltage. Second, the additional current through the resistance of the armature winding causes a further drop in output voltage.

The reduction in prime-mover speed is caused by the development of a countertorque. When a load is connected to the generator terminals, the current in the armature conductors sets up a magnetic field that interacts with the field of the rotating magnets. The mechanical force produced by this action is in opposition to the driving torque of the prime mover. This magnetic action is discussed in Sec. 1-16 and illustrated in Fig. 1-13*e*.

An elementary circuit diagram and its one-line diagram counterpart for a generator and distribution system are shown in Fig. 8-4. The generator feeds a distribution bus, which in turn supplies various connected loads. The millivoltmeter, calibrated in amperes, is used in conjunction with an ammeter shunt to measure high values of direct current. Changes in prime-mover speed caused by the application of load are corrected by an automatic speed governor that uses electronic or mechanical sensors to detect a change in speed and then automatically increases or decreases the energy input to the prime mover. Similarly, an automatic voltage regulator may be used to detect a change in voltage and then automatically adjust the resistance in the field circuit to raise or lower the voltage.

8-6 Compound Generator

Modification of the basic dc generator to obtain certain desirable voltage-current characteristics is generally accomplished by the addition of an auxiliary field winding called the *series field*. The series field coils are wound with heavy copper conductors on the same pole iron as the main field coils and are connected in series with the armature, as shown in Fig. 8-5*a*. Such machines are called *compound generators*. If the current in the series coils is in the same direction as the current in its respective

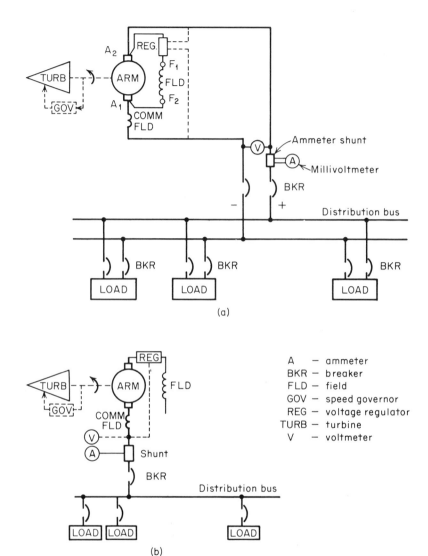

FIG. 8-4 Elementary circuit diagram and one-line counterpart for a dc generator and distribution system.

main field coil (cumulative compound), the flux contribution due to the series field adds to the main field flux; if the current in the series coils is in the opposite direction with respect to the current in the corresponding main field coil (differential compound), the series field flux subtracts from the main field flux.

Figure 8-5b illustrates the voltage-current characteristics of compound and shunt-connected generators for different amounts of compound-

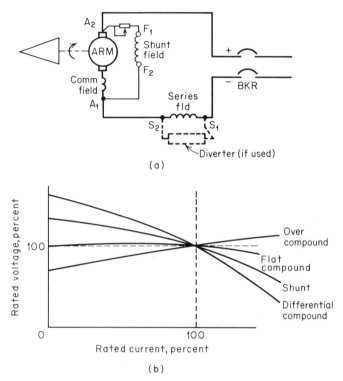

FIG. 8-5 Compound generator. (a) Circuit diagram; (b) voltage-current characteristics.

ing. The amount of compounding may be increased by adding more turns of copper conductor to each series field coil. Thus, with the same armature and shunt field winding, the overcompound machine has more turns of copper conductor in its series field than the flat compound machine, and the flat compound machine has more turns of copper conductor in its series field than the undercompound machine. The compounding effect of the series field can be reduced by connecting a low-resistance path, called a *diverter,* in parallel with the series field to bypass some of the armature current that would otherwise pass through the series field. The connections for the diverter are shown with dotted lines in Fig. 8-5a. Overcompound machines have limited applications, and are used principally for long-distance transmission. Flat compound machines, undercompound machines, and shunt machines with automatic voltage regulators are most commonly used to supply dc distribution systems. The differential compound machine has applications in electric hoist systems where it is desirable that the voltage decrease as the load on the hoist increases.

8-7 Armature Reaction and Compensating Windings

As the load on a dc generator is increased, the flux that was uniformly distributed across the pole face at no load is shifted in the direction of rotation, crowding into the trailing edge of the pole. This effect is called armature reaction. See Fig. 8-2*b* for the location of the trailing edge and leading edge of a pole. At full load there is a high concentration of flux in the trailing edge and a very small concentration in the leading edge. This shifting of the flux back and forth across the pole face as the load is increased and decreased generates a voltage in the armature coils. The magnitude of this voltage depends on the rate of change of the flux as it shifts through the window of the respective armature coil.

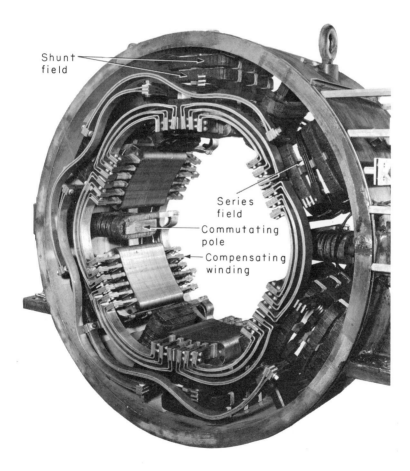

Shunt field

Series field

Commutating pole

Compensating winding

FIG. 8-6 Field structure of a dc generator equipped with a compensating winding. (*Marathon Electric Mfg. Corp.*)

A very rapid shift of flux across the pole face caused by a large and sudden increase in load or a severe short circuit can induce a voltage high enough to cause severe sparking and even a destructive flashover across the entire commutator.

In applications such as rolling mills, electrically propelled vessels,

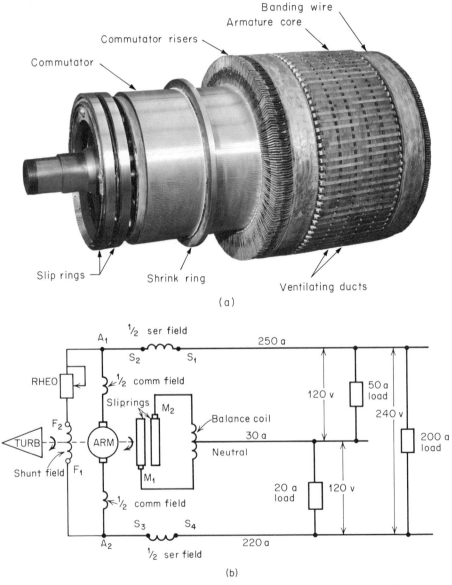

(a)

(b)

FIG. 8-7 Three-wire dc generator. (a) Armature (*Marathon Electric Mfg. Corp.*); (b) circuit diagram.

electric trains, etc., where the generator is subject to rapidly fluctuating loads of large magnitude, including reversing duty, compensating windings are added to the pole faces to prevent the flux from shifting across the pole face. Thus, the compensating windings eliminate armature reaction. Figure 8-6 shows the field structure of a dc generator equipped with a compensating winding, also called a *pole-face winding.* The compensating winding is connected in series with the armature, so that its strength is always proportional to the armature current. The conductors of the compensating winding are placed in slots in the pole face that are parallel to the slots in the armature. As load on the generator is increased, the armature reaction tends to shift the flux in the direction of rotation, and the compensating winding tends to shift it in the opposite direction. The result is a stalemate, and armature reaction is eliminated.

8-8 Three-wire DC Generator

A three-wire dc generator is used to provide two voltages for a three-wire dc distribution system—120 volts for lighting circuits and 240 volts for power.

The construction details of the three-wire generator differ from that of the standard two-wire machine by the addition of two slip rings, as shown in Fig. 8-7a (one, three, or four slip rings may also be used). The slip rings, also called *collector rings,* are connected to appropriate points in the armature winding. Carbon brushes riding on the slip rings connect to an autotransformer, called a *balance coil.* The connection diagram for the three-wire dc generator and some representative loads is shown in Fig. 8-7b. The center or common line is called the *neutral.* The current in the neutral line (unbalanced current) is the difference between the currents in the positive and negative lines. Alternating current appears in the balance coil, but because of the rectifying action of the commutator, the currents in the positive, negative, and neutral lines are direct.

8-9 Procedure for Single-generator Operation

Before starting the prime mover, be sure that the generator breaker is open, and then close the disconnect links located in the rear of the switchboard.[1] The purpose of the links is to allow the complete isolation of the breaker from the bus for test purposes. These links, shown in Fig. 4-10, must be bolted to the connection bars with an insulated wrench.

> **Warning.** The circuit breaker must always be opened before opening or closing the disconnect links.

[1] On old installations a disconnect switch is mounted on the front of the switchboard.

The following procedure may be used as a guide for single-generator operation:

1. Be sure that the generator breaker is open.
2. Bolt the disconnect links to the connection bars. (Use an insulated wrench.)
3. Switch the voltage regulator to "automatic."
4. Start the turbine or engine, and bring it up to speed.
5. Adjust the automatic voltage regulator to obtain rated voltage.
6. Close the circuit breaker manually or by turning the breaker switch to "close."

8-10 Procedure for Parallel Operation of DC Generators

1. Be sure that the breaker for the incoming machine is open.
2. Bolt the disconnect links of the incoming machine to the connection bars. (Use an insulated wrench.)
3. Switch the voltage regulator to "automatic."
4. Start the turbine or engine, and bring it up to speed.
5. Adjust the automatic voltage regulator so that the voltage of the incoming machine is equal to or slightly higher than the bus voltage. If the voltage of the incoming machine is less than that of the bus, it will be motorized at the instant it is paralleled. The energy to motorize the incoming machine comes from the other generators on the bus. If these machines are heavily loaded, the additional current required to motorize the incoming machine may be sufficient to trip them off the bus and black out the plant. Hence, when dc generators are to be paralleled, it is necessary that the voltage of the incoming machine be at least equal to the bus voltage.
6. Close the circuit breaker.
7. Adjust the automatic regulator control of the incoming machine in the raise-voltage direction, and the other machine in the lower-voltage direction, until the ammeters of both machines indicate the same. If on manual control, use the field rheostats instead of the automatic voltage control.

In factories, hotels, or large ships that require two or three dc generators in parallel to supply the connected load, it is good operating practice to balance the current between machines in proportion to their ampere ratings. By doing so, transient overcurrents and short circuits in the distribution system are shared by all generators, reducing the probability of one generator tripping. The tripping of one generator due to overcurrent will throw the entire load on the remaining unit or units, causing them to trip and blacking out the plant.

8-11 Removing a DC Generator from the Bus

To remove a dc generator from the bus, transfer the load to the other machines by adjustment of the field rheostats, or automatic voltage regulator, trip the circuit breaker, and then open the disconnect links. If the generator is separately excited, its field switch should be opened last; opening the field circuit causes the voltage to drop to a very low value, and if the machine is still on the bus, the other machines will feed into it. Severe damage to the commutator and brushes may result.

The generator circuit breaker provides protection against overcurrents, and through a reverse-current trip provides protection against motorization. Although motorization does not harm the machine, it does result in a wasteful expenditure of energy by acting as a load connected to the other generators.

8-12 Instrumentation and Control of DC Generators

Figure 8-8 is a functional diagram for a two-generator system, showing the minimum instruments, switches, and adjustable controls that an operating engineer must be acquainted with when operating dc generators singly or in parallel. The connecting lines indicate what each control device operates and where each meter gets its signal.

When the load on a dc system exceeds the amount that can be supplied by a single generator, additional machines must be connected to the system to supply the required energy. The incoming machine must be paralleled in a manner that enables each machine to supply its proper proportion of power to the common load.

8-13 Load-sharing Problems of DC Generators in Parallel

Compound generators in parallel. Stability of operation of two or more compound generators in parallel requires a third connection, called an *equalizer connection,* that parallels the series fields of all machines. This is illustrated in Fig. 8-9. The equalizer serves two functions. It ensures the division of load in accordance with the machine's ampere rating, and it prevents the reversal of current in the series field in the event that one machine becomes motorized. With the series fields in parallel, the current through each is in the inverse ratio of its respective resistance. Hence all incoming load cannot be "hogged" by one machine. In the case of two or more identical machines in parallel, the equalizer ensures an equal current through each series field, regardless of the division of load by the generator armatures. The equalizer connections of all machines are made to a common bus, which is referred to as the *equalizer bus.*

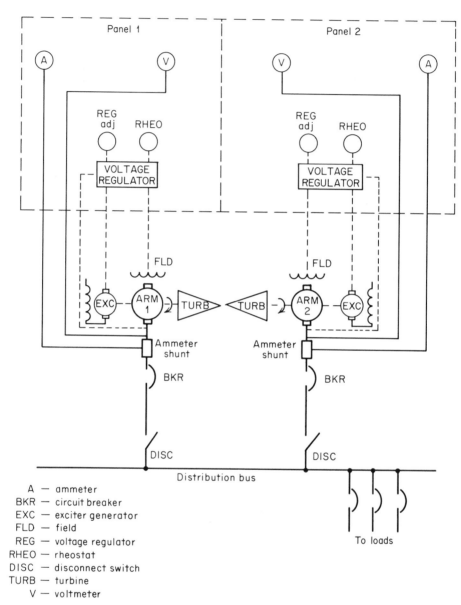

A — ammeter
BKR — circuit breaker
EXC — exciter generator
FLD — field
REG — voltage regulator
RHEO — rheostat
DISC — disconnect switch
TURB — turbine
V — voltmeter

FIG. 8-8 Functional diagram for a two-wire dc generator and distribution system.

The connections from each machine to the equalizer bus are made with relatively heavy copper conductors so as not to add appreciably to the resistance of the circuit.

If two identical compound generators are connected in parallel without an equalizer, any attempt to transfer load from one to the other will cause the machine with the greater excitation to take the entire load and drive the other as a motor. The machine that has its excitation in-

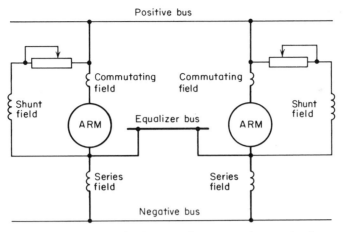

FIG. 8-9 Elementary diagram showing two-wire compound generators in parallel.

creased takes load from the other and, in doing so, effects an increase in its series field current. The machine that loses some of its load has its series field current reduced and thus suffers a net reduction in its series field strength. The result is an even greater internal voltage in the machine that gains the load and a further reduced internal voltage in the machine that loses some load; a rapid buildup of voltage occurs in one machine, accompanied by a simultaneous drop in the other. A vicious circle is initiated, resulting in the absorption of all load by one machine and motorization of the other.

When a compound generator without an equalizer is motorized, the direction of current in its series field is reversed. If the series field is sufficiently strong, it may reverse the residual magnetism of the iron poles, and the polarity of the machine will be reversed.

Effect of voltage droop on load transfer. The transfer of load from one dc machine to another is accomplished by adjustment of the field rheostats of the respective generators. The generator that is to accept more load should have its field excitation increased until one-half of the load that is to be transferred has been shifted, and the machine that is to lose some of its load should have its field excitation decreased until the remaining half has been shifted. If convenient, a simultaneous adjustment may be made on both machines. Unlike the ac system, the no-load speed setting of the governors is fixed; it is not varied to transfer load. The transfer of load is done by raising or lowering the no-load voltage setting of the machines. The load transfer between dc generators is observed on ammeters; wattmeters are rarely used, because the power factor is always unity.

For optimum performance when operating in parallel, dc generators should have the same or very similar load-voltage characteristics. Machines with identical characteristics share the load in proportion to their ampere rating; whereas machines with different characteristics, even with the same ampere rating, do not share any increase or decrease in bus load proportionately. The machine with the smallest percentage droop takes the greater portion of the increasing load.

Figure 8-10a illustrates a typical droop characteristic for a dc generator. A voltage droop of 1 to 3 percent is essential for stable operation when in parallel; the percent droop is the percent change in voltage that occurs when the load is changed from full load to no load. If two dc generators are in parallel and both characteristics are almost horizontal (essentially zero voltage droop), one machine will tend to hog the load and drive the other as a motor. Because the generator voltage is a function of both the speed of rotation and the field flux, the desired voltage droop can be obtained by governor droop or electrical design.

The point of intersection of the droop characteristic with the voltage axis is the no-load voltage of the machine for the given setting of the field rheostat. Turning the field rheostat in the "raise-voltage" or "lower-voltage" direction raises or lowers the no-load voltage setting of the generator but does not change the droop; as indicated by the broken lines in Fig. 8-10a, adjustment of the rheostat raises or lowers the entire characteristic. The droop remains unchanged.

For a given no-load voltage setting of the field rheostat, there is a fixed relationship between the generator voltage and the ampere load. Increasing the ampere load causes the output voltage to decrease; decreasing the ampere load causes the output voltage to increase. Thus assuming that the generator referred to in Fig. 8-10a is operating at its rated values of 1,000 amp and 240 volts, with a 3 percent voltage droop, when the load is removed, the voltage will rise to its no-load value of

$$240 + 0.03 \times 240 = 247.2 \text{ volts}$$

Figure 8-10b shows two dc generators with identical droop characteristics and the same no-load voltage settings operating in parallel. Note that increased bus load, indicated by the lowered system voltage, is divided equally between the two machines.

Figure 8-10c shows two dc generators with identical droop characteristics but with different no-load voltage settings. Note that the increases in bus load, indicated by the lowered system voltage, are divided almost equally between machines.

Figure 8-10d shows the effect of radically different voltage droop characteristics on the division of load current. Starting with an initial balanced condition of 500 amp per generator and a bus voltage of 240 volts, the lowered voltage caused by the increased bus load causes machine B to take more of the additional load than machine A. If, as loading continues, the ampere load on machine B causes its breaker to trip, all

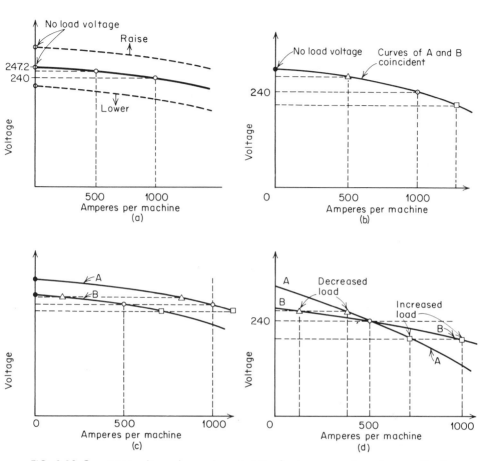

FIG. 8-10 Generator voltage-droop characteristics for two generator sets operating in parallel. (a) Typical voltage-current droop characteristic; (b) identical droop characteristics and same no-load voltage; (c) identical droop characteristics and different no-load voltages; (d) radically different droop characteristics.

the bus load will fall on machine A and it too will trip, blacking out the plant. Referring again to Fig. 8-10d, if the bus load is decreased, the system voltage will increase. For this condition machine B loses a greater amount of the overall reduction in bus load than machine A. A large reduction in bus load can cause machine B to lose all its load and then be motorized by machine A. If this occurs, a reverse-current relay will trip machine B off the bus.

   If after balancing the bus amperes between two generators with supposedly identical voltage droops, the switching of load on or off the bus causes an unbalance condition, it is an indication that the governor droops of the two prime movers are not the same. One or both require droop adjustment, which should be done by experienced personnel.

8-14 Reversed Polarity of DC Generators

Reversed polarity is indicated by a reverse reading of the voltmeter on the generator panel. It may be caused by the wrong direction of rotation, the reversal of the exciter connections, or the reversal of the residual magnetism in the iron-pole pieces of a self-excited machine.

In order for the direction of the residual magnetism of a self-excited machine to be corrected, it is necessary for current to be sent from another machine into the shunt field winding of the affected unit. If the machine is equipped for parallel operation, the correction is easily made. The machine should be stopped, the brushes insulated from the commutator by several thicknesses of heavy paper, and its field rheostat turned to the position of maximum resistance. The disconnect switch and circuit breaker should then be closed, the field rheostat turned to the all-out position and then back to the all-in position, the circuit breaker tripped, and the disconnect switch opened. The paper separating the brushes from the commutator may then be removed and the machine restarted. If the brushes of the affected machine are not insulated from the commutator while this correction is made, the machine on the bus will feed into the armature of the affected machine, cause severe damage to the commutator, and black out the plant. If the machine is not equipped for parallel operation, batteries or another generator may be connected to the shunt field to supply the correct magnetic flux.

8-15 Voltage Failure of DC Generators

The generated voltage of any machine is a function of two variables, speed and field strength. Hence, if the machine is revolving at rated speed, failure to build up voltage may be attributed to a failure of the magnetic field. In a separately excited machine, this can be caused only by an open somewhere in the shunt field circuit or the field rheostat, or a failure of the exciter generator. A voltage check of the exciter generator will determine if it is at fault. To test the rheostat, short-circuit its terminals, and observe the generator voltage. A buildup of voltage indicates an open rheostat. If the exciter is operating at rated voltage and the generator fails to build up when the rheostat is shorted, the trouble is in the field circuit or the wires connecting the exciter to the field.

In a self-excited machine, the trouble can be caused by low speed, the rheostat set at too high a resistance, poor contact between the burshes and the commutator, inadequate spring tension, an open somewhere in the shunt field circuit, or too low a value of residual magnetism. If the generator voltmeter indicates 10 volts or higher, the trouble is not a lack of residual magnetism but one or more of the other factors.

A relatively simple method of strengthening the residual magnetism of a self-excited cumulative compound generator is to short-circuit the armature and series field through a fuse, as shown in Fig. 8-11. The

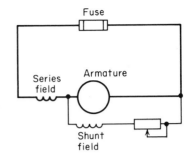

FIG. 8-11 Circuit for increasing the residual magnetism of a cumulative compound generator.

fuse should not have a greater current rating than the generator and should have adequate interrupting capacity. The small value of generator voltage causes current to flow through the armature and series field. The series field increases the total flux, which in turn increases the voltage, and there is more current. The higher current causes an even greater series field strength, greater voltage, and still higher current, etc., until the fuse blows. The momentary high value of series field strength should be adequate to restore the residual magnetism. All resistance should be cut out of the shunt field rheostat when this remedy is applied.

The residual magnetism of a shunt machine may also be restored by following the procedure outlined for reversed polarity.

Bibliography

Hubert, C. I.: "Operational Electricity," John Wiley & Sons, Inc., New York, 1961.

Questions

8-1. What causes a voltage to be induced in an armature coil when the coil is rotating within a magnetic field?

8-2. What is the function of the commutator in a dc generator?

8-3. What is the brush neutral position? Explain the undesirable effects that occur if the brushes are not in the neutral position.

8-4. What is the function of interpoles? Where are they located?'

8-5. What causes the voltage of a self-excited generator to build up from a low value, initiated by residual magnetism, to its rated value?

8-6. When a relatively heavy load is connected to a dc generator, the speed and voltage output of the machine are reduced. Explain why this happens. What equipment is provided to minimize this effect, and how does it work?

8-7. In what way does a compound generator differ from a shunt machine? Sketch the voltage versus current characteristics of both machines, and explain why they are different.

8-8. Explain how an overcompound machine may be given a flat-compound characteristic without changing the number of turns in the field windings and without changing the speed.

8-9. What is armature reaction; what are its adverse effects; and how may it be eliminated?

8-10. How does a three-wire dc generator differ from the standard dc generator? What is the function of the balance coil?

8-11. What is the purpose of generator disconnect links when a generator has a circuit breaker? Where are the links located, and how are they opened and closed? What cautions must be observed?

8-12. Outline the steps required for paralleling a dc generator with another already on the bus. Assume that the machine on the bus is operating at 250 volts and has a bus load of 1,200 amp. Include the correct procedure for dividing the bus current equally between machines.

8-13. Why is it good practice to balance the load currents in proportion to the generator kw ratings?

8-14. Assume that two identical dc generators are operating in parallel and sharing a small bus load. Outline the correct procedure for removing one of the generators from the bus.

8-15. What is motorization? Is it harmful? How may it be detected and corrected?

8-16. What damage may occur if a dc generator, in parallel with other machines, loses its field excitation?

8-17. For stability of operation in parallel, compound generators require an equalizer connection. Explain how the equalizer provides this stability.

8-18. What fault condition can cause a generator to reverse its polarity? Assume that the direction of rotation is not changed.

8-19. Describe a method for correcting reversed polarity of a dc generator. Assume that the only source of energy available is that of another generator on the bus.

8-20. What effect do different load-voltage characteristics have on the division of oncoming load between generators? Explain with the aid of a sketch.

8-21. Describe a method that may be used to strengthen the residual magnetism of a self-excited dc generator.

8-22. What fault conditions can prevent a self-excited generator from building up to rated voltage?

8-23. What fault conditions can prevent a separately excited machine from building up to rated voltage?

8-24. Describe a simple test which does not require instruments and which can determine whether or not a field rheostat is at fault when a generator fails to build up voltage.

nine

dc motor

The construction details of a direct-current motor are identical to that of a dc generator and are shown in Figs. 8-1, 8-6, and 9-7.

9-1 Principle of Motor Action

Because there is no essential difference between a dc motor and a dc generator, the principle of motor operation may be explained by examining a motorized generator. The sketches shown in Fig. 9-1 correspond to those of the dc generator shown in Fig. 8-2, except that its prime mover is not operating and it is motorized by a battery or another dc generator. When motorized, the current in each armature coil is reversed with respect to its direction when operating as a .generator; the polarity of the field poles does not change.

The development of motor torque requires an interaction of the magnetic field of the poles with the magnetic field produced by the current in the armature conductors. Figure 9-1a, b, and c illustrates the direction of current in each conductor and their relationship with respect to the field poles. The $\odot$ marks represent the electron current coming out of the conductor toward the reader, and the $\oplus$ marks indicate the current going into the conductor. The direction of the mechanical force produced by the interaction of the magnetic fields is clockwise, as shown by the heavy arrows in Fig. 9-1c, resulting in clockwise rotation of the armature (see Sec. 1-16).

When operated as a generator, the rotating commutator and associated carbon brushes served as a rotary switch that switched the internal alternating current to direct current for the

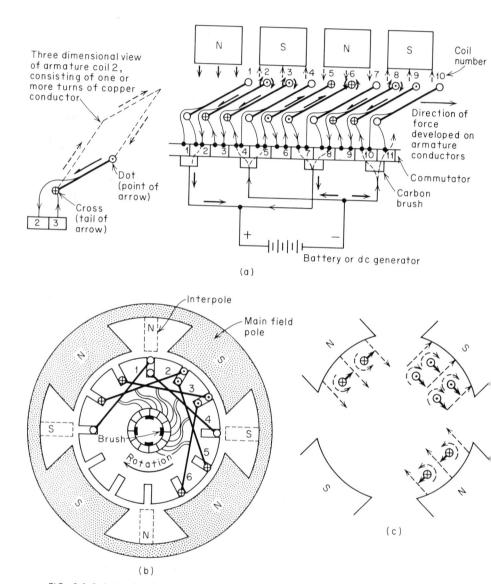

Three dimensional view of armature coil 2, consisting of one or more turns of copper conductor

Dot (point of arrow)

Cross (tail of arrow)

Coil number

Direction of force developed on armature conductors

Commutator

Carbon brush

Battery or dc generator

(a)

Interpole

Main field pole

Brush

Rotation

(b)

(c)

FIG. 9-1 Relationship between the current in the armature coils and the polarity of the poles in a dc motor. (a) Laid-out view of field poles and armature coils; (b) commutator-end view of armature and field structure; (c) direction of mechanical force produced by interaction of magnetic fields.

external circuit. If the same machine is operated as a motor, the commutator provides the same type of action but in reverse; although direct current is supplied to the armature, the switching action of the commutator and brushes reverses the current in each armature coil every time the center of a coil window passes the center of a main field pole. It is this reversal of armature current which causes the mechanical force on a given conductor to be in the same direction when under a north pole and when under a south pole, as shown in Fig. 9-1c.

As with the dc generator, commutating poles, called *interpoles,* are used to neutralize the inductive effect of the armature coils, which would otherwise cause destructive sparking during the commutation process (see Sec. 8-3).

Likewise, compensating windings (pole-face windings) are used to eliminate armature reaction in those motor applications where rapidly fluctuating loads are experienced (see Sec. 8-7). Armature reaction in a generator shifts the flux across the pole face in the direction of armature rotation; armature reaction in a motor shifts the flux across the pole face in a direction opposite to the armature rotation.

The magnitude of the mechanical force developed on the armature conductors depends on the amount of current in the conductors and the density of the pole flux in which the conductors are immersed. Expressed mathematically,[1]

$$F = BIl$$

where F = mechanical force developed on the armature conductor, newtons

B = density of pole flux in webers per sq meter

I = current in armature conductor, amp

l = length of that part of armature conductor which is immersed in pole flux (active length), meter

9-2 Reversing a DC Motor

To reverse the direction of rotation of a dc motor, it is only necessary to reverse the polarity of the field poles or to reverse the direction of the direct current supplied to the brushes by the battery, but not both. Reversing the polarity of the field poles and the current supplied to the brushes does not reverse the direction of rotation of the armature. This is illustrated in Fig. 9-2a for a representative pole and armature conductor. The heavy arrows indicate the direction of the mechanical force and hence the direction of rotation of the armature.

Except for some fractional-horsepower machines, all dc motors use electromagnets rather than permanent magnets in their field assemblies. The connection diagram for dc shunt motors, showing the standard terminal markings and connections for the two directions of rotation, is shown in Fig. 9-2b.[2] The direction of rotation is as viewed when facing the end opposite the drive shaft.

[1] Newtons $\times$ 0.00026 = pounds, meters $\times$ 3.281 = feet, webers per square meter $\times$ 10^4 = gauss (lines of force per square meter).
[2] Terminal Markings for Electrical Apparatus, USAS C6.1-1956.

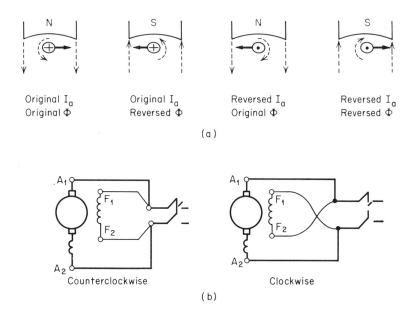

Original I_a
Original Φ

Original I_a
Reversed Φ

Reversed I_a
Original Φ

Reversed I_a
Reversed Φ

(a)

Counterclockwise

Clockwise

(b)

FIG. 9-2 (a) Effect of the polarity of the poles and the direction of armature current on the direction of rotation: I_a = current in the armature conductor, Φ = flux of field pole; (b) standard connections for dc motors, direction of rotation as viewed from end opposite the drive shaft.

9-3 Generator Action in a DC Motor

The application of a driving voltage to a dc motor causes the armature to revolve within the magnetic flux set up by the field poles. As the armature rotates, the flux through the respective windows of each armature coil increases, decreases, and reverses. This changing flux through the coil windows generates a voltage, which, in accordance with Lenz's law, is always in opposition to the driving voltage. The magnitude of this magnetically induced countervoltage depends on the rate of change of the flux through the coil window. This relationship, identical to that for the dc generator, is expressed by Faraday's law:

$$e = N\frac{d\phi}{dt}$$

where e = countervoltage generated in each armature coil

N = number of turns of wire in each armature coil

$\dfrac{d\phi}{dt}$ = rate of change of flux through window of each armature coil

The sum of the series-connected countervoltages generated in the armature adds up to an overall armature countervoltage E_a that is almost equal to the driving voltage V_T. Figure 9-3 illustrates the relative directions

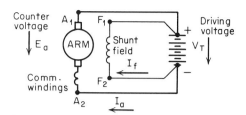

FIG. 9-3 Elementary diagram showing the relative directions of voltage and current in a dc motor.

of driving voltage, countervoltage, armature current, and field current in a dc motor.

When a driving voltage is first applied to a dc motor, the armature is at rest, and the countervoltage E_a is zero; the current to the armature is limited only by the combined resistance of the armature, interpoles, and connecting cables. Thus, using Ohm's law,

$$I_a = \frac{V_T}{R_a + R_{IP} + R_c}$$

where R_a = resistance of armature as measured between positive and negative brushes

R_{IP} = resistance of interpole windings

R_c = resistance of cables

Because the resistance of the armature, commutating winding, and cable is deliberately kept small, only a small fraction of an ohm, the armature current on starting is excessively high. To prevent this high current from damaging the machine, additional resistance, called *starting resistance,* is placed in series with the armature when starting.

As the armature accelerates, the armature countervoltage E_a builds up, causing the armature current to decrease from its high starting value to the normal value for the particular shaft load. When up to speed, the starting resistance is removed, and the armature current is expressed by

$$I_a = \frac{V_T - E_a}{R_a + R_{IP} + R_c}$$

The shunt field coils are wound with many turns (several hundred to several thousand) of relatively small-diameter wire, so that their resistance is fairly high, and the field current low. The field current as determined by Ohm's law is

$$I_f = \frac{V_T}{R_f}$$

where R_f = overall resistance of entire shunt field circuit

The many turns of wire around each field pole produce a relatively high flux with only a small current.

9-4 Starting and Speed Adjustment of DC Motors

The base speed or nameplate speed of a motor is its speed when operating with rated shaft load, at its rated temperature, with rated voltage applied, and no external resistors or rheostats connected in series with the armature or field.

Adding external resistance in series with the armature reduces the armature current; this in turn reduces the mechanical force developed on the armature conductor, and the motor slows down.

Adding external resistance to the shunt field circuit reduces the field flux. With less pole flux, the rate of change of flux through the windows of the armature coils is reduced, causing a lowered countervoltage. The reduced countervoltage permits a higher armature current and hence a higher motor speed. Figure 9-4 illustrates the two methods for adjusting the speed of the motor above or below its base speed. Rheostat adjustments should be made slowly; a sudden increase in field rheostat resistance or a sudden decrease in armature rheostat resistance causes excessive currents that may damage the commutator and brushes.

The speed of a dc motor is extremely sensitive to changes in the field flux. If the shunt field is overly weakened by too much field rheostat resistance or by an open in the shunt field circuit and the load on the motor is light, the machine may accelerate to destruction. Figure 9-5 shows the damage caused by overspeed when the shunt field circuit of a dc motor was accidentally opened.

Dc motors above ½ hp require current-limiting starters. A manually operated starter is shown in Fig. 9-6. Automatic starters for dc motors

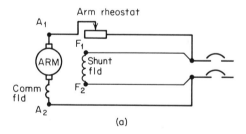

(a)

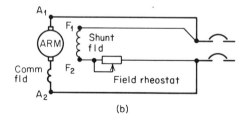

(b)

FIG. 9-4 Speed adjustment of a dc motor. (a) Rheostat control for speeds below the base speed; (b) rheostat control for speeds above the base speed.

FIG. 9-5 Damage to a dc motor caused by excessive speed. (*Westinghouse Electric* Corp.)

are discussed in Chap. 13. To operate this type of starter, the rheostat handle is slowly moved from the "stop" position up to the full-speed or "run" position, where it is held in place by an electromagnet. This type of starter should not be used for adjusting the speed of the motor. The starting lever should be moved just fast enough to provide a uniform and gradual acceleration. Moving the rheostat handle too rapidly may damage the armature or commutator by causing excessive current. Moving

FIG. 9-6 Manually operated starter for a dc motor. (*Cutler-Hammer, Inc.*)

it too slowly may damage the rheostat. Starting rheostats are designed for a short-time duty and will overheat if operated in an intermediate position for an extended period.

9-5 Compound and Series Motors

Modification of the basic dc motor to obtain certain desirable torque-speed characteristics is generally accomplished by the addition of an auxiliary winding called the *series field*. The series field coils are wound with heavy copper conductors on the same pole iron as are the shunt field coils, as shown in Fig. 9-7, and are connected in series with the armature.

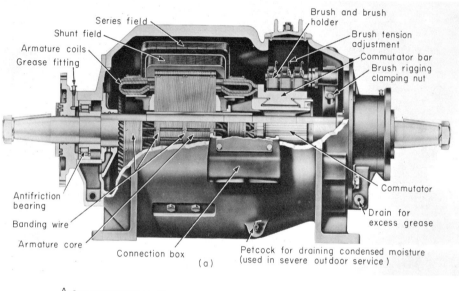

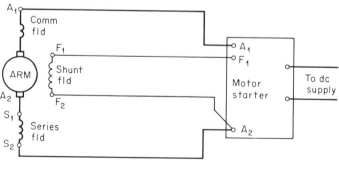

FIG. 9-7 (a) Compound motor for severe outdoor service (*Westinghouse Electric Corp.*); (b) compound motor connections.

Such machines are called compound motors. If the current in each series coil is in the same direction as the current in its respective shunt field coil (cumulative compound), the flux contribution due to the series field adds to the shunt field flux; if the current in each series coil is in the opposite direction with respect to the current in the corresponding shunt coil (differential compound), the series field flux subtracts from the shunt field flux.

However, because of ~~its~~ *their* instability, differential compound motors are not used in power applications. Although differentially connected compound generators do have some applications, particularly in electric hoist systems, a differential connection for a motor poses a hazard to life and property, and must be avoided. The behavior of a differential motor is both peculiar and frightening, especially the higher-horsepower machines; if connected to the power supply, it may start in the wrong direction, then quickly reverse and accelerate to destruction. Figure 9-8 illustrates the torque and speed characteristics of compound, shunt, and series motors. A compound motor with just one or two series field turns is

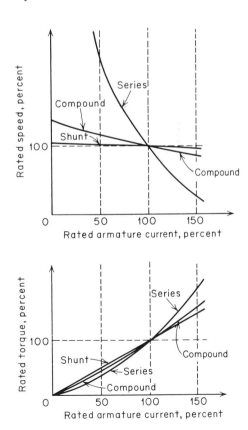

FIG. 9-8 Comparison of torque and speed characteristics of dc motors.

called a *stabilized shunt machine*. Motors with only a heavy series field winding and no shunt field have extremely high starting torques. However, they must be directly geared to the load to prevent loss of load and the resultant excessive speed. Shunt and stabilized shunt motors are generally used for centrifugal pumps and fans, compound motors for electric hoists and punch presses, and series motors for traction motors in railroad applications.

9-6 DC Motor Connections

The standard terminal markings[1] for dc motors and the proper connections for different directions of rotation are shown in Fig. 9-9; the ammeter and voltmeter should be inserted at the indicated points when checking current and voltage. The directions of rotation are as viewed from the end opposite the drive shaft. Unmarked terminals should be identified

[1] Terminal Markings for Electrical Apparatus, USAS C6.1-1956.

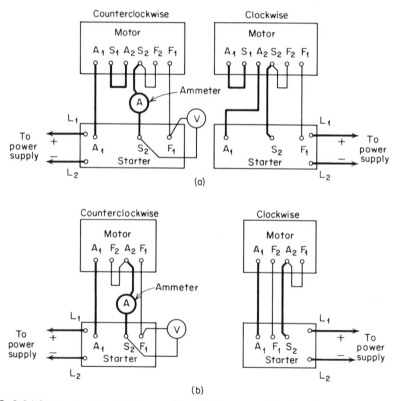

FIG. 9-9 USA standard terminal connections for dc motors. Direction of rotation as viewed from the end opposite the drive. (a) Compound or stabilized shunt motors; (b) shunt motors.

by test before the machine is connected to the starter; otherwise blown fuses or damage to the starter and motor may result.

Except for fractional-horsepower motors, the shunt field leads are generally smaller in diameter than the leads of the armature and series field, and are therefore easily identified. However, testing with a 60-watt 120-volt lamp connected in series with two test points and a 120-volt line will resolve any doubts. This is illustrated in Fig. 9-10a. One test point should be connected to a motor lead, and the other point should

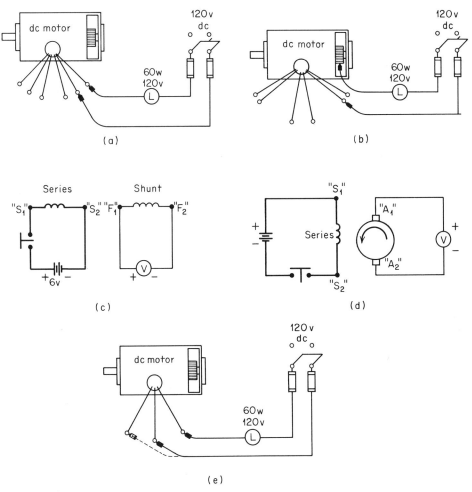

FIG. 9-10 Identification of unmarked motor terminals. (a) Pairing off the armature and field circuits; (b) identification of the armature circuit; (c) determination of the relative polarity of series and shunt fields; (d) determination of the relative polarity of the armature; (e) identification of motor leads for a three-terminal motor.

be used to probe the remaining wires until the lamp lights or a spark is obtained. The pair of motor leads that lights the lamp or causes a spark should be pushed to the side, and the remaining four should be tested in the same manner until the other two pairs are determined. The pair of motor leads that causes the lamp to burn at reduced brillancy is the shunt field, and should be so marked. The two armature leads can be identified by testing from the brush holders to each of the other four wires, as shown in Fig. 9-10b; the armature leads cause the lamp to burn. The remaining two wires are those of the series field.

Because it is extremely dangerous to start a compound motor that is differentially connected, the relative polarity of the series and shunt fields must be established. This can be determined in a safe manner with the simple circuit shown in Fig. 9-10c. Connect a six-volt battery consisting of four No. 6 dry cells in series with a pushbutton and the series field, and connect a 10/50 volt dc voltmeter across the shunt field. Make a momentary contact by depressing and releasing the button quickly. If the voltmeter reads upscale when the button is pushed and downscale when released, mark the shunt field lead that connects to the $+$ terminal of the voltmeter F_1, and the series field lead that connects to the $+$ terminal of the battery S_1; if the correct deflection is not indicated, interchange the two shunt field leads, and try again.

The circuit shown in Fig. 9-10d may be used to determine the terminal markings for the armature. Be sure to connect the positive terminal of the battery to the S_1 lead, and connect the voltmeter across the armature. Manually, or otherwise, rotate the armature in the counterclockwise (CCW) direction, as viewed from the end opposite the drive shaft. While rotating CCW, push the button, and note the voltmeter deflection. If the pointer deflects upscale, the armature lead that connects to the $+$ terminal of the voltmeter should be marked A_1; if the correct deflection is not indicated, interchange the two armature leads, and try again. Although the presence of residual magnetism in the iron pole pieces may cause a small voltage to be generated, it is the direction of the additional voltage generated when the button is depressed that is significant. After establishing the relative terminal markings, the machine may be connected in accordance with the circuits shown in Fig. 9-9, with the assurance that it is not differentially connected.

To identify the leads of a dc motor that has only three external wires, a test point should be connected to one of the three wires, and the remaining two should be probed with the other point, as shown in Fig. 9-10e. The F_1 wire will indicate a low brilliance when tested separately with each of the other two. To identify the S_2 wire, the brushes should be lifted and the test made from F_1 to the other two wires. Wire S_2 will cause the lamp to burn. The remaining wire is A_1.

Table 9-1 General Effect of Voltage Variation on DC Motor Characteristics

Voltage variation	Starting and max. run torque	Full-load speed	Efficiency			Full-load current	Temp. rise full load	Maximum overload capacity	Magnetic noise
			Full load	¾ load	½ load				
Shunt wound									
120% voltage	Increased 30%	110%	Slight increase	No change	Slight decrease	Decrease 17%	Main field increase. Commutating field and armature decrease	Increased 30%	Slight increase
110% voltage	Increased 15%	105%	Slight increase	No change	Slight decrease	Decrease 8.5%	Main field increase. Commutating field and armature decrease	Increased 15%	Slight increase
90% voltage	Decreased 16%	95%	Slight decrease	No change	Slight increase	Increase 11.5%	Main field decrease. Commutating field and armature increase	Decreased 16%	Slight decrease
Compound wound									
120% voltage	Increased 30%	112%	Slight increase	No change	Slight decrease	Decrease 17%	Main field increase. Commutating field and armature decrease	Increased 30%	Increased slightly
110% voltage	Increased 15%	106%	Slight increase	No change	Slight decrease	Decrease 8.5%	Main field increase. Commutating field and armature decrease	Increased 15%	Increased slightly
90% voltage	Decreased 16%	94%	Slight decrease	No change	Slight increase	Increase 11.5%	Main field decrease. Commutating field and armature increase	Decreased 16%	Decreased slightly

NOTE: This table shows general effects, which will vary somewhat for specific ratings.
SOURCE: From D. G. Fink and J.M. Carroll, "Standard Handbook for Electrical Engineers," sec. 18-21, McGraw-Hill Book Company, New York, 1968, with permission.

9-7 Checking Motor Performance

Overloading of dc motors and generators results in the generation of excessive heat; the higher temperatures shorten the life of insulation, cause sparking at the brushes of dc machines, and loosen the soldered banding wire and the commutator connections. If emergency conditions require a machine to supply more load than it can safely handle without excessive heating, the machine may be cooled with portable blowers and fans. If it appears that a machine is overheating, its temperature should be checked with a bulb thermometer, as shown in Fig. 5-16.

The behavior of dc motors when operating above or below rated nameplate voltage is indicated in Table 9-1. Note particularly the effect of voltage variations on the heating of armature and fields.

If the motor overheats and voltmeter and ammeter readings indicate rated voltage and rated or below-rated current respectively, the ventilating ducts may be clogged, or the ambient temperature may be above normal (see Sec. 11-13).

Bibliography

Hubert, C. I.: "Operational Electricity," John Wiley & Sons, Inc., New York, 1961.

Motor and Generator Standards, NEMA MG1-1963.

Shoults, D. R., C. J. Rife, and T. C. Johnson: "Electric Motors in Industry," John Wiley & Sons, Inc., New York, 1942.

Questions

9-1. Explain how motor torque is developed.

9-2. What is the function of the commutator in a dc motor?

9-3. What is the function of interpoles and compensating windings?

9-4. Explain why a running motor develops a counter voltage. What is the relative magnitude of this voltage compared with the driving voltage?

9-5. Why is additional resistance needed to limit the current when starting a dc motor?

9-6. How may a dc motor be reversed?

9-7. What is meant by the base speed of a dc motor? What methods are used to adjust the speed above and below the base speed?

9-8. What precautions should be observed when increasing the resistance of a rheostat in series with the shunt field? Explain.

9-9. What effect does an open in the shunt field circuit have on the performance of a lightly loaded motor?

9-10. State the correct procedure for starting a motor with a manually operated starter. Can improper operation of the control handle damage the motor and starter? Explain.

9-11. What is a compound motor?

9-12. Explain why a differentially connected compound motor poses a hazard to life and property.

9-13. What is a stabilized shunt motor?

9-14. Why must a series motor be geared directly to the load?

9-15. Describe a simple polarity test for determining the relative polarity of the series and shunt fields when the terminals are unmarked.

9-16. State how you would identify the unmarked terminals of a compound motor that has six external leads.

9-17. State how you would identify the unmarked terminals of a compound motor that has three external leads.

9-18. State how you would identify the unmarked terminals of a shunt motor with four external leads.

9-19. What adverse effect does continued overloading have on a dc motor? What protective action should be taken if emergency conditions require prolonged operation above rated load?

Problems

9-1. A 230-volt dc shunt motor has a combined armature and interpole resistance of 0.01 ohm. The shunt field resistance is 100 ohms. Each of the two cables connecting the armature to the 230-volt supply line has a resistance of 0.025 ohm, and each of the two cables connecting the shunt field to the supply line has a resistance of 0.05 ohm. (a) Sketch the circuit; (b) determine the starting current drawn by the armature if no starting resistance is used; (c) determine the shunt field current.

9-2. Assume that the motor in Prob. 9-1 was properly started and that it is running at rated speed. If the armature current is 100 amp, what is the countervoltage?

9-3. A 240-volt dc shunt motor has an armature resistance of 0.002 ohm, an interpole winding resistance of 0.005 ohm, and a shunt field resistance of 86 ohms. A starting resistance must be connected in series with the armature to limit the armature current to 150 amp. (a) Sketch the circuit; (b) assuming that the resistance of the connecting cables is negligible, determine the resistance of the starting resistor.

ten

commutator, slip-ring, and brush maintenance

10-1 Commutator Surface Film

During normal operation, commutators and slip rings acquire a shiny protective gloss that reduces wear and thereby lengthens service life. This surface film generally contains copper oxide and graphite in varying proportions and is formed after several days or weeks of operation. Water vapor adsorbed by this film plays an important role in providing the lubrication required for reduced brush and commutator wear. In fact, in areas of extremely low humidity, as at high altitudes or during extended dry periods of cold winter months, the surface film disappears, and the brushes, commutators, and slip rings wear severely. Operating conditions, atmospheric conditions, and brush grade affect the makeup and color of this film, which may range from a light straw color to jet black. However, the color most common to a good surface film is chocolate brown.

A good surface film has the added advantage of higher contact resistance between the brush face and the commutator than would be obtained with a raw copper surface. This is beneficial in reducing circulating currents that occur each time the brush short-circuits an armature coil. (Although under ideal conditions the short-circuit current is zero, under very heavy loads saturation of the interpole iron permits a short-circuit current.) A raw copper surface may be caused by the wrong brush grade, some mechanical or electrical fault, or low current density at the brush surface brought on by consistently low loads. In the latter case, a current density of about 40 amp per sq in. is often necessary if a good surface film is to be maintained, and may be accomplished by removing one or more brushes from each brush arm. The increased current in each of the remaining brushes raises the brush temperature, thereby reducing the friction between brush and commutator or brush and slip ring. Figure 10-1 illustrates the effect of brush-face temperature on the coefficient of friction for electrographitic brushes. The greatest friction and hence the greatest rate of brush wear occur at about 70°C; at temperatures between 85 and 115°C, the friction is minimal, increasing again for temperatures above 115°C. The low brush current of lightly loaded machines does not generate sufficient heat for efficient brush operation, and excessive brush wear occurs. A similar relationship exists with graphite and metal-graphite brushes.

Atmospheric contamination, like that produced by oil vapors, salt air, hydrogen sulphide gas, or silicone vapors, is also very destructive to the surface film. The silicone vapors from only a very small amount of silicone material, such as silicone tape, silicone varnish, or a silicone-rubber–insulated lead, cause very rapid brush wear; furthermore, the considerable amount of carbon dust, resulting from this rapid brush wear, is deposited on the windings, causing a lowering of the insulation resistance to ground. This adverse characteristic of silicone insulation is particularly bad in enclosed machines.

Oil contamination is generally indicated by a relatively thick black film that causes poor brush contact and excessive heating at the conducting spots. If the temperature of the commutator at the relatively few conducting spots beomes high enough, very small globules of copper will break away from the commutator or slip-ring surface. Some of these tiny globules embed themselves in the brush and cut threadlike lines in the commutator or slip rings.

The carbon brushes used on commutators and slip rings are composed of lampblack, coke, or graphite, to which various other elements may be added to obtain special characteristics. For example, copper and silver are added to some brushes to increase their current-carrying capacity, and silica is added to others to provide a cleaning action. Very soft brushes

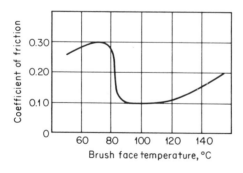

FIG. 10-1 Brush friction versus temperature for electrographitic brushes. (*Union Carbide Corp.*)

deposit graphite on the commutator, and hard abrasive brushes scour the commutator; each has its field of application, and no single brush grade serves well in all machines. The softer film-forming brushes are generally preferred for service at light loads, and the harder scouring grade for operation with heavy loads in an oily atmosphere. Brushes of different grades should never be mixed, and when replacement brushes are ordered, the specifications of the manufacturer should be followed.

10-2 Brush Sparking

Brush sparking is a symptom of a fault that may be of a mechanical, electrical, or operating nature and can be caused by a great variety of disorders. Hence, the cause of sparking is not always easily discerned. More often than not, the sparking is instigated by a dirty commutator or some mechanical fault.

Mechanical unbalance, machine misalignment, commutator eccentricity, incorrect positioning of brushes, wrong brush tension, incorrect brush grade, etc., are all detrimental to good commutation and may cause sparking.

10-3 Cleaning Commutators and Slip Rings

Commutators and slip rings may be cleaned with a wiper made of 16 layers of 6- or 8-oz hard-woven duck canvas fastened to the end of a flexible wooden stick, as shown in Fig. 10-2. The rivets or screws used to fasten the canvas should be countersunk to avoid scratching the commutator. The frequency of application is determined by the circumstances surrounding the machine, and should be often enough to prevent the formation of a smutty surface. In the case of open machines, an application once every 24 hr of operation will prolong the life of the commutator and extend the periods between shutdowns. Canvas is not abrasive and hence will not damage the surface film so necessary for good commutation. To be effective, the wiper should be applied with considerable pressure. It may be used edgewise on small machines or sideways, as shown in Fig. 10-2b and c. The use of solvents on the commutator should be avoided, because they may damage the surface film.

10-4 Commutator Surfacing

Before a commutator is surfaced, it should be checked for loose bars by gently tapping each one with a lightweight hammer. A clear bell-like ring indicates a tight commutator; a dull thud indicates a loose bar. Loose commutators should be tightened with a calibrated torque wrench, in accordance with the manufacturer's specification. Indiscriminate tightening of commutator nuts, without regard to torque limits, can result in a distorted commutator or buckled bars.

A commutator that is out of round, grooved, burned, possessed of flat spots, or otherwise in bad condition should be resurfaced by grinding

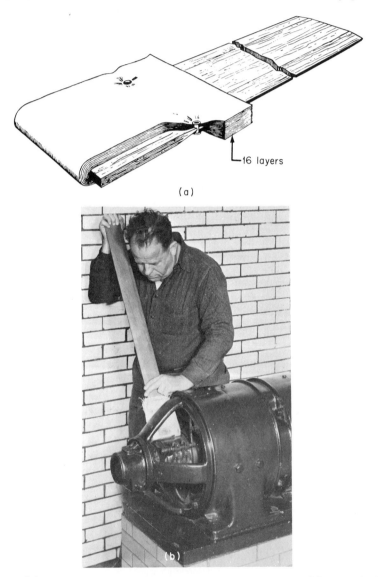

Fig. 10-2 (a) Canvas wiper for cleaning commutators and slip rings. (b) applied edgewise to the commutator of a small machine; (c) applied sidewise to the commutator of a large machine. (*Union Carbide Corp.*)

FIG. 10-2 (Continued)

FIG. 10-3 Grinding rig in use on a commutator. (*Martindale Electric Co.*)

or turning. However, excessive grinding or turning should be avoided, because a considerably reduced commutator diameter may cause the brushes to contact the commutator at an off-neutral point. The preferred method is grinding with a grinding rig, as shown in Fig. 10-3. The abrasive stone is moved slowly, back and forth across the commutator, with the machine revolving at rated speed. Operating the machine at rated speed allows centrifugal force to position any loose bars that may have escaped detection.

Several light cuts are preferred, because excessive pressure on the stone will make the grinding rig deflect and cause an inferior job. Some commutators and slip rings have a very hard glasslike surface that is difficult to cut without considerable pressure, particularly on slow-speed machines. Once the glaze is removed, however, the cutting proceeds with ease. Badly worn commutators may be ground with a coarse stone and then finished with a fine one. If only a light dressing of the commutator is necessary, a fine-grit stone is still preferred; sandpaper polishes the commutator but does not remove any incipient flat spots. Abrasive stones are also called commutator stones.

Any grease, oil, or dirt that may be present must be wiped off the commutator or rings before the stone is applied, because it will interfere with the cutting process. The armature windings and field coils should be protected against the entry of copper dust by using a vacuum cleaner and by shielding the windings with fiber or stiff cardboard.

Hand stones may be used in the absence of a grinding rig. A hand stone should be wide enough to cover the distance between adjacent sets of brushes, and its length should be less than the length of the commutator to allow for a side-to-side motion while grinding. However, it should be longer than the largest flat spot, if the flatness is to be removed. Hand stones come in many grades and shapes to suit different commutator conditions. Figure 10-4 illustrates the application of a hand stone to a commutator; and Fig. 10-5 illustrates some of the different forms in which stones are manufactured. Although hand stoning will remove flat spots, it will not remedy an eccentric commutator. Such correction must be made by a grinding rig or a diamond-shaped lathe tool.

Commutator stones are nonconductive and easy to apply. Hence, if necessary, they may be used while the machine is operating at rated load and voltage, provided that the voltage does not exceed 300 volts. However, for safety reasons, whenever practicable the commutator should be resurfaced with the machine deenergized. This is easily done with generators, because the prime mover does the driving. In the case of a motor, it is safer to accelerate the machine under its own power, then deenergize it and grind the surface while it is coasting, repeating as often as necessary. If the grinding must be done on a live circuit, the operator should use insulated gloves and stand on a rubber mat; if an adjustable circuit breaker is used, it should be reset after starting, to as low a value as possible. Because heavy concentrations of carbon and copper dust can cause a flashover between positive and negative brushes, very light cuts should be made when grinding live.

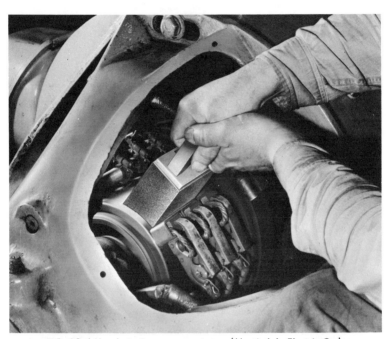

FIG. 10-4 Hand stoning a commutator. (*Martindale Electric* Co.)

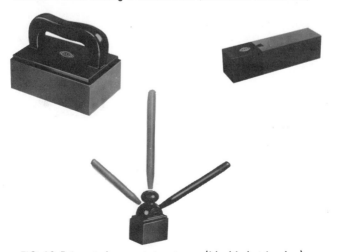

FIG. 10-5 Assorted commutator stones. (*Ideal Industries, Inc.*)

After resurfacing, if an exceptionally smooth finish is desired, the commutator may be burnished wtih a hardwood block applied manually with considerable pressure. Burnishing serves a twofold purpose: it removes the microscopic edges that may remain from the grinding process, and the heat generated serves to develop a thin oxide film, necessary to good commutation. The wood block should be applied with the end grain in contact with the commutator surface.

Freshly ground commutators are sometimes air-cured with dry compressed air at 30 to 40 psi. With the machine assembled and operated at no load, the compressed air is applied to the commutator at the leading edge of the brush, and the load is gradually applied. The compressed air compacts the residual dirt, composed of copper and carbon dust, and burns it out; the controlled burning action assists in forming a thin oxide film on the commutator.

Do not touch the freshly ground surface of a commutator or slip

(a)

FIG. 10-6 (a) Direct-drive mica-undercutter slotting mica with the brush rigging in place; (b) slotting file being used to smooth rough slot edges after the undercutting operation. (*Ideal Industries, Inc.*)

(b)

FIG. 10-6 (Continued)

rings; the oily film deposited on the fresh surface will prevent or delay the development of a normal surface film.

10-5 Mica Undercutting and Repair

Mica undercutting is essential to good commutation and should not be neglected. Mica is harder than copper and, if not undercut, will form ridges as the copper wears away. The mica ridge will cause sparking and chipping of brushes. Mica should be undercut to a depth approximating its thickness. This is easily done with an undercutting machine and slotting file, as shown in Fig. 10-6. Hacksaw blades may also be used for this purpose, provided that the sides of the blade are ground to remove the set (the protruding sides of the teeth).

During the process of undercutting, some or all of the bars may acquire a featheredge, or mica fins, as shown in Fig. 10-7. To remedy this, the copper bars may be side-cut with a knife, a tool fashioned from a power hacksaw blade, or a slotter that chamfers the corners of the

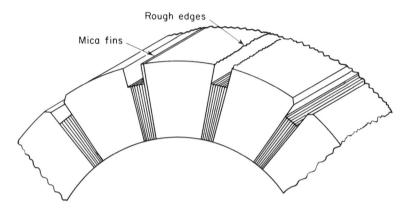

FIG. 10-7 Rough edges and mica fins that must be removed by side cutting.

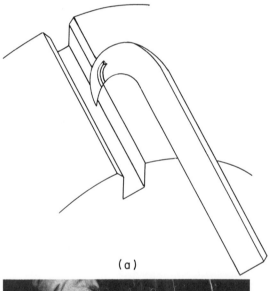

(a)

(b)

FIG. 10-8 (a) Side-cutting tool fashioned from a power hacksaw blade; (b) combination slotting and chamfering tool. (Ideal Industries, Inc.)

bars to make certain that the mica is well below the surface. This is a very important part of the undercutting process and must not be neglected. If allowed to remain, the featheredges of mica and copper will flake off, embed in the brush face, and score the commutator. Figure 10-8 illustrates the use of a side-cutting tool.

All carbonized mica should be dug out and the holes filled with commutator cement. Burned or carbonized mica, caused by the absorption of oil, moisture, and carbon dust, forms conducting paths between the bars and between the bars and ground.

After undercutting and side cutting, the slots should be insulated with a light coating of air-drying varnish or a sprayed-on acrylic plastic coat. This will prevent the absorption of oil by the mica and also reduce the possibility of shorts by insulating adjacent bars from accumulated carbon dust.

10-6 Brush Adjustment

The correct adjustment of the brushes is just as important as the maintenance of a good commutator surface. In fact, a good commutator surface is dependent to a large extent on correct brush adjustment. Figure 10-9 shows the effects of dirt and incorrect adjustment on the brush face.

When a machine is to be overhauled, the brushes should be removed and tagged to ensure their replacement in the same brush holders; the unavoidable small variations in brush-holder angle and brush clearance may cause faulty commutation and require reseating of all brushes. The brush holders should be thoroughly cleaned and inspected for sticky brushes or excessive movement in the holders. Excessive play permits unnecessary forward and side motion, resulting in partial seating of the brush in several positions. Partial seating increases the current density at the contact surface, causing overheating and sparking. On the other hand, insufficient play interferes with the free movement of the brush, causing it to stick in the holder and arc at the commutator. Brushes should be replaced when they have worn down to one-quarter of the useful length. Short brushes eventually cause burning of the commutator and brush pigtails, as shown in Fig. 10-10.

When replacing brushes, the whole set should be replaced at the same time. If this is not practicable, all the brushes on a single arm should be replaced; replacing only a few brushes on a multibrush arm may cause unequal current distributions and lead to commutation difficulties.

10-7 Brush Clearance

All brushes of a given machine should be accurately aligned with the commutator bars. Misalignment of brushes results in some brushes being

FIG. 10-9 Comparison of the effects of dirt and incorrect brush adjustment on the brushes. (a) Appearance of brushes requiring cleaning; (b) brushes damaged by sparking caused by incorrect and nonuniform brush-spring pressure; (c) effect of high mica and rough commutator surface on brushes; (d) brush appearance resulting from excessive clearance in holder. (*General Electric* Co.)

FIG. 10-10 Damage to motor caused by short brushes. Arcing under the brushes caused a flashover that burned the brush pigtails and the commutator. (*General Electric* Co.)

off neutral while others on the same stud are in the neutral position. Such conditions may give rise to serious sparking. The clearance between the brush holders and the commutator varies with different machines and may be $\frac{1}{16}$ in. in some machines and $\frac{3}{16}$ in. in others; all brush holders of a given machine should be set the same distance from the commutator. A gauge made from several thicknesses of fiber may be used on commutators to obtain the desired clearance; each brush-holder clamp is loosened, and the box is pressed firmly down on the fiber and then reclamped.

10-8 Brush Staggering

Staggering of the brush holders should be done in positive and negative pairs, as shown in Fig. 10-11. Each set of positive brushes must be in

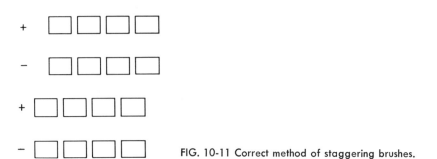

FIG. 10-11 Correct method of staggering brushes.

line with a negative set. If brushes of the same polarity track each other, the commutator will wear unevenly.

Figure 10-12 illustrates a commutator that had been operating with the brushes incorrectly staggered. The alternate bands of bright copper and dull brown or black are characteristic of incorrect staggering. For

FIG. 10-12 Characteristic light and dark bands caused by incorrect brush staggering. (*General Electric* Co.)

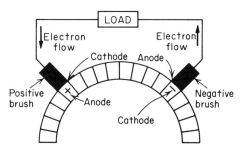

FIG. 10-13 Anodes and cathodes formed at contact surfaces between the brushes and commutator of a dc generator.

the same reason, slip rings that carry direct current should have their connections reversed every 6 months to provide for better brush and ring wear. This phenomenon may be explained with the aid of Fig. 10-13, where it is shown for the commutator of a dc generator. The moisture adsorbed by the commutator is decomposed by electrolysis; oxygen is freed at the anodes and hydrogen at the cathodes. At the negative brush, the oxygen formed at the anode causes the carbon surface to oxidize, forming carbon dioxide; the hydrogen formed at the cathode has a reducing action, thereby maintaining a bright copper surface. Hence both the negative brush and the commutator surface below it will wear relatively fast. At the positive brush, the oxygen formed at the anode causes a protective copper oxide film to form on the commutator; the hydrogen formed at the cathode acts as a reducing agent at the brush and lessens the brush wear. Hence in a generator the positive brush and the commutator surface below it will wear less than the brush and surface at the negative side. If the machine is operated as a motor, the cathodes become anodes and the anodes become cathodes. This will cause the positive brush and the commutator below it to wear relatively fast.

10-9 Brush Angle

The angle that the brush makes from the radial depends on the type of brush holder used. The two most common types of brush holders are the box type and the reaction type, shown in Fig. 10-14. The box type, shown in Fig 10-14a, b, and c, may be operated in the radial, trailing, or leading position; when used in the trailing position, the angle may be any value up to 20°; but when used in the leading position, the angle should be between a minimum of 30° and a maximum of 37½°. An angle of less than 30° in the leading position causes severe chattering. The radial position must be used for motors in reversing service; however, to prevent brush chatter and avoid commutation difficulties, the brush angle must not deviate from the radial by more than 4°. Box types of holders are designed with very little clearance to prevent excess side motion that would cause wear of both brush and holder. Unfortunately, this small clearance makes it subject to clogging with dirt, and it should therefore be checked periodically for sticky brushes; each brush should be checked by moving it up and down in the holder.

The reaction type of holder, shown in Fig. 10-14d, is positioned to hold its brush at an angle of 30° to 37½° from the radial and must be operated in the leading direction. Hence machines that have this type of holder should not be reversed. If set properly, the reaction type of holder offers the advantage of smooth performance with no brush chatter,

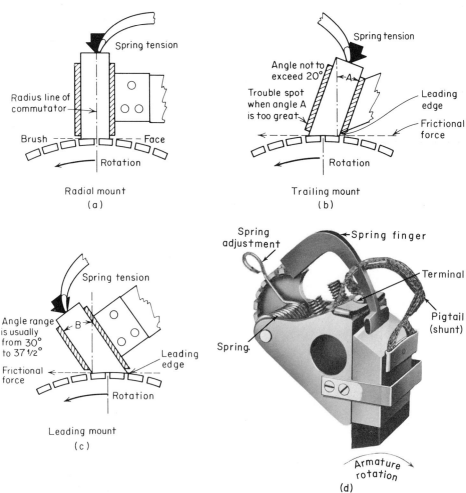

FIG. 10-14 Brush holders. (a), (b), (c) Box type; (d) reaction type. (*Union Carbide Corp.*)

and its open construction eliminates the possibility of sticky brushes. All brush holders on a given machine must have the same angle. Brushes set at the wrong angle do not make contact with the commutator at the correct place and cause severe sparking.

10-10 Brush Spacing

The spacing between positive and negative sets of brushes may be checked by wrapping a strip of paper around the commutator, as shown in Fig. 10-15. The paper is marked at the full circumference, then removed and carefully marked off in as many equal parts as there are brush arms. The paper is then replaced, and one of the marks is placed coincident with the edge of a brush and is held in position with Scotch tape. If all the brush arms are properly spaced, the brush edge of each will coincide with the respective markings on the paper strip. A deviation greater than $\frac{1}{64}$ in. should be corrected through adjustment of the brush arms.

10-11 Connecting Brush Pigtails

When installing brushes, the brush pigtail, also called *shunt,* must be properly bolted to the brush box, as shown in Fig. 10-14d. A loose shunt

FIG. 10-15 Checking brush spacing. (*United States Merchant Marine Academy.*)

terminal causes some of the current to be conducted through the brush box rather than through the pigtail, causing burning or rough spots on the inside of the box.

If two brushes are in parallel, as shown in Fig. 10-16, and the pigtail of one is not tightly bolted to its brush box, it will not carry its share of the current. The properly connected brush will carry a greater percentage of current and get hotter. Unfortunately, the resistance of carbon decreases with increasing temperature, causing the properly connected brush to take an even greater percentage of the total current. The high current density at this one brush may cause sparking at the commutator.

10-12 Brush Seating

The seating of all brushes to the exact curvature of the commutator and collector rings is a prerequisite to good brush performance. Improper seating of some brushes causes the others to carry a greater portion of the load. The net result is overheating and sparking, destructive to both the commutator surface film and the brushes.

Before new brushes are seated, it is essential that the holders be inspected and any roughness filed smooth. The pigtails should be checked for frayed wires and loose connections. Brushes with defective pigtails should not be used, because there will be current through the sides of the brush, damaging the holder. The brushes should be checked for freedom of movement in the holders without excess side play.

The actual fitting of brushes to the commutator should be done with sandpaper when the machine is not in operation, or with a brush-seating stone if the seating is to be done while the machine is running. Emery cloth or paper should never be used for seating brushes or polishing the commutator. Emery is very abrasive, and any particles that become embedded in the brush face will score the commutator.

When sandpaper is used for seating, it should be pulled back and forth under the brush with the sand side facing the brush. The brush tension should be adjusted for maximum pressure during the sanding operation, and the ends of the sandpaper should be pulled along the curvature of the commutator to prevent rounding the edges of the brushes. If the machine is nonreversing, the last few strokes should be made in the known direction of rotation. To facilitate the sanding operation, coarse sandpaper, grade $1\frac{1}{2}$, may be used for the initial cutting, followed by a fine grade, such as grade 0, for the final cut. Long sheets of sandpaper, which are purchased on a roll and are available to fit the entire circumference and length of the commutator, may be used for the simultaneous sandpapering of all brushes. The sandpaper should be wrapped around the commutator and held in position with Scotch tape; the brushes should be placed in the holders, and the armature should be revolved slowly until the brushes are shaped to the correct curvature. After the sanding operation is completed, the carbon dust should be removed with a vacuum cleaner or blown out with dry compressed air. You should avoid blowing the dust into the windings. Figure 10-16a presents an illustration of the proper method of sanding brushes.

Brush seating with a stone is illustrated in Fig. 10-16b. The stone is of fine-grain composition and is applied at the commutator close to the brush holder while the machine is in motion. The fine grains worn off the stone are carried under the brush, grinding it to the curvature of the commutator. The stone should be applied a little at a time, and the brush should be checked after each application until complete seating is obtained.

In addition to ease of operation, a brush-seating stone offers the advantage of grinding the brushes to the exact curvature of the commutator

FIG. 10-16 (a) Proper method of sanding brushes. (*Westinghouse Electric* Corp.)

FIG. 10-16 (b) Brush-seating stone applied to a commutator. (*Martindale Electric* Co.)

and, if not used to extremes, will not noticeably affect the commutator surface film.

Sandpaper has the disadvantage of grinding the brush surface to a radius equal to that of the commutator plus the thickness of the sandpaper; thus the true curvature will not be obtained until the machine has been in operation for some time. It is not advisable to operate a machine at full load until the brushes are worn to the true curvature. The procedure for seating brushes to a commutator also applies to seating brushes to collector rings. Machines should not be placed under load until at least 75 percent of the brush face is properly seated.

10-13 Brush Pressure

A curve showing the relationship between the rate of brush wear and brush pressure for electrographitic brushes is shown in Fig 10-17a; the best brush wear is obtained with a brush pressure of between 2 and 3

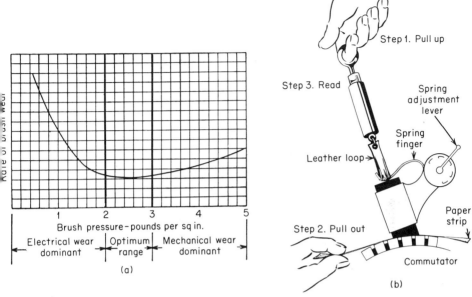

FIG. 10-17 (a) Relationship between brush pressure and rate of brush wear for lampblack base, electrographitic grades; (b) measuring total brush force with a spring balance. (*Union Carbide Corp.*)

psi. Insufficient pressure causes poor electrical contact and hence chattering and arcing between the commutator and the brush, rapidly wearing the brush. On the other hand, too much pressure causes excessive friction, also resulting in excessive brush wear. However, as indicated by the curve, pressures below the optimum value cause much greater brush wear than the correspondingly higher pressures. Similar effects occur with other grades of brushes, but the range of optimum performance may be different. The optimum brush pressure for other grades of brushes is given in Table 10-1.

The total force on the brush may be measured with a spring balance and a strip of paper, as shown in Fig. 10-17*b*. The leather loop must be placed under the end of the spring finger, where the finger touches the brush. The scale is read when the strip of paper placed between the commutator and the brush can be drawn out with very little effort. The amount of spring force necessary to provide the correct brush pressure may be obtained by multiplying the recommended pressure in pounds per square inch by the cross-sectional area of the brush face and then adjusting this figure, in the case of horizontal machines, by the weight

Table 10-1   Recommended Brush Pressure

Brush grade	Pressure, psi
Carbon	$1\frac{3}{4}-2\frac{1}{2}$
Carbon-graphite*	$1\frac{3}{4}-2\frac{1}{2}$
Graphite-carbon*	$1\frac{3}{4}-2\frac{1}{2}$
Electrographitic	2–3
Graphite	$1\frac{1}{4}-2$
Metal graphite	$2\frac{1}{2}-3\frac{1}{2}$
Fractional-horsepower motors	4–5

* A carbon-graphite brush has more coke type of carbon than graphite. A graphite-carbon brush has more graphite than coke type of carbon.
SOURCE: Union Carbide Co.

of the brush. Thus, brushes that ride the top of the commutator should have a lower spring force, and those which ride the bottom of the commutator should have a greater spring force than those which ride on the side.

BRUSH NEUTRAL

The neutral setting of the brush holders should not be adjusted unless there is reasonable evidence that the original installation was incorrect or the neutral was shifted during overhaul or commutator wear; a commutator whose diameter has been reduced by excessive wear may cause the brushes to make contact at an off-neutral position. The methods described for locating the neutral are based on the recommendations of the Institute of Electrical and Electronic Engineers.[1]

10-14 Physical-inspection Test

A chisel mark, covered with white paint for easy location, is often used to mark the neutral position. The mark is placed on the bracket that supports the brush rigging and on the end bell to which it is clamped. If off-neutral, the brush-rigging clamp is loosened and the rigging rotated until its chisel mark lines up with the one on the end bell. Some manufacturers of large machines mark two armature slots and three commutator bars to assist the operator in locating the proper brush position. To obtain the correct setting, the armature should be rotated so that the marked

[1] Test Code for Direct Current Machines, IEEE No. 113, 1962.

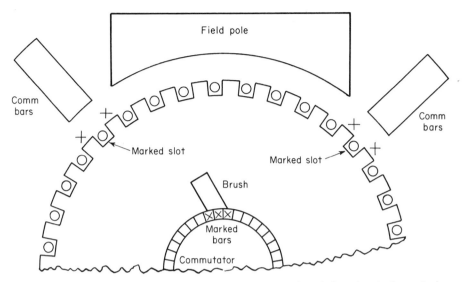

FIG. 10-18 Factory marks on armature slots and commutator bars help to locate the neutral position for the brushes.

slots are directly under the center of the interpoles. With the armature held in this position, the brush rigging should be adjusted so that one set of brushes is centered on the three marked commutator bars, as shown in Fig. 10-18. If no markings are present and the neutral position is in question, tests should be made to determine the correct position.

10-15 Inductive-kick Test (for Approximate Location of Neutral without Running Machine)

The inductive-kick test is a fairly accurate method for locating the no-load neutral and may be used when the brushes must be positioned prior to operation of the machine. It is equally effective with interpole and non-interpole machines. To make the test, the shunt field should be energized by a source of direct current in series with a rheostat and switch, as shown in Fig. 10-19a. All brushes should be removed, and two beveled brushes installed, one in a positive holder and the other in a negative holder, as illustrated in Fig. 10-19b. A dc voltmeter with a 0.5–1.5–15-volt range should have its 15-volt range connected across the two brushes of opposite polarity. The interpoles, series field, and compensating winding must not be used. With the rheostat adjusted to obtain a value of field current equal to 20 percent of its normal value, the switch is quickly opened and the momentary deflection of the voltmeter pointer is observed.

A knife switch with a quick-break blade, or a snap switch, provides the same decay rate for the field current each time the switch is opened. The brush rigging should then be shifted a few degrees, and the test repeated. If the magnitude of the voltmeter deflections increases, the brushes must be shifted in the opposite direction. The 15-volt connection should be used until the deflections fall within the range of the lower scales. The neutral position for the brush rigging is located when there is no deflection on the 0.5-volt range of the voltmeter as the field circuit is opened or closed.

The principle of the inductive-kick test is explained with the aid of Fig. 10-19c and d. Because the armature is stationary, the only voltages

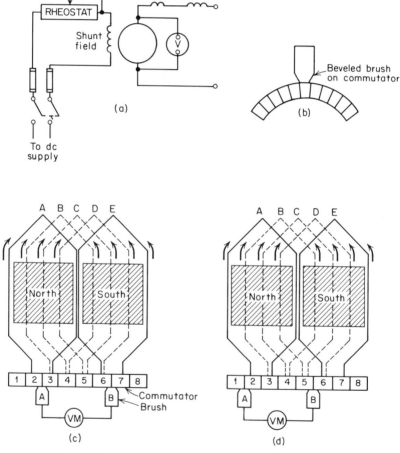

FIG. 10-19 Inductive-kick test for locating the brush neutral position. (a) Circuit diagram; (b) beveled brush resting on commutator; (c) brushes in neutral position; (d) brushes off neutral.

generated in the armature coils are those resulting from the "inductive kick" caused by making and breaking the field circuit. When the field circuit is closed, the poles establish north and south polarity as shown; when the field circuit is opened, the collapse of this flux through the windows of the armature coils generates voltages in the coils in a direction to oppose the collapse of flux (Lenz's law). See Sec. 8-2.

Flux from only the north pole passes through the window of coil A; hence the voltage induced in coil A, because of the collapse of flux, will be in the clockwise direction. Similarly, flux from only the south pole passes through the window of coil E; hence the voltage induced in coil E will be in the counterclockwise direction. However, flux from both poles passes through the windows of coils B, C, and D, causing voltages to be induced in each of these coils in both the clockwise and the counterclockwise directions. The direction of the induced voltages in each coil, as a result of opening the field circuit, is shown with arrows in Fig. 10-19c and d.

The relative magnitude of the voltage generated by each pole in each one of the five coils depends on the percentage of the total pole flux that passes through the respective coil window; all the north pole flux passes through the window of coil. A, three-quarters through coil B, one-half through coil C, one-quarter through coil D, and none through coil E. Assuming that the collapse of north pole flux causes the voltage in coil A to attain a peak value of 4 volts, then the corresponding voltages in the other coils, because of the action of this same pole, will be 3 volts in B, 2 volts in C, 1 volt in D, and zero volts in E. Similarly, the collapse of south pole flux will cause the generated voltages in coils E, D, C, B, and A to have peak values of 4, 3, 2, 1, and 0 volts respectively. The magnitude and direction of the voltages generated in each coil by the collapse of flux in the north and south poles are listed in Table 10-2. Counterclockwise voltages are assumed positive ($+$), clock-

Table 10-2 *Magnitude and Direction of Voltage in Armature Coils Due to Buildup of Flux in Main Field Poles*

Armature coil	North pole	South pole	Net voltage in each coil
A	-4 (CW)	0	-4
B	-3 (CW)	$+1$ (CCW)	-2
C	-2 (CW)	$+2$ (CCW)	0
D	-1 (CW)	$+3$ (CCW)	$+2$
E	0	$+4$ (CCW)	$+4$

wise voltages are assumed negative (−). The net voltage in each coil is listed in the right-hand column.

With the brushes in the neutral position (Fig. 10-19c), the voltmeter reads the voltage between commutator bars 3 and 6; coils A and B are each shorted by a brush, and therefore do not contribute to the total voltage. This voltage, which is the summation of the net voltages in coils B, C, and D, adds up to zero:

$$-2 + 0 + 2 = 0$$

With the brushes off neutral, a distance of only one commutator bar (Fig. 10-19d), the voltmeter reads the summation of voltages between commutator bars 2 and 5; discounting the coils shorted by the brushes, the voltmeter indicates the summation of the net voltages in coils A, B, and C. This voltage adds up to

$$-4 - 2 + 0 = -6 \text{ volts}$$

10-16 Commutating-pole Test (for Commutating-pole Generators and Motors above 1 Hp)

This test is very accurate and applies only to commutating-pole machines. It is particularly adaptable to generators but may be used on motors if some mechanical means is provided to drive them as generators. The test utilizes the voltage generated in the armature coils as they sweep past the commutating poles and requires a set of beveled brushes placed in brush holders of opposite polarity. The brush holders are connected to a low-reading dc voltmeter, as was described for the inductive-kick test. However, the beveled edge of the brushes must be slightly wider than the thickness of the mica. The commutating poles and compensating winding (if used) must be separately excited with about 3 percent of the rated current. The series field should not be used. Before the test is made, all traces of residual magnetism must be removed. This may be accomplished by placing a reversing switch in the shunt field circuit, and flashing the field in one direction or the other until all traces of residual magnetism are eliminated. These connections are illustrated in Fig. 10-20a.

To locate the neutral, start the prime mover in its normal direction of rotation, adjust it to rated speed, and flash the field with the reversing switch to remove the residual magnetism; this will be indicated by a zero reading on the voltmeter. Then, with the reversing switch open, energize the commutating poles, and shift the brush rigging a few degrees at a time until the voltmeter indicates zero. This is the neutral position. When the neutral is located, open the commutating-pole circuit, and again check for the presence of residual magnetism. If evidence of residual magnetism is indicated on the voltmeter, repeat the entire procedure.

The principle of the commutating-pole test is explained with the aid of Fig. 10-20b and c. Because the coils wound around the main field poles are not energized and all the residual magnetism in the iron of the main poles is eliminated, the only voltage generated in the rotating armature coils is due to the commutating-pole flux. Because a commutating pole is very narrow, its flux through the window of an armature coil changes from all-in to all-out in a very brief interval of time (see Sec.

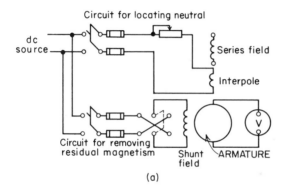

(a)

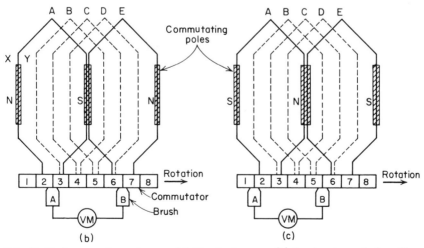

(b)   (c)

FIG. 10-20 Commutating-pole test. (a) Circuit diagram; (b) brushes in neutral position; (c) brushes off neutral.

8-3). This is shown in Fig. 10-20b. When the left side of coil A is in position X, all the flux passes through the window of coil A, but when in position Y, none of the flux passes through its window. Furthermore, because the interpole is narrow, it generates a voltage in only one or two armature coils at a time. Thus, for the instant shown in Fig. 10-20b and c, a voltage is generated in coils A and E, but none in coils B, C, and D.

If the brushes are in the neutral position (Fig. 10-20b), they will short-circuit coils A and E, and the voltage indicated on the voltmeter will be zero. If the brushes are off neutral (Fig. 10-20c), they will not short-circuit coils A and E, and the voltmeter will indicate the summation of all coil voltages between bars 2 and 5; and because the voltage is zero in coils B and C, the voltmeter will indicate the voltage generated in coil A.

10-17 Reverse-rotation Test (for Reversible Motors)

The speed of the motor should be accurately determined with a tachometer for both directions of rotation and with the same load and voltage. If the speed in both directions is not the same, the brushes should be shifted a few degrees, and the test repeated. The correct position for the brushes is the one that results in the same speed for both directions of rotation. Although the test is most accurate when made under full load, it may be done at any load or no load so long as it is the same for both directions. Adjustable-speed motors should be tested at the maximum rated speed. All brushes must be properly seated before this test is made.

10-18 Causes of Unsatisfactory Brush and Commutator Performance

Tables 10-3 and 10-4, along with Fig. 10-21, provide an effective means for determining the primary source of most brush and commutator troubles; the numbers listed in the third column of Table 10-3 are keyed to Table 10-4.

• *Example 10-1.* Suppose the brushes of a 10-pole dc motor are sparking, and examination of the commutator reveals bar marking at five regions equally spaced around the commutator. Because the bar marking is at pole pitch spacing (see Fig. 10-21b), Table 10-3 suggests that the primary source of trouble is listed under 25 and 37 in Table 10-4. These faults are (25) open or high resistance connection at the commutator, or (26) poor connection at the brush pigtail, or both.

Table 10-3 Trouble-shooting Chart for Commutators and Brushes

Indication appearing at brushes	Immediate cause	Key to primary sources of poor brush performance
Sparking	Commutator surface condition	1-2-3-43-44-45-46-49-59-60
	Overcommutation	7-12-31-33
	Undercommutation	7-12-30-32
	Too rapid reversal of current	7-12-30-32
	Faulty machine adjustment	8-9-11
	Mechanical fault in machine	6-14-15-16-17-18-19-20-21-22
	Electrical fault in machine	25-27-28-29
	Bad load condition	38-39-40-41-42
	Poorly equalized parallel operation	7-13-23-34
	Vibration	51-52
	Chattering of brushes	See "Chattering or noisy brushes"
	Wrong brush grade	55-57-59
	Fluctuating contact drop	50
Etched or burned bands on brush face	Overcommutation	7-12-31-33
	Undercommutation	7-12-30-32
	Too rapid reversal of current	7-12-30-32
Pitting of brush face	Glowing	See "Glowing at brush face"
	Embedded copper	See "Copper in brush face"
Rapid brush wear	Commutator surface condition	See specific surface fault in evidence; also 50
	Severe sparking	See "Sparking"
	Imperfect contact with commutator	11-14-15-16-51-52
	Wrong brush grade	54-58
Glowing at brush face	Embedded copper	See "Copper in brush face"
	Faulty machine adjustment	7-12
	Severe load condition	38-39-41-42
	Bad service condition	46-47
	Wrong brush grade	57-61-62
Copper in brush face	Commutator surface condition	2-3
	Bad service condition	43-46-47-48-49
	Wrong brush grade	59-61
Flashover at brushes	Machine condition	14-35
	Bad load condition	38-39-41-53
	Lack of attention	5-11
Chattering or noisy brushes	Commutator surface condition	See specific surface fault in evidence
	Looseness in machine	15-16-17
	Faulty machine adjustment	10-11
	High friction	6-43-45-49-52-58-59
	Wrong brush grade	55-58-59

Table 10-3   Trouble-shooting Chart for Commutators and Brushes (continued)

Indications appearing at commutator surface	Immediate cause	Key to primary source of poor commutator performance
Brush chipping or breakage	Commutator surface condition	See specific surface fault in evidence
	Looseness in machine	15-16-17
	Vibration	52
	Chattering	See "Chattering or noisy brushes"
	Sluggish brush movement	14
Rough or uneven surface		1-2-3-4-17
Dull or dirty surface		5-44-60
Eccentric surface		1-19-22-52
High commutator bar	Sparking	17
Low commutator bar	Sparking	2-25
Streaking or threading of surface (Fig. 10-21d and e)	Sparking	43-44-45-46-49-59
	Copper or foreign material in brush face	2-3-46-47-48-61
	Glowing	See "Glowing at brush face"
Bar etching or burning	Sparking	2-3-7-12-30-31-32-33
	Flashover	5-11-14-35-38-39-41-53
Bar marking at pole pitch spacing (Fig. 10-21b)*	Sparking	25-37
Bar marking at slot pitch spacing (Fig. 10-21c)†	Sparking	7-12-30-57-60
Flat spot	Sparking	19-23-25-41-42-53
	Flashover	5-11-14-35-38-39-41-53
	Lack of attention	1-5-11
Discoloration of surface	High temperature	See "Heating at commutator"
	Atmospheric condition	44-46
	Wrong brush grade	60

Table 10-3   Trouble-shooting Chart for Commutators and Brushes (continued)

Indication appearing at commutator surface	Immediate cause	Key to primary sources of poor commutator performance
Raw copper surface	Embedded copper Bad service condition Wrong brush grade	See "Copper in brush face" 43-45-47-49 59-61
Rapid commutator wear with blackened surface	Burning Severe sparking	2-3-11-14 See "Sparking"
Rapid commutator wear with bright surface (Fig. 10-21f) Copper dragging (Fig. 10-21a)	Foreign material in brush face Wrong brush grade Brush vibration	43-45-47-49 61 39-52-58-59

Indication appearing as heating	Immediate cause	Key to primary sources of overheating
Heating in windings	Severe load condition Unbalanced magnetic field Unbalanced armature currents Poorly equalized parallel operation Lack of ventilation	38-41-42-53 18-19-20-21-27-28-29 8-19-22-25-27-28-29-37 7-13-23-34 6
Heating at commutator	Severe load condition Severe sparking High friction Poor commutator surface Depreciation High contact resistance	38-41-42 7-8-9-12-20-33-45-57 10-11-36-43-45-49-58-59 See specific surface fault in evidence 6-24 56
Heating at brushes	Severe load condition Faulty machine adjustment Severe sparking Raw streaks on commutator surface Embedded copper Wrong brush grade	38-41-42 7-10-11-12-26 See "Sparking" See "Streaking or threading of surface" See "Copper in brush face" 57-58-59-61-62

* Bar marking at pole pitch spacing is indicated by discoloration or etching of the commutator in groups of bars equally spaced around the commutator. The number of discolored areas is related to one-half the number of main field poles. For example, a six-pole machine may have three discolored areas equally spaced around the commutator (see Fig. 10-21b).

† Bar marking at slot pitch spacing is indicated by a film discoloration or mark every second, third, or fourth commutator bar. This is related to the number of coils contained in one of armature slots (see Fig. 10-21c).

SOURCE: Union Carbide Co.

Table 10-4   Primary Sources of Poor Brush and Commutator Performance

Numbers are keyed to Table 10-3

Preparation and Care of Machine
1. Poor preparation of commutator surface
2. High mica
3. Featheredge mica
4. Bar edges not chamfered after undercutting
5. Need for periodic cleaning
6. Clogged ventilating ducts

Machine Adjustment
7. Brushes in wrong position
8. Unequal brush spacing
9. Poor alignment of brush holders
10. Incorrect brush angle
11. Incorrect spring tension
12. Interpoles improperly adjusted
13. Series field improperly adjusted

Mechanical Fault in Machine
14. Brushes tight in holders
15. Brushes too loose in holders
16. Brush holders loose at mounting
17. Commutator loose
18. Loose pole pieces or pole-face shoes
19. Loose or worn bearings
20. Unequal air gaps
21. Unequal pole spacing
22. Dynamic unbalance
23. Variable angular velocity
24. Commutator too small

Electrical Fault in Machine
25. Open or high resistance connection at commutator
26. Poor connection at pigtail
27. Short circuit in field or armature winding
28. Ground in field or armature winding
29. Reversed polarity on main pole or interpole

Machine Design
30. Commutating zone too narrow *

31. Commutating zone too wide*
32. Brushes too thin
33. Brushes too thick
34. Magnetic saturation of interpoles
35. High bar-to-bar voltage
36. High ratio of brush contact to commutator surface area
37. Insufficient cross connection of armature coils

Load or Service Condition
38. Overload
39. Rapid change of load
40. Reversing operation of non-interpole machine
41. Plugging
42. Dynamic braking
43. Low average current density in brushes
44. Contaminated atmosphere
45. "Contact poisons"
46. Oil on commutator or oil mist in air
47. Abrasive dust in air
48. Humidity too high
49. Humidity too low
50. Silicone contamination

Disturbing External Condition
51. Loose or unstable foundation
52. External source of vibration
53. External short circuit or very heavy load surge

Wrong Brush Grade
54. "Commutation factor" too high§
55. "Commutation factor" too low†
56. Contact drop of brushes too high
57. Contact drop of brushes too low
58. Coefficient of friction too high
59. Lack of film forming properties in brush
60. Lack of polishing action in brush
61. Brushes too abrasive
62. Lack of carrying capacity

* The commutating zone is the neutral region between adjacent main field poles. Any flux in this region is caused by the commutating poles (interpoles). The size of the commutating zone depends on the number of commutator bars spanned by a brush.
† Commutation factor is a measure of the shock-absorbing characteristic of the carbon brush. The factor depends on the brush grade.

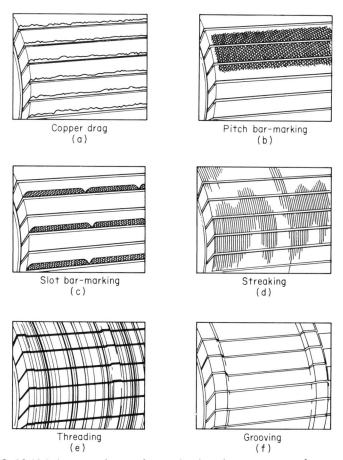

Copper drag	Pitch bar-marking
(a)	(b)
Slot bar-marking	Streaking
(c)	(d)
Threading	Grooving
(e)	(f)

FIG. 10-21 Indications of unsatisfactory brush and commutator performance.

Bibliography

Kalb, W. C., and F. K. Lutz: "Carbon Brushes for Electrical Equipment," Union Carbide Corp., Carbon Products Division, 1966.

Lynn, C., and H. Elsey: Effects of Commutator Surface Film Conditions on Commutation, *Trans. Proc. AIEE* (1929), vol. 68, 1949. ?

Van Brunt, C., and R. Savage: Carbon Brush Contact Films, Part I, *Gen. Elec. Rev.,* 47, 16–19, July, 1944.

Test Code for Direct Current Machines, IEEE No. 113, 1962.

Questions

10-1. What is the composition of the surface film that forms on commutators and slip rings during normal operation?

10-2. What is the advantage of maintaining a good surface film on a commutator?

10-3. What causes a raw copper surface?

10-4. How does the brush temperature affect brush wear?

10-5. What adverse effect does silicone vapor have on the performance of dc machines?

10-6. What causes threading on commutators and slip rings?

10-7. State the operating conditions that require (*a*) very soft brushes and (*b*) very hard brushes.

10-8. What are some of the more common causes of brush sparking?

10-9. How should a commutator be cleaned?

10-10. How may a loose commutator bar be detected? What is the correct procedure for tightening a commutator?

10-11. What is the best method of resurfacing a commutator? Describe it in detail.

10-12. Why is sandpaper undesirable for resurfacing commutators, even if only a light dressing is needed?

10-13. What method should be used to remedy an eccentric commutator?

10-14. Explain why excessive grinding of a commutator may result in sparking.

10-15. What is the one commutator problem that hand stoning cannot correct?

10-16. Is it safe to use a commutator stone while the machine is carrying load? Explain.

10-17. What is the advantage of burnishing a commutator, and how is it done?

10-18. What is air-curing as applied to a commutator?

10-19. Why should one avoid touching a freshly ground commutator?

10-20. Describe several methods for undercutting mica. When should it be done, and why is it necessary?

10-21. What can be done to prevent the mica from absorbing oil and moisture?

10-22. If the brushes are to be removed during overhaul, why is it important that they be tagged for return to the same holder?

10-23. Make a sketch showing the correct way to stagger the brushes of a six-pole machine.

10-24. What bad effect does excessive brush play in a brush holder have on the operation of the machine?

10-25. When should brushes be replaced? Why should all brushes on a multibrush arm be replaced at the same time?

10-26. Explain why slip rings that carry direct current should have their connections reversed every 6 months.

10-27. What is the essential difference between the reaction type and the box type of brush holder? What are the advantages and disadvantages of each?

10-28. Describe the entire procedure, in proper sequence, for installing and seating new brushes.

10-29. What is a brush-seating stone? How is it used?

10-30. How much of the brush-contact area should be seated before full load may be applied?

10-31. What adverse reactions will be instigated by too high or too low values of brush-spring pressure?

10-32. Should the neutral setting of the brushes be adjusted every 6 months? Explain.

10-33. Describe the physical-inspection test for locating the brush neutral setting.

10-34. Describe the inductive-kick test for locating the brush neutral setting of a dc machine.

10-35. Describe the commutating-pole test for locating the brush neutral setting of a dc generator.

10-36. Describe the reverse-rotation test for locating the brush neutral setting of a reversible motor.

eleven

trouble shooting and emergency repairs of dc machinery

This chapter is devoted to logical methods for determining some of the common troubles that occur in dc machinery, and to suggestions for emergency repairs that will keep the equipment in operation until time is available for a major overhaul. Difficulties of a mechanical nature, such as brush position, brush tension, commutator irregularities, bearing wear, and vibration, are covered in other chapters.

Electrical apparatus itself does not reverse the connections to its field, armature, interpole, etc. Hence, unless there is reasonable suspicion that the connections were changed, it may be assumed that they are correct. The nameplate data of the machine should always be checked against the actual operating values of voltage, current, temperature, speed, etc. Poor performance may often be traced to operating above or below the rated nameplate values. Figure 11-1 illustrates some of the more visible effects that overload can inflict on electrical machinery. High temperature caused the commutator and armature banding wire to throw solder, grounding the field coils and the armature. A major overhaul was necessary to put the machine back in service.

11-1 Trouble-shooting a DC Motor

When a motor fails to start, the branch-circuit fuses (or molded-case breaker), the thermal overload relay, and the small fuses on the motor control panel should be the first things checked (see Secs. 14-3, 14-9, and 14-12). If the fuses are good and the breaker and relays are not tripped, the control panel should be checked (see Sec. 13-5).

To check the motor, disconnect the armature and field leads from the controller, and make continuity and ground tests of the shunt field circuit, the series field circuit, and the armature circuit, as shown in Fig. 11-2. The ground test is made by connecting one terminal of the megohmmeter to the shaft or an unpainted part of the motor frame and connecting the other terminal to the F leads, A leads, and S leads in turn. A zero reading indicates a grounded circuit.

The continuity test (open-circuit test) is made by connecting the two megohmmeter terminals to the corresponding circuit leads F_1F_2, A_1A_2, and S_1S_2, as shown in Fig. 11-2b. A zero reading indicates a complete circuit. A high reading on the megohmmeter indicates an open circuit.

A listing of some of the common troubles experienced with dc motors and their possible causes is given in Sec. 11-13.

11-2 Locating Shorted Shunt Field Coils

To check for short circuits in the shunt field coils of a dc motor or generator, a 120- or 240-volt 60-hertz voltage should be applied to the shunt field circuit, and the voltage drop measured across each field coil,

(a)

FIG. 11-1 (a) Overheating causes commutator and armature bands to throw solder. Operator should be cautioned to avoid overloading. (b) Overloading causes solder thrown from armature bands to be deposited on field and interpole coils, resulting in grounds and shorts. (c) Banding wire loosened by overheating damages the field coils. (General Electric Co.)

(b)

(c)

FIG. 11-1 (Continued)

FIG. 11-1 (d) Overheating of the armature causes the solder from the banding wire to run into the winding, resulting in a burnout. (*General Electric Co.*)

as illustrated in Fig. 11-3. Because each field coil has the same number of turns, the voltage across each should be the same.

• *Example 11-1.* For the six-pole field shown in Fig. 11-3, if the voltage applied to the shunt field is 240 volts, the average voltage across each coil should be

$$\frac{240}{6} = 40 \text{ volts}$$

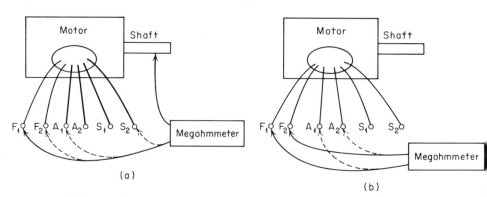

FIG. 11-2 (a) Ground test of motor circuits; (b) continuity test of motor circuits.

See also p.173

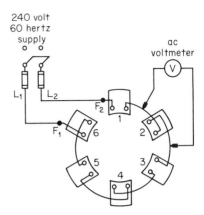

FIG. 11-3 Testing for shorted shunt field coils.

If the voltages across the coils are equal, it can be assumed that all six coils are free of shorted turns. A shorted turn in one coil causes its voltage and that of the adjacent coils to be noticeably lower than the average value. If the voltage drops across any coil is 10 percent less than the average, one or more turns may be shorted.

In Example 11-1, the average voltage, determined by dividing the impressed voltage by the number of field poles, is 40 volts. Hence, a voltage of less than 36 volts $(40 - 4)$ across a field coil is indicative of shorted turns. A tabulation of representative voltage drops in Table 11-1, for a machine with a shorted field coil, indicates that coil 4 has shorted turns.

Table 11-1 Voltage Drops for Shunt Field Test

Coil	Voltage
1	42
2	42
3	41
4	32*
5	41
6	42

* Shorted turn or turns.

Faulty field coils should be removed, and the coils unwound to note the general condition of the insulation and to locate the fault. If the insulation is very brittle and cracked, the other coils may be similarly deteriorated, and though not shorted, should be replaced.

If an ac test voltage is not available, the test may be performed with direct current. However, the dc test is not so sensitive as the ac test and will not detect a short involving one or two turns. Nevertheless, the dc test will detect a shorted coil if 10 percent or more of its turns are shorted.

If each coil in Example 11-1 has 200 turns, one shorted turn will result in a resistance change of only one two-hundredth of the coil resistance. This very small change in resistance will have no noticeable effect on the dc voltage measured across the coil. However, with ac applied, the transformer action of the shorted turn will lower the impedance of the affected coil and the adjacent coils appreciably, resulting in noticeably lower voltages (see Sec. 7-2).

11-3 Locating an Open Shunt Field Coil

An open shunt field coil may be located by connecting the field circuit to its rated voltage or less and testing with a voltmeter from one line terminal to successive coil leads, as shown in Fig. 11-4. The voltmeter will indicate zero until it passes the coil with the open; then full line voltage will be indicated. The test voltage may be either ac or dc. An open coil may also be located with an ohmmeter or megohmmeter test across each one; a closed coil will indicate a very low or zero reading, and an open coil will indicate infinity. An emergency repair of an open field coil can be made by unwinding the coil to expose the break, and then soldering the broken ends together.

11-4 Locating a Grounded Shunt Field Coil

A grounded coil may be located by connecting the field circuit to its rated voltage or less and measuring the voltage drop between each coil and ground, as shown in Fig. 11-5. A grounded coil will indicate very low or no voltage when tested from its terminals to ground.

A grounded coil may be located with a megohmmeter, but each

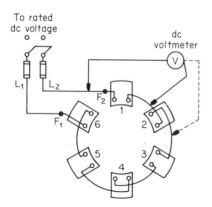

FIG. 11-4. Locating open shunt field coils.

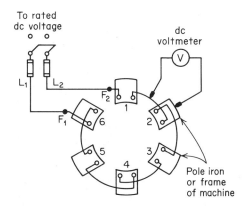

FIG. 11-5 Locating grounded shunt field coils.

coil must be disconnected from the others and tested to ground individually. A grounded coil will indicate zero. The megohmmeter test is recommended for shipboard use, because the voltmeter test requires the field frame to be insulated from the ship's hull or floor plates, and this is often difficult to do.

A ground caused by dirt and carbon dust can generally be cleared by cleaning and revarnishing. However, a coil that is grounded by chafing of insulation against the pole iron should be replaced if the copper shows signs of having worn away.

11-5 Testing for a Reversed Shunt Field Coil

A reversed-field coil in a generator or motor may be detected by testing with a magnetic compass. To make the test, the field should be connected to a source of direct current no higher than the rated voltage, and the polarity of each pole should be tested with a compass. The poles should test alternately north and south around the frame. The compass should not be held too close to the field, or its own polarity may be reversed. An alternate method for testing magnetic polarity makes use of the short steel bolts bridged across the gap between adjacent field poles; correct polarity will cause a strong attraction between the bolts, but a reversed coil will cause the ends to repel. These tests are illustrated in Fig. 11-6. Because of the very strong magnetic field at the pole faces, testing with a single bolt may give misleading results. Two bolts should be used.

11-6 Reversed Commutating Winding (Interpoles)

Because the function of a commutating winding is to reduce or eliminate sparking, its reversal will result in worse sparking than if no commutating winding were used. Hence, if a rebuilt or repaired machine sparks severely

(a)

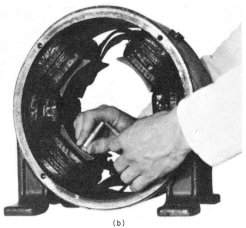

(b)

FIG. 11-6 Polarity test of field poles (a) using a compass and (b) using two steel bolts. (*United States Merchant Marine Academy.*)

when under load and when starting, the commutating poles may have been reversed. Shorting the interpole connections with a heavy copper jumper while the machine is under load will confirm or deny the suspicion. If the sparking is reduced when the winding is shorted, it indicates that the commutating winding is connected in reverse and should be corrected.

11-7 Location of Open Armature Coils

An open in the armature of a dc machine can generally be detected by a visual inspection of the commutator. The bars that connect to the defective coils will be partly burned or discolored by excessive heating and arcing, as shown in Fig. 11-7a. Many opens originate by overheating due to overload or inadequate ventilation, which causes the solder to

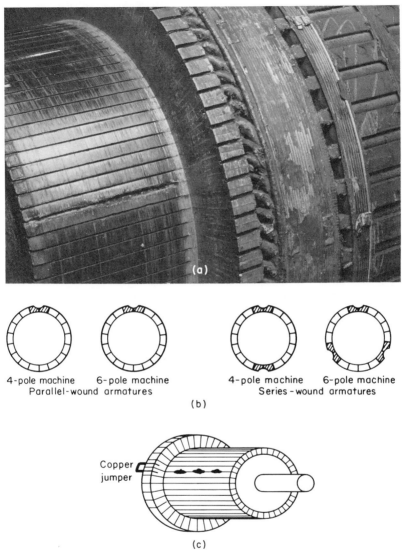

FIG. 11-7 (a) Open armature coil indicated by burned commutator bars (*General Electric Co.*); (b) relationship between burn spots and type of armature winding; (c) emergency repair of an open armature coil.

soften and be thrown from the coil connections at the commutator risers. Inspection of the risers will reveal this condition. The risers and coil connections should be thoroughly cleaned and resoldered with a hard solder, and the machine should be thoroughly cleaned to clear any clogged ventilating ducts. A bar-to-bar test of the armature will reveal any opens or high-resistance connections not detected by visual inspection. When the machine is put back in service, the load current should be checked and compared with the rated nameplate value. A higher than normal current is an indication of overload. If the open is not in the risers, identification of the type of winding and the required emergency repair may be determined by the number of burned spots and their geometric location on the commutator.

11-8 Emergency Repair of an Open Armature Coil

Figure 11-7b illustrates the number and location of burned spots caused by a single open in the two principal types of windings, using four-pole and six-pole machines as examples.

Parallel-wound armature. A parallel-wound armature will have only one burned spot on the commutator for each open, regardless of the number of poles. An emergency repair may be made by bridging over the affected bars, as illustrated in Fig. 11-7c. The jumper should be of the same size of armature wire and must be soldered in place. If the open is in a fractional-horsepower machine, the repair may be made with a lump of solder bridging the affected risers.

Series-wound armature. In machines with four or more poles, if the burned spots are symmetrically spaced around the commutator and the number of burned spots is equal to the number of poles divided by 2, the armature has a series winding, and only one (any one will do) of the burned spots should be bridged over. Bridging over the others will cause circulating currents in two or more coils and cause overheating.

11-9 Locating Shorted Armature Coils

A shorted armature coil can usually be detected by sparking at the brushes, overheating discoloration, and burning of its insulation. If a shorted coil is allowed to operate, the heat it generates may loosen the soldered connections to the commutator. A growler test or a bar-to-bar test may be used to locate a short that is not discernible by visual inspection; the very low coil resistance of integral-horsepower armatures makes it quite difficult, if not impossible, to check for shorted coils with a conventional ohmmeter.

Growler test. The growler test, shown in Fig. 11-8, represents a quick and reliable method for locating short-circuited armature coils. The growler is energized from the 120-volt 60 hertz supply voltage,

FIG. 11-8 Testing armature for short circuits with a portable growler. (*United States Merchant Marine Academy.*)

and a feeler fashioned from a hacksaw blade is "wiped" across the armature from slot to slot, as shown. The feeler will vibrate with a loud growling noise when directly over a slot containing the shorted coil. Feeler-test all slots with the growler set in different positions around the armature; shift the growler two or three slots at a time.

The principle of growler action may be explained by reference to Fig. 6-2, where the alternating magnetic flux set up by the growler is shown passing through the window of an armature coil. The changing flux through the coil window generates an alternating voltage in the armature coil. If the coil is shorted, the circuit is complete, and an alternating current will appear; the resultant magnetic field encircling the armature conductors will attract the hacksaw blade, causing it to vibrate in synchronism with the alternating current. A vibrating noise (growling) indicates that the coil is shorted. The size of the armature that can be tested in this manner depends on the size of the growler used. The portable growler shown in Fig. 11-8 is most effective in the range of $\frac{1}{10}$ to 15 hp. However, with careful observations and a sensitive feeler, it may be extended to 50 hp.

Bar-to-bar test. A bar-to-bar test is made by passing about 10 percent of the rated current through the armature and then measuring the voltage from bar to bar with a millivoltmeter. The current may be supplied by a battery or generator connected in series with a rheostat or bank of lamps, as shown in Fig. 11-9. The current may be conducted to the armature through two strips of copper secured to the commutator with cord or electrician's tape at points that are one pole apart. A millivoltmeter may be improvised by using an external shunt ammeter without its shunt.

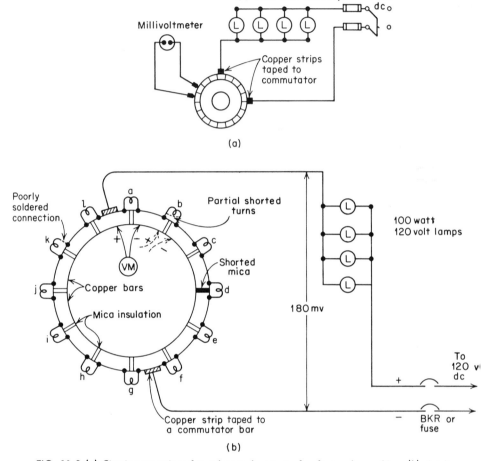

FIG. 11-9 (a) Circuit connections for a bar-to-bar test of a four-pole machine; (b) circuit connections for a two-pole machine (three different types of faults are indicated).

If the voltmeter deflection for a given bar-to-bar measurement is less than the average bar-to-bar value, a shorted coil is indicated. A zero reading indicates a dead short, and a very high or off-scale deflection indicates a high-resistance connection of the soldered joints at the risers or an open in the coil.

> **Warning.** (1) Before starting this test, the rheostat or the number of lamps in parallel should be adjusted to obtain no more than a half-scale deflection on the millivoltmeter for a few sample bar-to-bar readings. This will prevent overvoltage, caused by high-resistance armature connections, from damaging the millivoltmeter. (2) When making the bar-to-bar test, do not span more than one mica segment

at a time; spanning two or three mica segments doubles or triples the voltage across the millivoltmeter and may damage it.

The principle of the bar-to-bar test is explained with the aid of Fig. 11-9b, which shows the armature coil connections for a two-pole machine. The lamp bank serves to lower the 120-volt dc supply to 180 millivolts (0.180 volt) across the commutator. The armature current divides into two parallel paths consisting of six coils each: coils a, b, c, d, e, and f and coils l, k, j, i, h, and g. If there are no defective connections or defective coils, the voltage across each coil will be the same and equal to

$$\frac{180}{6} = 30 \text{ mv}$$

However, for the faults shown in Fig. 11-9b, the voltages across the different coils will not be the same. Shorted coils will offer less opposition to the current, and high-resistance connections more opposition to the current, causing the voltage drops to be respectively lower and higher than the average values. A tabulation of representative voltage drops for the fault conditions shown in Fig. 11-9b is given in Table 11-2.

Table 11-2   Voltage Drops for Bar-to-bar Test

Coil	Millivolts
a	39
b	24*
c	39
d	0†
e	39
f	39
g	27
h	27
i	27
j	27
k	45‡
l	27

* Partially shorted turns.
† Shorted mica.
‡ Poorly soldered connection.

Lamp test for fractional-horsepower armatures. A lamp test may be used to check for shorted coils in the armatures of fractional-horsepower machines, such as drills, vacuum cleaners, mixers, etc. The circuit hookup shown in Fig. 11-10 is very simple. The choice of lamp is by trial and error; the correct size lamp will burn at reduced brilliance when connected

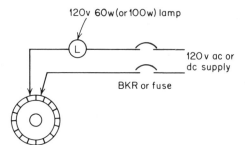

FIG. 11-10 Lamp test of fractional-horsepower armatures.

across a normal coil, but will burn brightly when connected across a shorted coil. Lamps with ratings of 25w, 40w, 60w, and 100 watts (all 120-volt) should be kept handy for such tests.

11-10 Emergency Repair of a Shorted Armature Coil

An emergency repair may be made by cutting the shorted coil in half at the back end of the armature. This may be done with a sharp chisel, hacksaw, or diagonal-cutting pliers, as shown in Fig. 11-11. Care must be taken to prevent damage to the adjacent coils. The open ends should be taped and secured to the armature. After the coil is opened, its commutator bars must be bridged over; use the procedure outlined for the emergency repair of an open armature coil, as illustrated in Fig. 11-7c.

Shorts caused by the absorption of carbon dust, oil, dirt, and moisture in the mica insulation between the commutator bars should be cleared by digging out and filling with commutator cement.

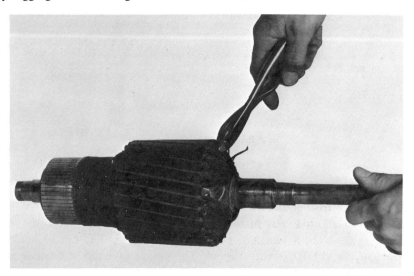

FIG. 11-11 Cutting an armature coil as part of an emergency repair. (*United States Merchant Marine Academy.*)

11-11 Locating Grounded Armature Coils

A grounded armature coil may be located with a bar-to-ground test. To make this test, about 10 percent of the rated armature current should be passed through the armature, and the voltage should be measured from bar to ground with a millivoltmeter. The current may be supplied by a battery or by a generator connected in series with a rheostat or a bank of lamps, as illustrated in Fig. 11-12. When this test is made aboard ship, the shaft and armature core must be insulated from the steel deck. The current may be conducted to the armature through two strips of copper conductor secured to the commutator with cord or electrician's tape at points that are one pole apart. One terminal of the milli-

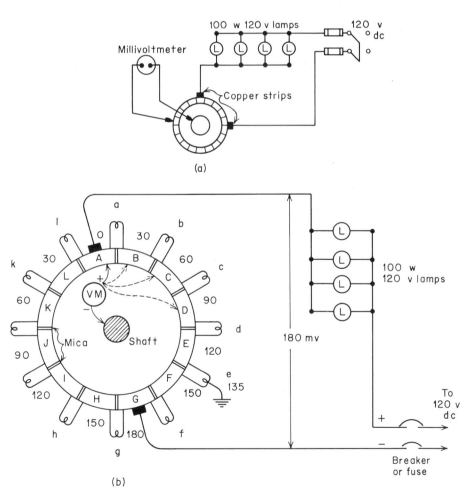

FIG. 11-12 Bar-to-ground test. (a) Circuit connections for a four-pole machine; (b) circuit connections for a two-pole machine (a ground fault is indicated).

voltmeter should be connected to the shaft, and the other should be moved from bar to bar along the commutator surface. Every bar should be tested, and those which indicate zero volts should be marked with chalk.[1] To complete the test, both copper conductors must be shifted several bars in the same direction, and the bar-to-ground test must be reapplied to the chalk-marked bars only. Those which indicate zero on the second test are the true grounds. The bars that indicate zero on the first test, but not on the second, are called *phantom* or *sympathetic grounds.*

Phantom grounds are indicated at points in the armature that have the same voltage as the grounded coil. Because all armature windings have at least two parallel paths, there will always be twice as many ground indications as actual grounds. Moving the power connections several bars shifts the phantom grounds, but the true grounds remain unchanged.

The principle of the bar-to-ground test is explained with the aid of Fig. 11-12b, where coil e is shown with a ground at its center. Assuming no opens, shorts, or high-resistance connections, the voltage across each coil will be 30 mv. To locate the grounded coil, connect one terminal of the millivoltmeter to the armature shaft. This connects it electrically to the center of coil e, which is at a potential of 135 mv. Then probe the commutator with the other test point, bar by bar, watching for the lowest meter readings. Because a voltmeter (or millivoltmeter) indicates the difference in potential (voltage) across its terminals, a zero or mini-

[1] An ungrounded armature will indicate zero volts between every bar and ground. Hence, the bar-to-ground test should be used only to determine the location of a grounded coil in an armature that is known to be grounded.

Table 11-3 Millivoltmeter Readings for Bar-to-ground Test

Bar	Millivolts
A	135
B	105
C	75
D	45
E	15*
F	15*
G	45
H	15†
I	15†
J	45
K	75
L	105

* Grounded coil.
† Phantom ground.

mum reading in the bar-to-ground test indicates that the "roving" test point is electrically connected to the grounded coil.

A tabulation of the millivoltmeter indications for the bar-to-ground test of Fig. 11-12b is given in Table 11-3; the lowest readings occur at the actual ground and at the phantom ground.

11-12 Emergency Repair of a Grounded Armature

The coil connections to the marked bars should be unsoldered, and each wire should be individually tested for grounds with an insulation-resistance tester, as shown in Fig. 11-13. The wires that indicate zero are grounded and should not be reconnected; they should be insulated from the others and secured to the armature with stout cord to prevent shifting by centrifugal force. The remaining wires may then be reconnected and an emergency repair effected for the now open armature circuit, using the procedure described for open armature coils (see Fig. 11-7c).

However, if the coil connections test clear of grounds, the fault lies in the commutator bars. A megohmmeter ground test of the marked bars will identify the faulty one. The most likely place for the ground to occur is at the string band that holds the mica in position, as illustrated in Fig. 12-3c; absorption of carbon dust, oil, dirt, and moisture may in time cause carbonized paths to occur between the commutator bars and the steel clamping ring, resulting in a ground. The ground may be cleared by digging out the "carbonized" mica and filling with commutator cement.

11-13 DC Motor Trouble-shooting Chart

The trouble-shooting chart in Table 11-4 lists the most common dc motor troubles, their possible cause, and the pertinent book section.

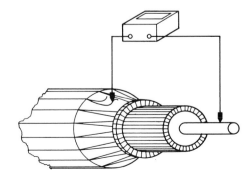

FIG. 11-13 Ohmmeter connections for the ground test of individual coils after disconnecting from the commutator.

Table 11-4   Trouble-shooting Chart

Dc motors

Trouble	Possible cause	Section
Fails to start	1. Blown fuses	14-3
	2. Frozen shaft	12-3
	3. Brushes not in contact with commutator	10-6
	4. Open in shunt field	11-3
	5. Open in armature circuit	11-7
Runs hot	1. Overload	9-7
	2. Shorted armature coils	11-9, 11-10
	3. Clogged ventilating ducts	2-18
	4. Brushes off neutral	10-14, 10-15, 10-16, 10-17
	5. Voltage too high or too low	9-7
	6. Shorted field coils	11-2
	7. High ambient temperature	
Sparks at brushes	1. Commutator and brushes dirty	10-1, 10-3
	2. Wrong brush grade	10-1
	3. Wrong brush adjustment	10-6
	4. Brushes off neutral	10-14, 10-15, 10-16, 10-17
	5. Open armature coil	11-7, 11-8
	6. Commutator eccentric	10-4
	7. High mica or high commutator bars	10-5
	8. Vibration	12-6
	9. Shorted or reversed commutating pole	11-6
Runs fast	1. Voltage too high	9-7
	2. Series field bucking shunt field	11-5
	3. Open in shunt field of a lightly loaded compound motor	11-3
	4. Shunt field rheostat resistance set too high	9-4
	5. Shunt field coil shorted	
Runs slow	1. Low voltage	9-7
	2. Overload	9-7
	3. Short circuit in armature	11-9, 11-10
	4. Brushes off neutral	10-14, 10-15, 10-16, 10-17
	5. Starting resistance not cut out	13-6

Bibliography

Annett, F. A., and A. C. Roe: "Connecting and Testing Direct Current Machines," McGraw-Hill Book Company, New York, 1955.

Rosenberg, R.: "Electric Motor Repair," Holt, Rinehart and Winston, Inc., New York, 1960.

Stafford, H. E.: "Troubles of Electrical Equipment," McGraw-Hill Book Company, New York, 1947.

Terminal Markings for Electric Apparatus, USAS C6.1-1956.

Veinott, C. G.: "Fractional Horsepower Electric Motors," McGraw-Hill Book Company, New York, 1948.

Questions

11-1. State the routine procedure that should be followed when a motor fails to start.

11-2. Describe a continuity test applicable to the shunt field of a motor.

11-3. Explain why alternating current is better than direct current when testing the shunt field coils for short circuits. How should this test be made?

11-4. Describe a test for locating an open shunt field coil. State the procedure for making an emergency repair.

11-5. Describe a test for locating a grounded shunt field coil. Can a grounded coil be repaired? Explain.

11-6. Describe a test procedure (that does not require a compass) for checking the polarity of the shunt field poles.

11-7. What test procedure may be used to determine whether or not the interpole circuit is reversed?

11-8. What commutator bar markings are indicative of an open armature coil?

11-9. State the correct procedure for making an emergency repair of (a) a series-wound armature and (b) a parallel-wound armature.

11-10. Describe a growler test for locating a shorted armature coil.

11-11. Sketch the circuit and state the correct procedure for making a bar-to-bar test of an armature. What precautions should be observed? What are the test indications for a high-resistance connection at the commutator?

11-12. Describe the lamp test for fractional-horsepower armatures.

11-13. State the correct procedure for making an emergency repair of a shorted armature coil.

11-14. Describe a test for locating a grounded armature coil.

11-15. State the correct emergency procedure for repairing a grounded armature coil.

11-16. What causes commutators to become grounded? Can grounded commutators be repaired?

twelve

mechanical maintenance of electrical machinery

12-1 Disassembly of Machines

The disassembly of machines is an important phase of repair work that must be done without damage to the component parts. The stator end shields, bearing housing and clamping plate, and other removable parts should be carefully punch-marked with respect to one another to facilitate proper positioning when they are reassembled. To prevent loss, all small parts should be stored in a box and tagged for identification; bearings should be wrapped in clean lintless cloth.

Machines equipped with bearing housings that slide through the end shield (cartridge type) should not have the bearing cap removed unless the bearing is to be changed. Bearings that are locked to both the shaft and the end shield must have the locknut removed before the machine can be disassembled. Bearings that contain oil should be drained before disassembly. Before disassembly, all leads to the brush rigging should be disconnected and marked. The brushes should be removed, and thin fiber or several layers of heavy paper should be tied or taped in place around the commutator to prevent scratching of the surface film. If the air gap between the rotor and stator is large, several thicknesses of sheet fiber placed there will relieve the strain on the remaining bearing when one end shield is removed. If necessary, remove keys, and file any burrs on the keyway and shaft.

The end shields are generally closely fitted to the frame of the machine with a rabbet type of joint. To remove the end shield, take out the tie bolts, and then tap the lip or knock-off lugs with a mallet, or a hammer on

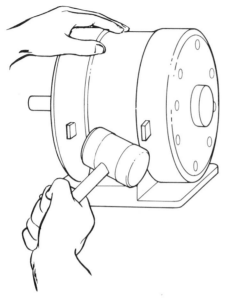

FIG. 12-1 Removing the end shield by tapping knock-off lugs with a mallet.

hardwood blocks, as shown in Fig. 12-1. Some ball-bearing and roller-bearing machines have very tight bearing fits and require the use of a *wheel puller* to assist in disassembly. The use of pry bars should be avoided because of possible damage to the mating edges and to the motor windings if the pry bar slips or is inserted too far.

The disassembly of explosion-proof machines requires exceptional care and attention to the manufacturer's instructions. Hammers and prying tools must not be used on the machined surfaces that form the flame-path surfaces of the enclosure. The tie bolts used to secure the end shields to the motor body are made of high-tensile-strength material in order to withstand the high pressures that may occur during an explosion. Hence, they must never be replaced with conventional low-tensile-strength bolts. These special bolts should be placed in a separate bag and marked with the motor serial number and the words "high-strength bolts."

The rotors of heavy machines should be lifted and removed in the manner shown in Fig. 12-2. A spreader bar is used to prevent the cable from cutting into the commutator or windings, and a length of pipe is used as an extension for short shafts. The rotor should not be lifted by the commutator, slip rings, or windings or allowed to rest on them. The armature should not be rolled on the floor, because the coils and banding wire are easily injured.

12-2 Reassembly of Machines

Before reassembling a machine, inspect all the mating surfaces for burrs and corrosion. Clean any corroded surfaces, file the burrs smooth, and

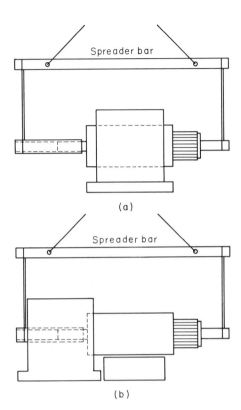

FIG. 12-2 Use of spreader bar to remove rotor from machine. (a) Position prior to lifting; (b) lowering to a platform.

apply an extremely light coating of oil or grease to the mating surfaces to aid in assembly. When reassembling, line up the holes in the end shield with the corresponding holes in the stator; square the mating parts by supporting the weight of the end shield; insert the bolts and then gently tap the end shield in position. It is generally easier to insert the side bolts first and the top and bottom bolts last. Do not force the parts together. Tap gently on the top, bottom, and sides, checking to see that the end shield is not cocked. The bolts should be tightened evenly and alternately, until the end shield is properly secured. If an inner bearing cap is used, an unusually long stud screwed into one of its mounting holes will facilitate alignment with the corresponding holes in the end shield.

When reassembling a machine with sleeve bearings, lift the oil ring and hold it in place with a piece of wire while the shaft is inserted. Do not hold the ring with your finger, or you may leave a piece of it in the bearing.

When driving pulleys, couplings, or fans on the motor shaft, the opposite end of the shaft should be supported against a stop or backed up with a heavy piece of brass or steel to absorb the blow and prevent damage to the bearing.

12-3 Bearings

The proper lubrication of electrical machinery extends the useful life of the bearings and prevents burned-out motors caused by a frozen shaft. Use the correct amount of lubrication. Excessive oiling or greasing forces the lubricant into the machine, where it may come into contact with the windings and damage the insulation. Figure 12-3 illustrates the damage that can be done by overlubrication. On the other hand, underlubrication causes excessive bearing wear; the rotor may rub on the stator, or the shaft may wipe the bearing and freeze.

When lubricating bearings, it is wise to check the operating conditions; belts that are too loose cause excessive pounding, and those which are too tight cause wear by overloading. Hot bearings indicate trouble. The safe operating temperature for most bearings, operating under normal conditions, is 40°C rise above the ambient temperature.

12-4 Care of Sleeve Bearings

Sleeve bearings are generally provided with oil reservoirs and settling chambers for collecting accumulated dirt. To clean this type of bearing, drain the old oil, flush with light mineral oil heated to about 75°C (an approved safety solvent may be used for flushing), and then refill with

FIG. 12-3 Example of damage that may result from overlubrication. (a) Overlubricated motor. This motor should be scheduled out of service and cleaned. (b) Interior of main generator frame with overlubricated commutator end bearing. (*General Electric Co.*)

FIG. 12-3 (Continued)

a good grade of mineral oil. SAE 20 or SAE 10 turbine oil is generally recommended, depending on the surrounding temperatures.

The correct oil level for various types of oil gauges is shown in Fig. 12-4. The oiling of sleeve bearings while the motor is in operation should be avoided, because it may result in overfilling. To drain the oil from machines not equipped with drain plugs, remove the bolts and revolve the end shield so that the filler cap is upside down.

The amount of bearing wear can be determined by measuring the air gap between the rotor and stator and comparing with previous measurements. Measuring on both sides and the bottom with a tapered, long-blade, feeler gauge is the accepted method.

Figure 12-5a illustrates a split arbor for removing sleeve-bearing liners. The section with the projection is inserted in the liner, and the projection is fitted into the oil-ring slot. The other section, or driving head, is then placed in the liner, and an arbor press is used to remove the bearing. Figure 12-5b illustrates a bearing-lining remover fashioned from a piece of steel key stock and a short length of pipe slightly smaller in diameter than the outer hole in the bearing housing. The key stock should be thin enough to enter the housing at an angle and just long enough to bear on the end of the steel shell of the liner. The key should

FIG. 12-3 (c) Grease and dirt damaged the string band and exposed the mica cone. Always keep string band and air inlet holes well cleaned. (*General Electric* Co.)

be properly positioned on the liner, the pipe placed over the key, and the lining pressed out with an arbor press or driven out with a hammer. If the bearing contains an oil ring, it should be held clear with a piece of wire while the lining is driven out.

Felt washers that are used to seal the ends of the bearing housing

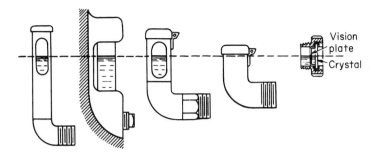

FIG. 12-4 Correct level for various types of oil gauges. (*General Electric Co.*)

should be replaced whenever new bearings are installed. The felt washer prevents airborne dust from contaminating the oil and prevents oil vapors from being drawn into the machine enclosure, where it can damage the insulation.

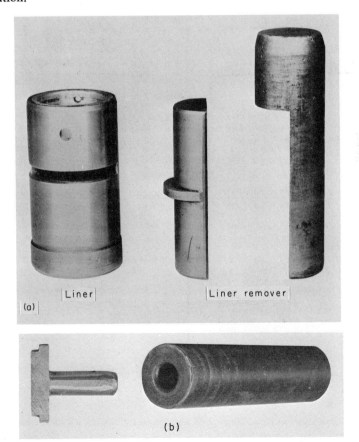

FIG. 12-5 Bearing-lining removers. (*a*) Split-arbor lining removers; (*b*) improvised lining remover. (*General Electric Co.*)

12-5 Care of Ball Bearings

Bearing housings equipped with pressure fittings and relief plugs may be cleaned and relubricated without ever removing the bearings from the housings. When this method of cleaning is used, both the bearing and housing are simultaneously purged of old grease. The cleaning agent may be either a light mineral oil heated to about 75°C or an approved safety solvent. If the latter is used, the bearing should be rinsed with a light mineral oil to remove all traces of solvent before regreasing. Figure 12-6a, b, c, and d illustrates the recommended method for cleaning horizontal motors that are equipped with pressure fittings and relief plugs. The bearing housings and relief fittings should be wiped clean to prevent the entry of dirt. The grease fittings and pressure-relief plugs, located respectively

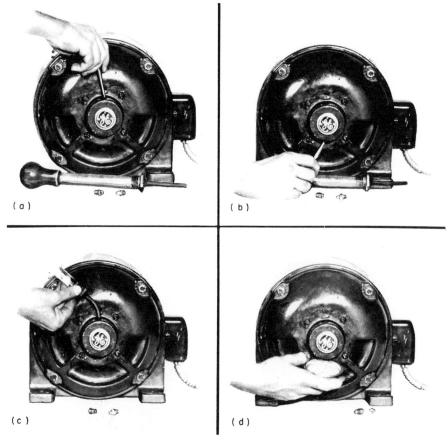

(a)   (b)

(c)   (d)

FIG. 12-6 Cleaning and lubricating a ball-bearing motor. (a), (b), (c), (d) Cleaning; (e), (f), (g), (h) lubricating. (*General Electric Co.*)

at the top and bottom of the bearing housings, should be removed, and a screw driver should be used to free the openings of hardened grease. Then with the motor running, an approved type of grease solvent should be injected into the bearing housing. This is easily done with a syringe, injecting the solvent into the bearing through the top hole. As the grease becomes thinned by the solvent, it drains out through the relief hole. Solvent should be added in small quantities until it drains out reasonably clear. The relief plug should then be replaced, and a small amount of solvent should be added and allowed to churn for a few minutes. The relief plug may then be removed, and the solvent allowed to drain. If the drainage is not clear, replace the relief plug, add more solvent, and allow it to churn a few minutes more. This should be repeated until the solvent drains clear. The bearing must then be flushed out with light mineral oil to remove all traces of solvent. The cleaning of vertical machines or machines not equipped with pressure-relief systems must be

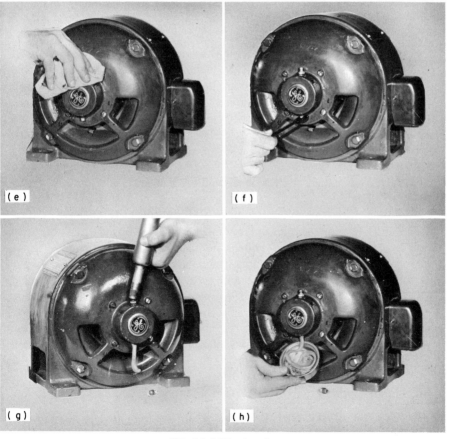

FIG. 12-6 (Continued)

done by disassembling the bearing housing or completely disassembling the machine.

The recommended procedure for regreasing machines that are equipped with pressure fittings and relief plugs is illustrated in Fig. 12-6e, f, g, and h. The bearing housings, pressure plugs, relief fittings, and grease gun should be wiped clean. The relief plugs should be removed, and the openings cleaned of hardened grease with a small screw driver. This permits the expulsion of old grease and prevents the buildup of excessive pressure in the bearing housing that might rupture the bearing seals. Then with the motor running, grease should be added with a hand-operated grease gun until it begins to flow from the relief hole. The motor should be allowed to run long enough to permit the bearing to expel all excess grease from the housing. The machine should then be stopped, the relief plug replaced, and the housing wiped clean. If it proves dangerous to lubricate the motor while it is running, follow the same procedure with the motor at standstill. Then start the machine, and allow it to run until all excess grease is expelled. Only grease recommended for electric motors should be used.

The greasing of vertical machines or machines that are not equipped with pressure-relief systems may be done by disassembling the bearing housings. The bearing and housing should be washed with an approved safety solvent, rinsed with light mineral oil, and then packed with grease. Only the lower half of the bearing housing and the space between the balls should be packed with grease. Do not fill the entire housing with grease; it may overheat and build up excessive pressure, thus forcing the grease into the motor housing.

A recommended regreasing schedule for electric motors is given in Table 12-1.

Ball bearings are precision-made and are adversely affected by dirt. Hence ball bearings should not be unnecessarily removed from the shaft and housing. New bearings should not be removed from their original wrapper until ready for immediate installation. A defective bearing should be replaced with the same size and type as the original. The defective bearing should be removed from the shaft with a bearing puller or arbor press by applying pressure against the inner race. The replacement bearing should be pressed on or tapped on, using a hammer and a clean metal tube or pipe that fits evenly against the inner race. Do not tap the outer race, because it may damage the bearing.[1]

If preheated to 200°F, the bearing will expand sufficiently to slip on the shaft with little or no driving. Heating of bearings is done best

---

[1] Tapping or pressing on the outer race is done only when the bearing is being pressed into a housing.

almost
same
till in
NFPA
70B
minimi
1982

Table 12-1   Recommended Regreasing Schedule

Type of service	Typical examples	Horsepower		
		½–7½	10–40	50–200
Easy	Motor operating infrequently (1 hr/day)	10 yr	7 yr	5 yr
Standard	Machine tools, fans, pumps, textile machinery	7 yr	5 yr	3 yr
Severe	Motors for continuous operations in key locations subject to severe vibration; steelmill service; coal and mining machinery	4 yr	2 yr	1 yr
Very severe	Dirty and vibrating applications where end of shaft is hot; high ambient	9 mo	4 mo	4 mo

SOURCE: General Electric Co.

in a temperature-controlled oven. If an oven is not available, bearings may be heated in a hot oil bath. However, heating in hot oil may cause contamination and deterioration of the grease.

12-6 Mechanical Vibration

Periodic vibration measurements of vital machinery can provide useful preventive-maintenance data on its mechanical condition. Although a sudden and significant increase in the amplitude of vibration is a very apparent indicator that something is wrong, a gradual increase in vibration may not be noticed until damage has occurred. In addition to damaging the machine, excessive vibration can cause damage to nearby equipment and to the buildings and ships in which it is installed; it also increases worker fatigue and reduces efficiency. Excessive vibration can be caused by misalignment between motor and driven equipment, loose mounting bolts, badly worn bearings, mechanically unbalanced rotor, bent or cracked shaft, or an excessively pulsating load.

Figure 12-7 illustrates two types of vibration-displacement indicators useful in vibration trouble shooting and preventive maintenance. The dial type of vibration indicator, shown in Fig. 12-7a, consists of a dial indicator mounted in a weighted case that rests on a coil spring. A plunger from the dial extends through the spring to make contact with the vibrating

(a)

(b)

FIG. 12-7 Vibration-displacement indicators. (a) Dial type;
(b) light-beam type. (*General
Electric* Co.)

surface. When the instrument is placed on a vibrating surface, the inertia
of the weighted case causes it to remain stationary while the plunger
is actuated by the vibrating surface. The vibrations cause the instrument
pointer to oscillate between two positions on the dial. The difference between the two points is the peak-to-peak displacement in mils (one mil
is 0.001 inch).

The light-beam type of vibration indicator, shown in Fig. 12-7b,
uses a stylus to transmit the vibrations to a small mirror. The mirror
reflects a light beam upon a frosted-glass scale marked off in mils. The
oscillation of the light beam across the scale gives the appearance of
a white line on the frosted glass. The length of the line indicates the
peak-to-peak displacement in mils. A flashlight battery and a miniature
lamp within the instrument supply the light.

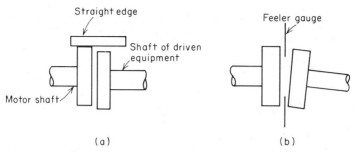

FIG. 12-8 (a) Checking horizontal and vertical alignment; (b) checking axial alignment.

More sophisticated equipment, not shown, can provide information on the frequency of vibration as well as the displacement. A change in the frequency of vibration of a rotating machine may signal a need for lubrication, thus providing a warning against bearing failure.

When trouble-shooting excessive vibration in rotating machinery, first check the mounting bolts, couplings, and foundations of the motor and the driven equipment. A loose mounting bolt can sometimes be detected by placing a finger at the parting line where the motor feet are bolted to the bedplate or deck. If this fails to indicate the trouble, and the bearings do not have excess play, uncouple the motor from the driven equipment and run a vibration test on the motor alone; if the motor vibrates it is probably out of balance.[1] If the motor does not vibrate when running alone, the vibration may be in the driven equipment or caused by shaft misalignment. Figure 12-8 illustrates an emergency method for checking shaft alignment using only a straightedge and a feeler gauge.

12-7 Rebanding Rotating Apparatus

Loosened banding wire on the rotors of ac and dc machines may be caused by shrinkage of the insulation and slot wedges, by rubbing of the rotor against the stator, or by softening of the solder from overheating. Whatever the cause, defective banding wire should be replaced, or the coils may be damaged by centrifugal force.

Sectional bands are easily replaced with a special clamping device shown in Fig. 12-9. When the device is pulled into position, a key is inserted into the band eyes, and the clamp is removed.

Figure 12-10 illustrates a simple method for obtaining the correct tension when nonsectional bands are applied. The rotor is placed in a lathe, and the banding wire is wound with little tension, over and beyond

[1] Electrical faults in the windings of motors or generators may also cause vibration (see Sec. 6-14).

FIG. 12-9 Sectional bands about to be clamped into position. (*General Electric Co.*)

the width of the regular band. Soldering clips are inserted under the band as it is wound and will serve to secure it when the correct tension has been obtained. Heavy insulation paper should be placed on top of the coils to prevent damage. The wire should be close-wound and, when past the end of the regular band width, should be wound over a small 2-by 4-in. wood block to provide slack for insertion of a pulley. The last few turns of banding wire should be wound over themselves and soldered in place to prevent loosening. The wood block should then be removed and the pulley slipped into position.

Then, using the arrangement shown in Fig. 12-10, the correct tension may be applied. The barrel should be filled with the required volume of water to obtain the proper tension, and the tackle should be adjusted to keep the beam level. With the lathe operating slowly, first in one direction and then in the other, the pulley moves back and forth across the band until all the slack is taken up. The pulley rises higher as the slack is given up, and it may be necessary to adjust the tackle to keep the beam level. If the angle θ of the wire over the sheave changes appreciably during the operation, an adjustment of the water in the barrel may be necessary.

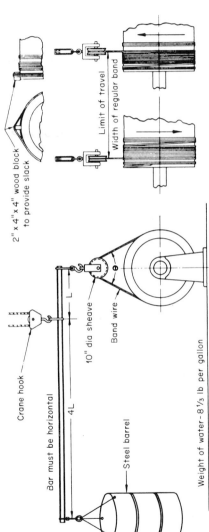

Total Load at End of Beam Including Weight of Barrel and 1/3 Weight of Beam.

Band Wire			Steel							Phosphor Bronze						
			Tension in Wire		Load On Beam — Lb.					Tension in Wire		Load On Beam — Lb.				
					Angle Θ of Wire Over Sheave							Angle Θ of Wire Over Sheave				
B. & S. Gage	Dia. In.	Area Sq. In.	Force Lb.	Stress Lb./Sq. In.	0°	30°	60°	90°	120°	Force Lb.	Stress Lb./Sq. In.	0°	30°	60°	90°	120°
8	.1285	.01295	850	66,000	425	410	368	300	212	600	46,000	300	290	260	212	150
10	.1019	.00817	600	73,500	300	290	260	212	150	400	49,000	200	193	173	141	100
12	.0808	.00512	400	78,000	200	193	173	141	100	250	49,000	112	108	97	79*	56
14	.0641	.00322	250	78,000	125	121	108	88	63	160	50,000	80	77	69	56	40
16	.0508	.00203	160	78,000	80	77	69	57	40	100	50,000	50	48	43	35	25

Weight of water—8 1/3 lb per gallon

Crane hook

Bar must be horizontal

4L

L

10" dia sheave

Band wire

Steel barrel

2" x 4" x 4" wood block to provide slack

Limit of travel

Width of regular band

FIG. 12-10 Method of banding and table of proper wire tension for different types and sizes of banding wire. (Allis-Chalmers Mfg. Co.)

When the proper tension is obtained, as indicated by no more slack, the soldering clips are bent over and soldered to the banding wire, as shown in Fig. 12-11. The surplus wire on either end of the band may then be cut off, and the entire band soldered. Preheating the armature to about 85°C prior to banding softens the insulation and thus permits pressing the coils into position as the band is applied.

When rebanding, use wire of the same size and material as originally supplied by the manufacturer. If heavier wire and clips are used, the added thickness may cause the bands and clips to rub on the pole faces, resulting in damage to armature and fields. After rebanding, the armature should be ground-tested with a megohmmeter.

Some manufacturers and repair facilities use resin and polyester-filled glass-fiber bands in both new construction and for replacement of metal bands in repair work. Properly applied, the glass bands have the same or higher tensile strength as steel, without the disadvantage of increased creepage paths to ground and heating due to induced currents that are common to steel bands. Furthermore, because glass bands can withstand

FIG. 12-11 Soldering the finished band of a motor armature. (*United States Merchant Marine Academy.*)

higher temperatures than conventional soldered steel bands and are non-conductive, their use results in less damage to a machine when a serious fault involving the bands occurs.

12-8 Shaft Currents

Machines with very strong magnetic fields often generate a voltage in the shaft that causes a current through the bearings. The resultant electrolytic action will in time cause bearing failure and possible sludging of the lubricating oil. To eliminate shaft currents, either the outboard bearing shell is insulated from the housing, or one outboard bearing pedestal is insulated from the bedplate as shown in Fig. 12-12. Only one bearing shell or one bearing pedestal is insulated. Insulating both bearings would permit the buildup of electrical charges (static electricity); the resultant high voltage would be hazardous to operating personnel and could cause a breakdown of the machine insulation. All lube-oil piping connections, exciter bearings, etc., that make connection to the shaft or bearing in question must also be insulated. The shell or pedestal insulation should be kept clean and should never be painted with metallic or other conducting paints.

The pedestal insulation of a disassembled machine may be tested with a 250- or a 500-volt megohmmeter. An insulation resistance of 20,000 ohms or greater is an indication of satisfactory insulation.

The pedestal-insulation resistance of an assembled machine cannot be checked with an ohmmeter or megohmmeter, because the parallel path provided by the uninsulated pedestal will result in an erroneous indication. However, the condition of the pedestal insulation may be checked while the machine is running by measuring the voltage between the shaft and

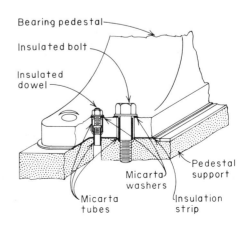

FIG. 12-12 Insulated bearing pedestal. (*Westinghouse Electric Corp.*)

Bearing pedestal

Insulated bolt

Insulated dowel

Micarta washers

Micarta tubes

Pedestal support

Insulation strip

the bedplate, as shown in Fig. 12-13.[1] Two readings are taken, one with a jumper connecting the shaft to the pedestal and one without the jumper. If the insulation is good, both readings will be alike; if defective, the reading in Fig. 12-13a will be higher than the reading in Fig. 12-13b.

An equivalent circuit showing the path of shaft current through the resistance of the oil film R_{of} and the pedestal resistance (if shorted) is shown in Fig. 12-13c. If the pedestal insulation is in good condition, the circuit formed by the shaft, bearings, bedplate, and pedestal insulation

[1] Guide for Operation and Maintenance of Turbine Generators, IEEE No. 67, 1963.

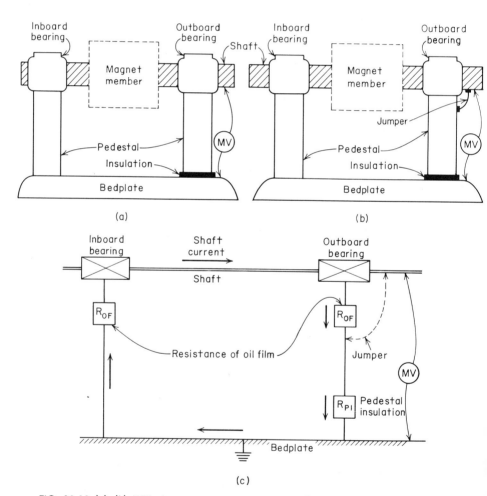

FIG. 12-13 (a), (b) Millivoltmeter measurements required to determine the condition of the bearing insulation of an assembled machine; (c) equivalent circuit showing the path of current through the oil film and through or across the defective pedestal insulation.

will be an equivalent open circuit; the very high insulation resistance will not permit any significant leakage current. With no shaft current present, the voltage drop across the oil film will be zero, and the full shaft voltage will appear across the insulation. Hence, the millivolt readings as measured in Fig. 12-13a and b will be the same. If the pedestal insulation is shorted, the resultant shaft current will cause a voltage drop across the resistance of the oil film, but there will be very little or no voltage across the shorted insulation. Hence, the millivoltmeter indication for the test in Fig. 12-13a will be higher than that for the test in Fig. 12-13b.

Bibliography

Morrow, L. C.: "Maintenance Engineering Handbook," McGraw-Hill Book Company, New York, 1966.

Questions

12-1. Describe the correct procedure for disassembling a dc machine equipped with the cartridge type of bearing housings.
12-2. What special care should be taken when disassembling explosion-proof machines?
12-3. What is the purpose of a spreader bar?
12-4. State the correct procedure for reassembling an overhauled induction motor.
12-5. Outline the correct procedure for cleaning and lubricating ball bearings on motors equipped with pressure-relief systems.
12-6. State the correct procedure for cleaning and lubricating sleeve bearings.
12-7. How may sleeve-bearing wear be detected without removing the bearing?
12-8. State the correct procedure for replacing worn sleeve bearings. What is the purpose of the felt washer?
12-9. State the correct procedure for removing a defective ball bearing and replacing it with a new one.
12-10. What are the principal causes of excessive vibration in electrical machinery? State the procedure that should be followed when trouble-shooting vibration.
12-11. Describe a simple method that may be used for rebanding a dc armature.
12-12. What causes shaft currents, and why are they harmful? Outline a procedure for checking bearing insulation.

thirteen

industrial control

Industrial-control systems use manual, magnetic, and electronic controls to start and stop motors and to provide reversing service, speed control, and other specialized features necessary to the automated operation of factories and power plants.

13-1 Reading and Interpreting Control Diagrams

Figure 13-1 illustrates two types of diagrams supplied by manufacturers for use with their equipment. The connection diagram, shown in Fig. 13-1*a,* shows the actual physical location of the component devices, their relative size, and actual wiring to each part. The heavy lines represent the power circuit, which carries the current to the motor; the light lines represent the wiring of the control circuit, which is used to start and stop the motor and to provide the necessary overload and low-voltage protection.

The elementary or schematic diagram, shown in Fig. 13-1*b,* is a simplified drawing that shows how the system works. The power circuit, drawn with heavy lines, shows the path of current to the motor via contacts, overload heaters, disconnect switch, etc; the control circuit shows the sequence of control operation and includes push buttons, operating coils, overload contacts, etc., that provide the logic and commands for the operation of the power circuit.

To facilitate the understanding of controller operation, the power and control components of each device are generally separated on the elementary diagram, as shown in Fig. 13-1*b,* where they become parts of separate power and control circuits. However, all com-

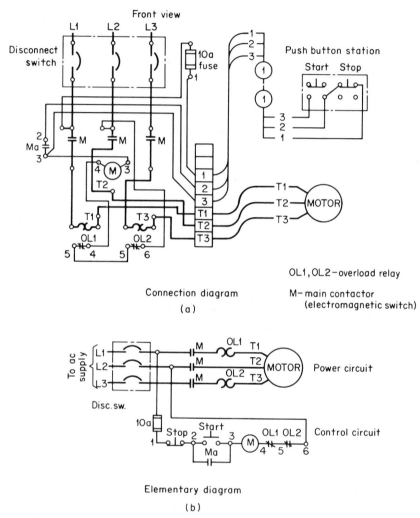

FIG. 13-1 Control diagrams. (a) Connection diagram; (b) elementary diagram.

ponents with the same letter designation are part of the same control device. Thus, the three *M* contacts in the power circuit, the M_a contact in the control circuit, and the *M* coil are all components of the magnetic contactor *M*.

A typical clapper type of magnetic contactor is shown in Fig. 13-2. Energizing the operating coil sets up a magnetic field that attracts the armature, causing the movable contacts, which are all mounted on a common insulated arm, to close. The pole shader, also called shading coil, is used with ac contactors to reduce contact noise and vibration that would otherwise be produced by the alternating current in the operating

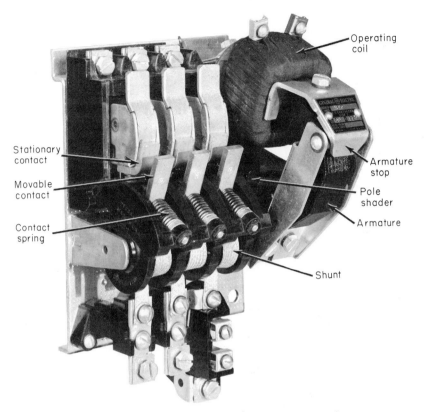

FIG. 13-2 Ac magnetic contactor. (*General Electric* Co.)

coil. The shading coil consists of a copper ring that surrounds a section
of the iron core. The function of the pole shader is to delay the buildup
and decay of flux in its section of iron; thus when the alternating flux
in the unshaded portion goes through zero, the flux in the shaded area
holds the armature in the sealed position until the flux in the unshaded
area builds up again in the opposite direction.

Correct reading and interpreting of control diagrams are essential
to effective maintenance and trouble shooting and require complete
familiarity with the standard graphical symbols for control components,
shown in Fig. 13-8. Figure 13-3 illustrates some of the typical circuits
used in magnetic controllers. The circuit in Fig. 13-3a provides low-voltage
release (LVR) and thermal-overload protection (OL). Pushing the start
button energizes operating coil *M*, causing the normally open contacts
M to close, and the motor starts. If a power failure occurs, coil *M* is
deenergized, and the motor stops; when power is restored, the motor
will automatically restart. It is not necessary to repush the start button,
because the snap-action switch remains in the start position after pushing.

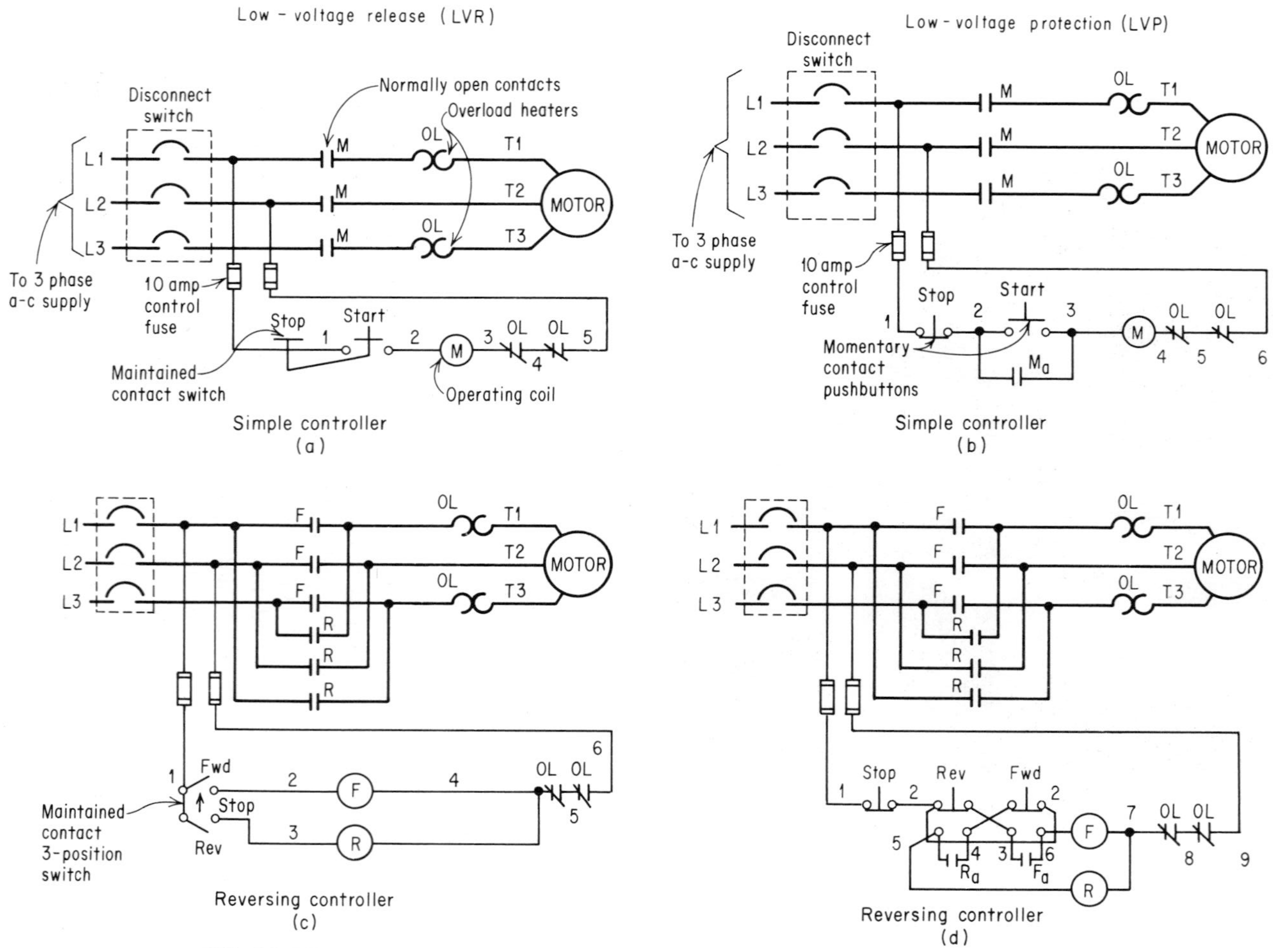

Low - voltage release (LVR)

Normally open contacts
Overload heaters

MOTOR
T1 T2 T3
OL OL
M M M

Disconnect switch
L1 L2 L3
10 amp control fuse
To 3 phase a-c supply

Stop Start
1
Maintained contact switch

2 3 OL OL 5
M 4
Operating coil

Simple controller
(a)

Low - voltage protection (LVP)

MOTOR
T1 T2 T3
OL OL
M M M

Disconnect switch
L1 L2 L3
10 amp control fuse
To 3 phase a-c supply

Stop Start
1 2 3
Momentary contact pushbuttons
Ma

M 4 OL OL 5
6

Simple controller
(b)

MOTOR
T1 T2 T3
OL OL

L1 L2 L3

F F F
R R R

Fwd Stop Rev
1 2 3
Maintained contact 3-position switch
F 4 R
OL OL 5
6

Reversing controller
(c)

MOTOR
T1 T2 T3
OL OL

L1 L2 L3

F F F
R R R

Stop Rev Fwd
1 2 5
Ra 4 3 Fa 6
F 7 OL OL 8
R 9

Reversing controller
(d)

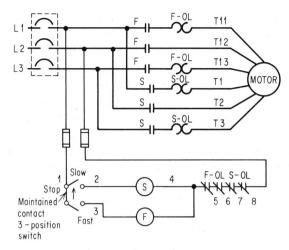

Two speed controller
(e)

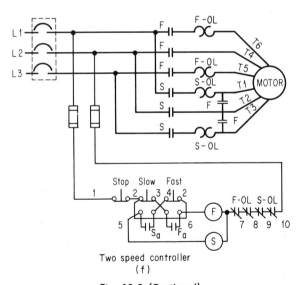

Two speed controller
(f)

Fig. 13-3 (Continued)

If the motor is overloaded, excessive current through the overload heaters causes the OL contacts to open (see Sec. 14-9), deenergizing the operating coil; this causes contacts *M* to fall to the normally open position, and the motor stops. To restart the motor, it is necessary for the OL contacts to be manually reset by pushing a reset button. (Automatic resetting is provided in some applications where vital equipment must be operated in remote locations.)

The elementary diagram shown in Fig. 13-3*b* has the same power circuit as in Fig. 13-3*a,* but its control circuit provides low-voltage protection (LVP) instead of low-voltage release. The basic difference is in the type of push button used. The LVP circuit uses momentary-contact buttons with spring return to the normal position. The "start" button is held normally open, and the "stop" button normally closed, by the respective springs. Pushing the "start" button energizes coil M, which closes contacts M in the power circuit and auxiliary contact M_a in the control circuit. Closing of contact M_a (the maintaining contact) provides a bypass around the start button, so that releasing the "start" button does not stop the motor. If coil M is severely weakened by low voltage, contacts M and M_a open, and the motor stops; when power is restored, the motor does not restart. It is necessary to push the "start" button to restart the motor. A starter with low-voltage protection protects against automatic restarting.

If low voltage or voltage failure stops the operation of all motors in a plant, their simultaneous starting when voltage is restored may overload the generator or distribution system, blacking out the plant. Hence, whenever possible, motors that are not vital to the operation of the plant should have their control circuits arranged for low-voltage protection. Low-voltage release circuits should be used in those applications where it is imperative that restarting be automatic.

Low-voltage release and low-voltage protection guard the motor against overheating caused by low voltage. For the effect of low voltage on motor performance, see Secs. 5-6 and 9-7.

The elementary circuit diagrams for reversing and two-speed controllers are shown in Fig. 13-3*c, d, e,* and *f.* In Fig. 13-3*d* the "rev" button has a set of normally closed contacts (2–3) and directly below it a set of normally open contacts (5–4). Pushing the "rev" button opens the normally closed contacts (2–3) and closes the normally open contacts (5–4). Do not confuse the push-button contacts (5–4) with auxiliary contact R_a (5–4); push-button contacts are always shown as small circles. Contact R_a (5–4) closes only when coil R is energized. The double push-button arrangement for each of the "rev" and "fwd" push buttons provides protective interlocking to prevent energizing both the R and F coils at the same time; such action would short-circuit the connecting lines between the disconnect switch and the motor. In addition, a mechanical interlock is generally provided at the contactors to ensure that both contactors cannot be manually closed at the same time.

13-2 Autotransformer Starting

Figure 13-4 illustrates the application of a three-phase wye-connected autotransformer (see Sec. 7-3) for reduced-voltage starting of a three-

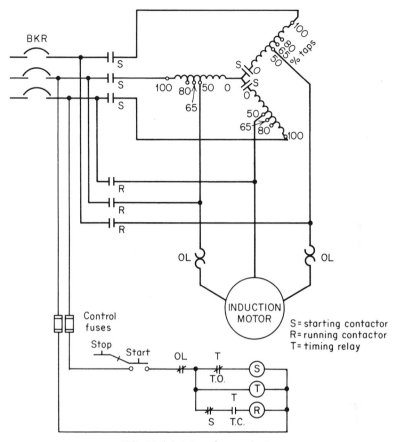

BKR

100 80 50 0
65

50
65 80 100

100
80
65
50
50 % taps
0
0

S
S
S
S
S

R
R
R

OL OL

INDUCTION
MOTOR

S = starting contactor
R = running contactor
T = timing relay

Control
fuses

Stop Start OL T
T.O.
S
T
T
R
S T.C.

FIG. 13-4 Autotransformer starter.

phase induction motor. Relay T is a pneumatic timing relay that is adjust-able from 0.2 sec to 3 min. Starting contactor S operates five normally open contacts in the power circuit and one normally closed sequence-inter-lock contact in series with the R coil. Pushing the start button energizes operating coils S and T. The closing of power contacts S energizes the autotransformer, which then supplies reduced voltage to the motor. After a time delay, contact T_{TO} opens and contact T_{TC} closes. This causes contactor S to assume its normally deenergized state, which disconnects the autotransformer from the line; contactor R is energized, connecting the motor across full line voltage.

13-3 AC Hoist Control

Figure 13-5 illustrates a reversing magnetic controller for a wound-rotor induction motor used to drive a hoist. A cam-operated master switch

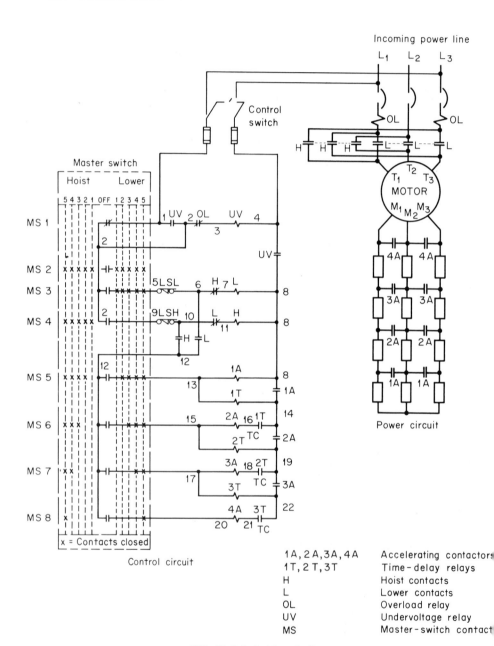

FIG. 13-5 Ac hoist controller.

1A, 2A, 3A, 4A	Accelerating contactors
1T, 2T, 3T	Time-delay relays
H	Hoist contacts
L	Lower contacts
OL	Overload relay
UV	Undervoltage relay
MS	Master-switch contact

energizes relays and contactors in a prescribed manner to provide five hoisting and five lowering speeds. Changes in speed are accomplished by changing the amount of resistance in the rotor circuit (secondary resistors). See Sec. 5-5.

The X marks on the circuit diagram for the master switch designate the cam-operated contacts that are closed for that particular position of the master switch. For example, in position 2 "hoist," cam-operated contacts MS_2, MS_4, and MS_5 are closed; in position 4 "lower," contacts MS_2, MS_3, MS_5, MS_6, and MS_7 are closed. Cam-operated contact MS_2 is provided for brake operation, if used.

With the master switch in the "off" position, there will be a path of current from L_3 through the control switch, cam-operated contact MS_1, overload relay OL, and undervoltage relay coil UV; the energized UV coil causes contacts UV (1–2) and UV (4–8) to close. Moving the master switch to position 1 "hoist" establishes a path of current from L_3 through contact UV (1–2), contact MS_4, hoist limit switch LSH, normally closed contact L (10–11), and the hoist coil H; the energized H coil closes contacts H (10–12) in the control circuit and the H power contacts in the power circuit. Closing of the H power contacts causes the motor to run in the hoist direction.

The relays $1T$, $2T$, and $3T$ are time-delay closing relays; when the respective operating coils are energized, the relay contacts close after a predetermined time has elapsed. The time-delay relays protect the motor against such poor operating practices as rapidly throwing the master-switch lever from "off" to 5 "hoist," or from 5 "lower" to 5 "hoist." Throwing the master switch rapidly from "off" to 5 "hoist" closes cam-operated contacts MS_2, MS_4, MS_5, MS_6, MS_7, and MS_8. Cam contact MS_4 energizes coil H, which establishes the direction of rotation, and also closes H (10–12) to start the following accelerating sequence:

1. The $1A$ coil is energized, closing the two $1A$ power contacts at the secondary resistor and closing contact $1A$ (8–14), energizing coil $1T$. The closing of the $1A$ power contacts causes the motor to attain a higher speed.
2. After a time delay, contact $1T$ (16–14) closes, energizing coil $2A$. Coil $2A$ closes the two $2A$ power contacts at the secondary resistor, shorting out another block of resistance and causing additional acceleration. It also closes contact $2A$ (14–19), energizing coil $2T$.
3. After a second time delay, contact $2T$ (18–19) closes, energizing coil $3A$. Coil $3A$ closes contact $3A$ (19–22), energizing coil $3T$, and also closes the $3A$ power contacts at the secondary resistors, causing additional acceleration.
4. After a third time delay, contact $3T$ (21–22) closes, energizing coil

$4A$. Coil $4A$ closes the $4A$ power contacts, shorting the remaining resistance, and the motor then accelerates to its full speed.

13-4 DC Reversing Controller

The schematic diagram and drum control for a manually operated reversing controller is shown in Fig. 13-6. The OL relay and M contacts are operated by the respective coils. All other contacts are manually operated by the drum controller. The complete assembly of cylindrically shaped contact segments gives it the appearance of a cylinder or drum. Moving the handle of the controller rotates the segments against the respective stationary contacts (fingers).

The drum development, shown in Fig. 13-6a, is used to indicate the contacts that are closed for each position of the operating handle. Drum contact "reset" (7–8) is closed only in the "off" position; it is used in conjunction with line contactor M to provide low-voltage protection. Figure 13-6b illustrates how the rotary drum segments make contact with the stationary fingers.

Referring to the schematic diagram (Fig. 13-6c), the energizing of coil M in the "off" position of the drum controller closes contact M in the power circuit and contact M_a in the control circuit. Maintaining contact M_a serves as a bypass contact that maintains current in the M coil when the drum controller is moved away from the "off" position. The blowout coils mounted in the drum controller are used to aid in extinguishing the electric arc when the contacts are opened, thus reducing contact wear. Each blowout coil is situated close to its respective contacts and sets up a magnetic field perpendicular to the path of contact motion. The stream of electrons that arcs between the contacts is driven upward by the magnetic field, elongating, cooling, and extinguishing the arc. The polarity of the magnetic field and the direction of electron flow in the arc are similar to that shown in Fig. 1-13e for a conductor in a magnetic field, except that the polarity of the magnetic field would be reversed to cause an upward motion of the arc.

The drum contacts that are closed for different drum-handle positions are indicated by the intersection of the contacts (solid black) with the lines drawn downward from the "fwd" and "rev" position numbers. Thus in position 1 "rev," contacts $1A$, $2R$, L, $1R$, and F are closed, resulting in the armature circuit shown in Fig. 13-7a; contact M is closed by the operating coil M. Similarly, Fig. 13-7b, c, and d shows the power circuits for drum-controller positions 4 "rev," 1 "fwd," and 3 "fwd" respectively.

When moving from one position to another, the operator should always pause long enough in each speed position to allow time for the motor to accelerate. Moving the controller too fast, when going from

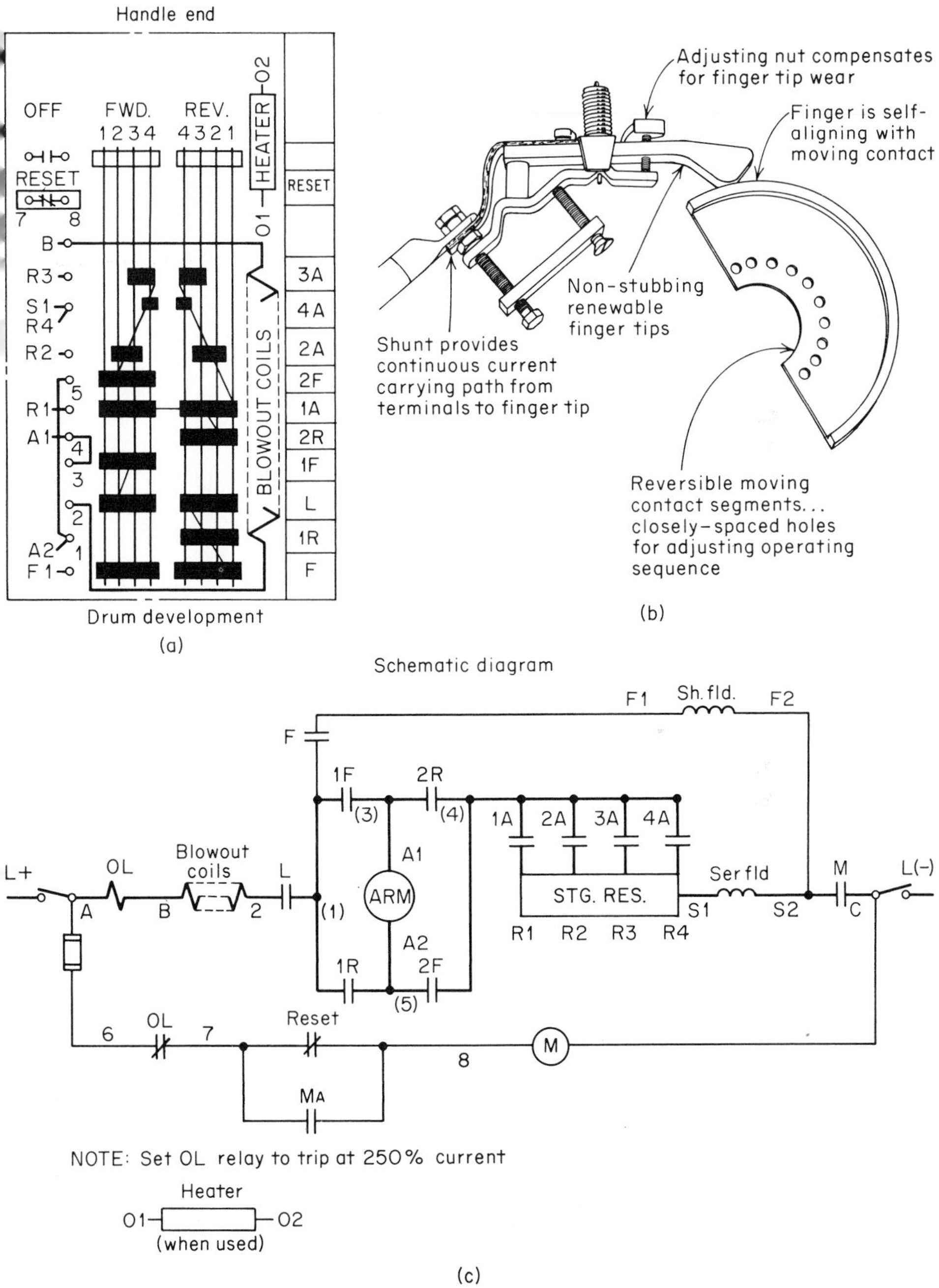

FIG. 13-6 Drum controller. (a) Drum development; (b) drum segment; (c) schematic diagram. (*Westinghouse Electric Corp.*)

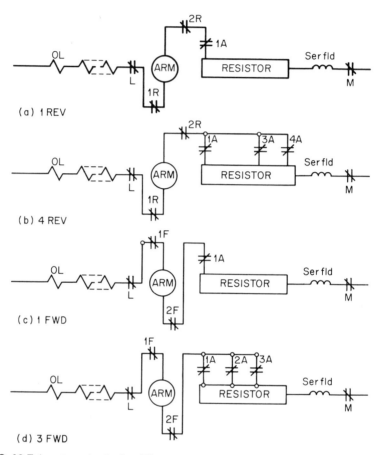

FIG. 13-7 Armature circuits for different positions of the drum controller in Fig. 13-6.

position 1 to position 3 or 4, without pausing in the intermediate positions, causes the motor to draw excessive current that may trip the overload relay OL. Each movement of the controller, from 1 to 2, 2 to 3, etc., should be made firmly and with a snap action to minimize arcing and burning of the contacts.

Low voltage or tripping of the OL relay will cause contactor *M* to open, and the motor will stop. The motor cannot be restarted until the drum controller is moved to the "off" position; this recloses the "reset" contact, reenergizing the *M* contactor.

The heater, shown in Fig. 13-7c, is used in regions of high humidity, such as outdoor or marine applications, to prevent condensation of moisture on the coils and contacts. Moisture condensing on contacts causes corrosion, and moisture absorbed by the insulation may cause short circuits.

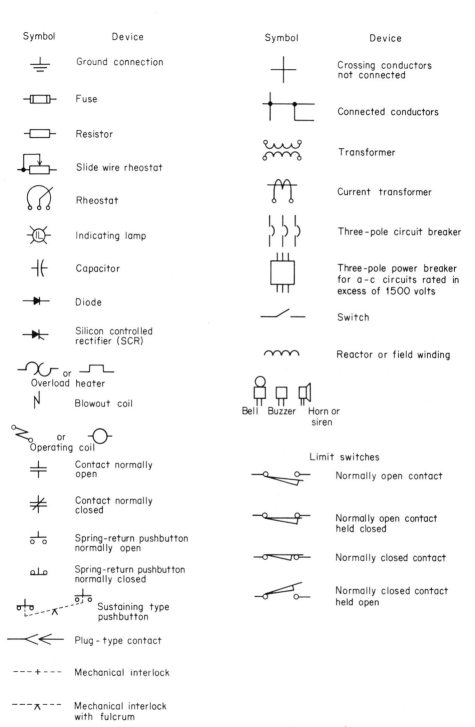

FIG. 13-8 Standard symbols for control components.

13-5 Motor-controller Troubles

When a motor fails to operate, the main line fuses, circuit breaker, and control fuses should be tested for continuity. Opens do not generally occur in the wiring of a control panel, and any break in a control circuit may usually be traced to opens in the operating coils or corroded or dirty contacts in the master switch or other contacts and push buttons in series with the operating coil. To check the master switch and operating coils, the test link should be removed or the motor disconnected, and the master switch should be operated in all positions. If each contactor and relay operates properly for its respective master-switch position, as indicated on the manufacturer's wiring diagram, the trouble is in the motor or resistor bank. The resistor bank should be checked for signs of overheating, loose connections, or broken resistors. The motor should be checked for opens in the field and armature circuits.

If a contactor or relay does not close when the position of the master switch indicates that it should, the trouble may be with either the master-switch contacts, the coil itself, or an auxiliary contact in series with the coil. The voltage across the coil should be checked with a voltmeter. If it has rated voltage but no magnetic pull, the coil is open. However, if the voltmeter registers zero, it indicates that the master-switch contacts or the auxiliary contacts in series with the coil are open or have a high-resistance connection.

Shorted coils are often indicated by discolored or charred insulation and by blown fuses in the control circuit. Motor controllers are seldom troubled with grounds, because the contactors and relays are mounted on a nonconducting panel.

13-6 Procedure for Trouble-shooting a Magnetic Motor Controller

Regardless of the apparent complexity of a schematic diagram, reading and interpreting it is essential to speedy trouble shooting and repair. The following suggestions should prove helpful in trouble-shooting the more complex problems posed by multispeed controllers that use a master switch along with contactors and relays, as shown in Fig. 13-9:

1. Familiarize yourself with the elementary diagram by locating the contacts that are associated with each operating coil.
2. From the symptoms of motor behavior, determine which positions of the master switch are involved in the fault.
3. Using the elementary diagram as a guide, make a check list of the operating coils and contacts, including the master-switch contacts, that

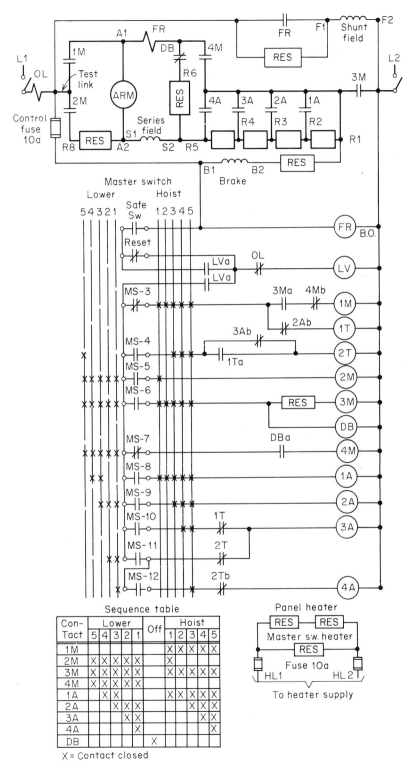

FIG. 13-9 Elementary diagram for a dc hoist controller. (*Westinghouse Electric Corp.*)

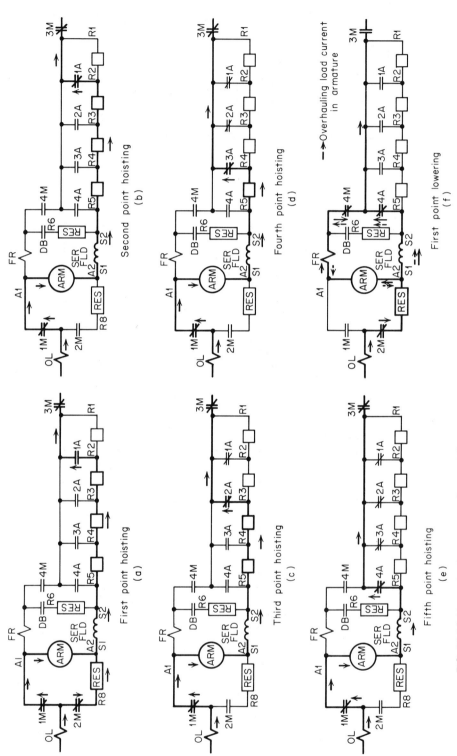

FIG. 13-10 Motor circuits corresponding to different master-switch positions in Fig. 13-9. (Westinghouse *Electric Corp.*)

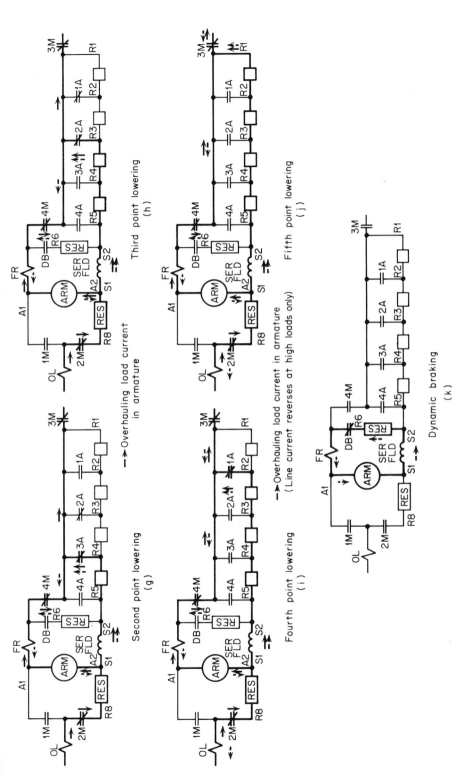

Second point lowering
(g)

Third point lowering
(h)

→ Overhauling load current
in armature

Fourth point lowering
(i)

Fifth point lowering
(j)

→ Overhauling load current in armature
(Line current reverses at high loads only)

Dynamic braking
(k)

FIG. 13-10 (Continued)

should be energized for each master-switch position in question, and then sketch the power circuit for each of these positions. The power circuit for each of the master-switch positions for the hoist control in Fig. 13-9 is drawn with heavy lines in Fig. 13-10.

4. Using the check list as a reference, operate the master switch, and determine whether these contactors or relays are functioning properly. Correct or replace the faulty ones. When testing the control, it is best that the motor be disconnected. This can be done by disconnecting the motor at the control terminals, removing the test links, or by opening a separate motor disconnect switch if one is provided; the hoist system in Fig. 13-5 has separate switches for the motor and the control.

5. If all contactors and relays appear to function normally and the supply voltage is correct, the trouble must be in the resistor bank or in the motor.

For the control circuit shown in Fig. 13-9, determine the probable sources of trouble in the following examples.

• *Example 13-1.* The motor will not run in any position of the master switch. However, clacking of contactors occurs when moving the master switch from "off" to 1 "hoist" or "off" to 1 "lower."

• *Solution.* Because all hoist and lower master-switch positions are involved, it is only necessary to check the diagram for those contacts or relays that could block the current to all hoisting and lowering coils. The common path of current to the operating coils for hoisting and lowering is from L_1, through the fuse, safety switch SW, "reset" contact, and LV_a contact. Because some clacking of contacts occurs, the fuse must be good. Hence, the fault must be with any one or combination of the following:

a. The SW or "reset" contacts in the master switch may be corroded, badly pitted, have defective springs, or loose connections.

b. Operating coil LV may be open, have a loose connection, or the OL contact in series with it may be corroded, have a weak spring, or a loose connection.

c. Contact LV_a may be defective.

• *Example 13-2.* The motor lowers but will not hoist in any position of the master switch.

• *Solution.* The MS contacts common to all hoisting positions are MS_3, MS_6, and MS_8. These contacts serve to energize coils $1M, 3M, DB,$ and $1A$. A sketch of the armature circuit that includes these contactors is shown in Fig. 13-11. The contactors that would prevent operation of the motor

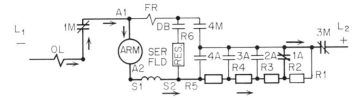

FIG. 13-11 Circuit diagram for Example 13-2.

are $1M$ and $3M$. However, because the motor does operate satisfactorily in the lower direction, contactor $3M$ must be in satisfactory condition. Hence, the fault must be with any one or combination of the following:

a. Master switch contact MS_3 may be corroded, badly pitted, have defective springs, or loose connections.
b. Coil $1M$ may be open or have a loose connection.
c. Contacts $3M_a$ or $4M_b$ may be corroded, badly pitted, have defective springs, or loose connections.
d. Contact $1M$ in the power circuit may be badly pitted, have a weak spring, or may be binding in the open position.

13-7 Maintenance of Magnetic Controllers

Control devices are essential to the successful control of electric motors but unfortunately are too often neglected. Failure of the control may result in poor performance, failure to start, and in many instances destruction of the machine itself. Failure of an overload trip may burn out the motor; failure of a shunt field contactor may accelerate a machine to destruction; failure of a brake contactor may cause the load on an electric hoist to fall; failure of the main propulsion control of an electric-drive ship may place the ship at the mercy of the elements, etc.

Figure 13-12 illustrates the extensive damage that can occur when protective devices fail. The motor shown was a 500-hp 600-volt 600-rpm compound motor driving a strip mill in a steel plant. The circuit breaker trip coil developed trouble and was disconnected to avoid production loss; it was to be replaced at the next scheduled shutdown. Because the control for this motor was equipped with an overspeed trip and a shunt-field failure relay, both of which were connected in the breaker-trip circuit, disconnecting the breaker-trip coil invalidated these protective features. Unfortunately, before the scheduled shutdown occurred, the shunt field lost its excitation, and the machine accelerated to destruction. The motor had to be replaced, and the mill was out of production for an extended period.

FIG. 13-12 Dc motor destroyed by overspeeding when a control circuit failed. (*Mutual Boiler and Machinery Insurance* Co.)

Circuit breaker trips, overspeed trips, loss-of-field relays, and other protective devices that are vital to equipment operation and safety should be tested periodically. If found defective, they should be repaired or replaced and properly adjusted for the specific application. Under no circumstances should protective equipment be bypassed or made inoperative; to do so is to invite disaster.

To provide for the proper maintenance of control equipment, a schedule should be arranged for periodic inspections. The frequency of inspection should be dictated by usage. Controllers should be checked for loose or worn contacts, weak spring pressure, displaced or burned arc shields, defective coils, low insulation resistance to ground, corrosion, etc. Figure 13-13 illustrates some of the preventive measures that are part of a well-planned maintenance program. An adequate supply of spare parts, as recommended by the manufacturer, should always be on hand for emergencies and preventive maintenance. See Sec. 16-4.

(a)

(b)

FIG. 13-13 Preventive measures that are part of a well-planned maintenance program. (a) Changing and tightening the contacts; (b) Changing a contactor coil that shows evidence of overheating. (*Westinghouse Electric Corp.*)

FIG. 13-13 (c) proper positioning of arc shields for good operation. (*Westinghouse Electric Corporation.*)

Control equipment should be kept clean, dry, and in proper working condition. Electric strip heaters used to prevent condensation of moisture should always be connected to ensure against insulation failure and corrosion. Excessive vibration should be avoided.

Cleaning contacts. Dirt and grease on contact surfaces increase the contact resistance and should be removed. The dust is easily removed with a vacuum cleaner, and the greasy film may be removed with a clean, dry cloth. Compressed air should be avoided, because it may force metallic dust, caused by contact wear, into the coil insulation. Clean delicate mechanical parts with a small stiff brush and an approved safety solvent.

Before work on control equipment is started, the circuit should be disconnected from the power line, and the fuses removed. A "Man Working on Line" sign should be fastened to the switch to prevent accidental closing.

Filing contacts. Heavy copper contacts and cadmium-plated contacts should be inspected regularly and filed when they become badly pitted. A rough-

FIG. 13-14 The two contacts on the right are badly worn and should be replaced. The two contacts on the left are still in good condition. (*United States Merchant Marine Academy.*)

ened butt contact can generally carry as much current as a smooth one and should be filed only when large projections caused by excessive arcing are evidenced. File carefully, with a smooth mill file, so that the contacts retain their original shape. Figure 13-14 illustrates badly worn contacts that require replacement. Contacts should never be allowed to wear to this extent.

Solid-silver contacts, used in relays and auxiliary control circuits, should not be filed unless sharp projections extend beyond the surface. The black silver oxide that forms on the surface should not be removed, because it is almost as good a conductor as the silver. When necessary, dress silver contacts with a fine-cut file or grade 0000 sandpaper fastened to a flat stick. In some cases, fine sandpaper may be drawn between the contacts while they are held together with moderate pressure.

Silver-plated contacts should be dressed carefully with grade 0000 sandpaper and should be replaced if badly pitted. Filing or burnishing of silver-plated contacts should be avoided, because the plating may be damaged.

Hard-alloy contacts which are pitted or corroded should be dressed with a burnishing tool instead of a file.

After filing, burnishing, or sandpapering, the contacts should be wiped with a clean cloth moistened with cleaning fluid, and then polished with a dry cloth.

Controller lubrication. The unnecessary use of lubricant may impair the effective operation of the controller. Dust in the air adheres to the lubricant and forms a gum that may cause the contactor or relay to fail. Butt-type contacts should never be lubricated. Sliding contacts, such as those used on rheostats and drum controllers, may be lubricated with a very thin film of petroleum jelly or special electric-contact lubricant. The bearings of contactors and relays are generally designed for operation without a lubricant. Hence they should not be lubricated unless specifically recommended by the manufacturer.

Welding of contacts. The welding of contacts is generally due to low contact pressure caused by low voltage or weak springs. Low operating voltage causes a weak magnetic pull; the contacts do not seal properly, and a relatively high resistance occurs between the connecting surfaces. If the contact resistance is considerably greater than the resistance of the rest of the circuit, the heat generated at the contact surfaces may cause the contacts to weld together. Low spring pressure produces the same effect; though the magnetic pull of the coil may be very strong, if the spring is weak or improperly adjusted, the high contact resistance may weld the contacts. Welding of contacts may also be caused by excessively high inrush currents at the instant the contacts close.

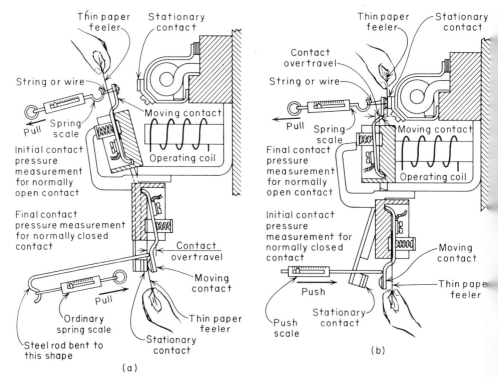

FIG. 13-15 Measuring contact spring pressure and contact overtravel (a) with armature open and (b) with armature closed. (*Westinghouse Electric Corp.*)

The spring pressure may be determined with a spring balance and a thin strip of paper, as shown in Fig. 13-15; the scale should be read at the instant the paper is released. For reasons of safety, it is best that the contactor be closed mechanically when the test is made. The contact overtravel or "wipe" provides a means for measuring contact wear. Contacts should be replaced when the overtravel falls below the minimum recommended by the manufacturer.

Resistors. The failure of resistors may be caused by poor connections, excessive vibration, inadequate ventilation, overload, defective insulation, and corrosion. Connections should be checked for proper bonding, because loose connections cause excessive heating and burning at junctions. Excessive vibration of cast grid types of resistors causes fractures to develop. Inadequate ventilation and overloads may result in burned-out units.

Where resistor banks composed of a number of individual series-connected units have worn thin by corrosion or burning, it may be advisable to replace the entire bank rather than renew a few units. Because of the reduction in cross-sectional area, badly worn units have a higher resistance than new ones. Hence placing new ones in series with the worn ones causes the higher-resistance units (worn ones) to take a greater load than the renewals, thus hastening breakdown.

13-8 Trouble-shooting Chart for Magnetic Controllers

The trouble shooting chart in Table 13-1 lists the most common troubles and their possible causes.

13-9 Maintenance of Electronic Control

The maintenance of industrial electronic apparatus calls for good housekeeping and carefully kept records if continuity of service is to be maintained. Apparatus such as photoelectric devices, electronic speed controllers for motors, electronic battery chargers, electronic welding control, and electronic heating, although different in circuitry, respond well to a carefully planned maintenance program. Electronic equipment should always be deenergized before servicing, because dangerously high voltages inherent in such apparatus may cause serious injury or death. All capacitors should be discharged, and the circuits grounded to the chassis or metal housing.

Electronic equipment should be cleaned by blowing out with clean, dry compressed air of about 30 psi. This removes dust that might otherwise cause overheating by reducing the heat-transfer characteristics of the components. Dust collected on high-voltage equipment may also result in short circuits. The wiring should be inspected for loose or broken connections;

Table 13-1   Trouble-shooting Chart for Magnetic Controllers

Trouble	Possible cause
Contact buzz (loud buzzing noise)	1. Broken shading coil 2. Misalignment of magnet faces 3. Dirt on magnet faces 4. Low voltage
Contact chatter (make-and-break action, staccato noise)	1. Poor contact in series with operating coil 2. Fluttering of a relay in the control circuit
Excessive burning of contacts	1. Excessive current (overload) 2. Weak spring pressure 3. Oxidized contacts 4. Poorly bolted connection 5. Arcing horn needs replacement
Welding of contacts	1. Excessive and rapid jogging 2. Low spring pressure 3. Excessive current (overload) 4. Low voltage on operating coil 5. Bouncing of contacts
Frequent tripping of OL relay	1. Overload 2. Wrong size of trip coil or heater (see Sec. 14-9) 3. Excessive ambient temperature (thermal relays)
Burned-out operating coil	1. Overvoltage 2. High ambient temperature 3. Shorted turns 4. Undervoltage (ac) 5. Dirt on magnet faces (ac) 6. Excessive jogging (ac)
Failure of contactor or relay to pick up	1. Low voltage 2. Open coil 3. Excessive magnet gap 4. Mechanical binding 5. Open circuit in series with operating coil
Failure of contactor or relay to drop out	1. Welded contacts 2. Improper adjustment 3. Accumulation of gum and other foreign matter 4. Misalignment 5. Voltage maintained at coil because of failure of contacts in series with operating coil

gently poking the wires with a clean, dry wood or plastic stick assists in detecting such faults. Lenses, mirrors, and light sources of photoelectric apparatus should be cleaned as frequently as local conditions require.

Resistors and rheostats should be inspected for signs of overheating, and discolored ones replaced. Capacitors which start to drip wax or fluid or which are discolored by excess heat should be replaced. Transformers that overheat, smoke, or have melted sealing compound should be replaced. Relays and contactors should be checked for defective components, and the procedure outlined for the maintenance of controllers should be carefully adhered to.

Water jackets used for cooling rectifier tubes should be flushed out at least once a year to remove any accumulation of rust or other sediment. Flushing with a 2 percent solution (by weight) of sodium carbonate, trisodium phosphate, or borax assists the cleaning process.

The first thing to do when checking faulty apparatus is to test all fuses and measure the line voltage. A volt-ohmmeter with a sensitivity of 1,000 ohms per volt or a vacuum-tube volt-ohmmeter aids considerably in diagnosing troubles. The ohmmeter measurements of resistance should be checked against the values stipulated by the manufacturer in his wiring diagram or specification sheet. All questionable resistors should be replaced. The color code for small tubular resistors is tabulated in Appendix 11 and will be of assistance in the absence of manufacturer's specifications.

Capacitors may be tested with an ohmmeter or megohmmeter, as described in Sec. 1-18.

The filaments of vacuum tubes may be checked by visual observation when the power is on; a good filament will light a glass tube and heat a metal tube. The filaments may also be tested with an ohmmeter by removing the tube and testing across the filament connections. A very high or infinite reading indicates a broken filament. If a tube tester is not available, the only way to check the characteristics of the tube is to replace it with another that is known to be good. If the replacement tube improves the performance of the apparatus, the old tube should be discarded. Tube caps should also be checked for tightness.

The filament or heater terminals of electronic tubes should be checked for low voltage. Insufficient voltage at the heater terminals reduces the electron emission and prevents proper operation of the electronic circuits. If low voltage is a problem, voltage-regulating transformers may be necessary.

Although the preventive-maintenance methods outlined above will reduce accidental shutdowns, they are not cure-alls. Electronic equipment is highly complex, and trouble can develop that may not respond to the recommended tests. In such cases more sophisticated test procedures must be used. Unfortunately a complete discussion of industrial electronics and its maintenance is outside the scope of this text. The reader is referred to the Bibliography for more detailed information and help in this area.

Bibliography

Control

Annett, F. A.: "Elevators," McGraw-Hill Book Company, New York, 1960.

Heumann, G .W.: "Magnetic Control of Industrial Motors," 3 vols., John Wiley & Sons, Inc., New York, 1961.

Hubert, C. I.: "Operational Electricity," John Wiley & Sons, Inc., New York, 1961.

Industrial Control, NEMA IC1-1965.

James, H. D., and L. E. Markle: "Controllers for Electric Motors," McGraw-Hill Book Company, New York, 1952.

McIntyre, R. L.: "Electric Motor Control Fundamentals," McGraw-Hill Book Company, New York, 1966.

Siskind, C. S.: "Electrical Control Systems in Industry," McGraw-Hill Book Company, New York, 1963.

Electronic Control

Chute, G. M.: "Electronics in Industry," McGraw-Hill Book Company, New York, 1965.

Kloeffler, R. G.: "Industrial Electronics and Control," John Wiley & Sons, Inc., New York, 1960.

Lytel, A.: "Industrial Electronics," McGraw-Hill Book Company, New York, 1962.

Platt, S.: "Magnetic Amplifiers: Theory and Applications," Prentice-Hall, Inc., Englewood Cliffs, N.J., 1958.

Wellman, W. R.: "Elementary Industrial Electronics," D. Van Nostrand Company, Inc., Princeton, N.J., 1957.

Questions

13-1. What is the function of a shading coil?

13-2. Explain the basic difference in operation between LVR and LVP circuits. State an application for each.

13-3. Sketch an autotransformer starting circuit, and state the sequence of operation when the start button is pushed.

13-4. Explain the sequence of operation of the controller, shown in

Fig. 13-5, when the master-switch lever is thrown rapidly from position 1 "hoist" to position 4 "lower." What is the function of the accelerating relays?

13-5. Referring to Fig. 13-6, sketch the armature circuits for drum-controller positions 2 "fwd" and 2 "rev." What is the function of the blowout coils?

13-6. When inspecting motor controllers, what items in particular should be checked?

13-7. What are some of the disastrous consequences that may result from an inadequate controller-maintenance program?

13-8. Outline the general approach that you would follow when troubleshooting a motor controller.

13-9. State the correct procedure for cleaning contacts. What controller parts require lubrication?

13-10. State the correct procedure for dressing (*a*) heavy butt types of contacts, (*b*) solid-silver contacts, (*c*) silver-plated contacts, and (*d*) hard-alloy contacts.

13-11. What may cause contact surfaces to weld together?

13-12. Describe a method for determining the spring pressure on a normally closed butt contact. What is contact wipe?

13-13. What are some of the factors that may cause failure of resistors?

13-14. Can the renewal of one or more resistor units in a bank of resistors cause some of the remaining units in the same bank to burn out? Explain.

13-15. Outline a preventive-maintenance program for industrial electronic controllers.

13-16. How may water-cooling jackets for rectifer tubes be cleaned?

13-17. How may shorted capacitors be detected?

13-18. How may the filaments of a vacuum tube be tested?

13-19. How may the characteristics of a vacuum tube be tested if a tube tester is not available?

fourteen

overcurrent protection of generators, distribution systems, motors, and other electrical apparatus

Excessive current can destroy insulation, melt conductors, cause explosions and fire, kill and maim operating personnel, and in general raise havoc with the immediate surroundings. Furthermore, the man-hours and production time lost as a result of this type of failure can never be replaced. Protection against damaging overcurrent can be provided by correctly applied and properly maintained fuses, circuit breakers, and relays.

14-1 Fuses

A fuse is an overcurrent protective device with a circuit-opening fusible link directly heated by the passage of overcurrent through it. Fuses are selected on the basis of operating voltage, normal or rated current, and interrupting capacity. Fuses should be able to carry 110 percent of their rated current without any externally visible soldered connection melting and without the fuse tube charring.[1] The interrupting capacity of a fuse is a rating based upon the highest direct current or rms alternating current it will successfully interrupt. If the short-circuit current exceeds the interrupting capacity of a fuse, it may explode violently, causing an electrical fire and severe damage to the equipment it was supposed to protect.

Fuses of the correct type and size provide good dependability and protection. They have no latches, coil, bi-metallic strips, or other moving devices common to circuit breakers and hence require little maintenance. The selec-

[1] Low Voltage Cartridge Fuses, NEMA FU-1-1966.

1974 — No longer exists

tion of fuses must be done with consideration for the type of circuit involved. Lighting and heating circuits need only the ordinary type of fuse protection, whereas motor circuits require a time-lag fuse to permit the existence of the high starting current. A time-lag fuse is designed to permit a transient starting current of 200 to 300 percent of the rated motor current and yet to provide protection against a sustained overload equal to 125 percent of the rated current. This is accomplished through a dual-element fuse, shown in Fig. 14-1. When a short circuit occurs, the high current causes the fuse link to blow. With a sustained overload, the fuse link remains intact, but the heat generated by the overload current softens the low-melting-point solder, which connects the link to the center piece of copper. The spring pulls the connector away, and the circuit is opened. Where thermal trip breakers are used for protection against sustained overloads, ordinary fuses or magnetic breakers are generally provided for protection against short circuits.

The maximum opening time for standard NEC fuses carrying overloads of 135 to 200 percent of rated current and the minimum opening time for dual-element fuses at 500 percent of rated current are given in Table 14-1.

When fuses of the correct size and type are found to blow frequently, look for trouble within the circuit. Do not overfuse. A fuse larger than necessary endangers the apparatus it is supposed to protect.

Common causes for blown fuses, other than short circuits and sustained overloads, are loose fuse clips, which heat because of high contact resistance, poor contact within the fuse, improper location of the fuse in extremely hot surroundings, and excessive vibration. Fuses that blow because of poor contact can generally be detected by discolorations or

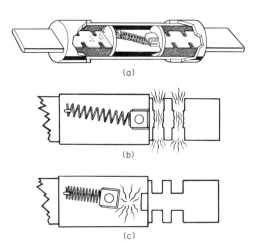

(a)

(b)

(c)

FIG. 14-1 Dual-element fuse. (a) Cutaway view; (b) fuse blowing at short circuit; (c) fuse blowing with overload. (*Bussmann Mfg. Co.*)

Table 14-1   Opening Time for Standard Low-voltage Cartridge Fuses

Fuse rating, amp	Opening time		
	Maximum, min		Minimum, sec, for dual-element fuses at 500 percent rating
	On 135 percent rating	On 200 percent rating	
0–30	60	2*	10
31–60	60	4*	10
61–100	120	6*	10
101–200	120	8	10
201–400	120	10	10
401–600	120	12	10

* May be 8 min for 600-volt fuses marked "For use only on motor circuits."
SOURCE: Low Voltage Cartridge Fuses, NEMA FU1-1966.

charring at the defective region. When replacing blown fuses, be sure that the fuse clips are straight, tight, and in good contact with the fuse. If the spring clips have lost their grip, they should be replaced, or clip clamps should be used, as shown in Fig. 14-2.

The maximum temperature rise above an ambient of 18 to 32°C when carrying 110 percent rated fuse current should not exceed the values shown in Table 14-2.

14-2   Current-limiting Fuses (CLF)

A current-limiting fuse differs from those having only a high interrupting capacity in that the melting and arcing time is so brief that the actual let-through current is much less than the maximum available short-circuit current. Whereas the ordinary high-interrupting-capacity fuses let through several half-cycles of short-circuit current before circuit interruption oc-

FIG. 14-2 Clip clamps for securing fuses to weak fuse clips. (*Ideal Industries, Inc.*)

Table 14-2

Fuse rating, amp	Maximum temperature rise above ambient temperature, °C	
	On fuse tube	On terminals
0–30	50	50
31–60	50	50
61–100	50	50
101–200	50	60
201–400	50	65
401–600	50	75

SOURCE: Low Voltage Cartridge Fuses, NEMA FU1-1966.

curs, the current-limiting fuse clears the fault in a small fraction of a cycle. This is illustrated in Fig. 14-3a. The broken line shows the fault current that would occur if the current-limiting fuse were not in the circuit. The triangular-shaped line shows the actual short-circuit current and length of time required for the CLF to clear the short circuit. The peak of the triangle represents the peak let-through current. A comparison of the current-limiting action of a 200-amp 600-volt CLF and standard National Electrical Code (NEC) fuses under short-circuit test conditions is shown in Fig. 14-3b. The CLF shows far greater current-limiting action than does the standard fuse when subject to high currents. For example, with an available short-circuit current of 50,000 rms amp, the CLF allows only 13,500 maximum peak amp, but the standard fuse allows 26,000 maximum peak amp. The standard NEC fuse provides effective short-circuit protection within its relatively low interrupting rating of approximately 10,000 rms amp. The current-limiting fuse shown in Fig. 14-3c, called an *amp trap,* has many silver links connected in parallel between heavy copper end blocks. The end blocks act as a heat sink to absorb, radiate, and conduct away the heat generated in the links. The links are embedded in pure quartz sand inside a heavy synthetic-resin-glass cloth-laminated tube that spaces the end blocks and seals it against the emission of gas or flames. Although the silver links have a high normal current-conducting ability, excessive current, such as that caused by a short circuit, almost instantaneously raises the temperature of the links to the melting point. The resultant arc formed in the surrounding quartz sand transforms the sand into glass. Because the silver vapor formed during the arc is practically nonconducting and the quartz sand gives off no gas, the arc is extinguished. The silver-melting process on a high-magnitude fault is so

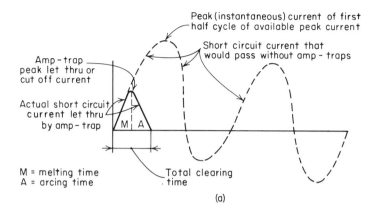

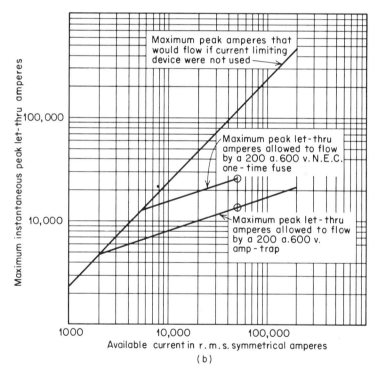

FIG. 14-3 Current-limiting fuse (amp trap). (a) Comparison of available short-circuit current and actual short-circuit current passed by current-limiting fuse; (b) current-limiting action in rms amperes for a 200 amp 600 volt amp trap and for a 200-amp 600-volt NEC one-time fuse. (Chase-Shawmut Co.)

(c)

FIG. 14-3 (c) Current-limiting fuse with outer case and sand filler removed, showing the silver links and the heavy copper end blocks. (*Chase-Shawmut Co.*)

rapid that the fault current does not have time to reach its maximum possible peak value but is "limited" to a relatively low value.

14-3 Fuse Testing

Fuse testing should be done with the circuit dead. If the circuit is dead, the fuses may be removed and tested with an ohmmeter, megohmmeter, magneto, or battery and buzzer test set. However, if the testing must be done on an energized circuit, a voltmeter or mechanical voltage tester of the correct voltage rating should be used. Light bulbs should be avoided, because there is always the danger of breakage from striking or overvoltage. The mechanical voltage tester, shown in Fig. 14-4, has the additional advantage of differentiating between alternating and direct current; when applied across an ac circuit, a slight vibration may be felt. Some mechanical voltage testers are equipped with a small neon light to determine polarity on dc systems. *Do not use your fingers to test fuses.*

To check for blown fuses in a single-phase or dc system, the switch should be closed, and the voltage tester applied across the top of both fuses, as shown in Fig. 14-4a, and then across the bottom of both fuses. If voltage is indicated across the tops but not across the bottoms, or vice versa, one or both of the fuses are blown. To locate the faulty fuse, the leads of the tester should be placed across the "hot" ends of both fuses; then one test point should be moved down the fuse to the other end, as in Fig. 14-4b. If the tester indicates voltage, the fuse is good.

This test may also be used for checking the three line fuses of a three-phase system, provided that the load is disconnected. If the load consists of a three-phase motor, the starter must be in the "stop" position; a lightly loaded three-phase motor will continue to operate with one fuse blown and will generate a voltage for the blown fuse, thereby preventing detection by a voltage tester. A continuity tester, such as a bell and battery, megohmmeter, or ohmmeter, provides a foolproof method for testing fuses in a three-phase system; the circuit must be deenergized, and the tester applied across each fuse, or the fuses must be removed and then tested. When fuses are removed or replaced, the circuit should be deenergized or an insulated fuse puller used.

14-4 Molded-case Circuit Breaker

A molded-case circuit breaker, shown in Fig. 14-5, is a nonadjustable air breaker assembled as an integral unit in a supporting and enclosing housing of insulating materials. It is used for circuit protection in low-voltage distribution systems. An air circuit breaker interrupts the current in air as contrasted with oil or gas interruption. Breakers of this type are able to provide overload protection for conductors as well as short-circuit protection for motors, control equipment, lighting circuits, and heating circuits.

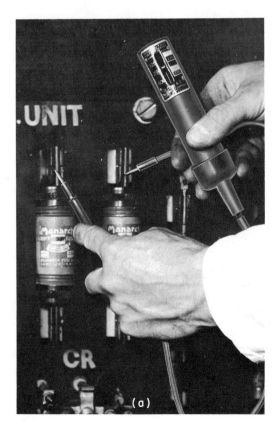

FIG. 14-4 Method of testing for blown fuses. (a) Testing across the fuse tops. (*United States Merchant Marine Academy.*)

FIG. 14-4 (b) Testing diago-
nally. (*United States Merchant
Marine Academy.*)

Molded-case breakers are classified by frame size, ampere rating, and interrupting capacity. The frame size, expressed in amperes, is the largest ampere rating available in the group. The ampere rating is the value of current that the breaker will carry continuously in an ambient temperature of 40°C without either tripping or exceeding the permissible temperature rise. A listing of standard frame sizes and ampere ratings for molded case breakers is given in Table 14-3 on page 356.

Protection against sustained overloads, as differentiated from a short circuit, is accomplished by a thermal-acting trip. It consists of a bimetal thermal element connected as shown in Fig. 14-6a and b. The bimetal element consists of two bonded strips of metal having different rates of thermal expansion. Excessive current passing through the bimetal element will cause it to deflect. This trips the latch that opens the contacts. In Fig. 14-6, the left strip of the bimetal has a greater rate of linear expansion than the right strip, causing the element to deflect to the right when heated. Thermal-acting trips have an inverse-time characteristic,

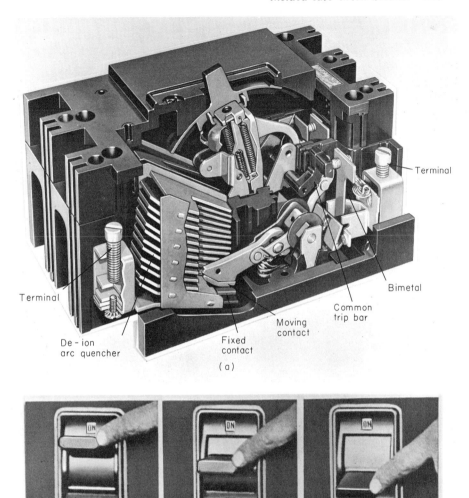

FIG. 14-5 (a) Cutaway view of a three-pole molded-case circuit breaker; (b) position of handle indicates status of breaker: "on," "tripped," or "off." (*Westinghouse Electric Corp.*)

providing a long time delay before tripping when a light overload occurs and a faster response for heavy overloads.

Protection against short circuits is accomplished by an electromagnet, as shown in Fig. 14-6c and d. The line current passes through the magnet element, causing a magnetic pull on a movable iron slug called the arma-

Table 14-3   Standard Frame Sizes and Continuous
Ampere Ratings for Molded-case Breakers

Frame size, amp							
100	200	225	400	600	800	1,000	1,200
15	125	70	200	300	300	600	700
20	150	100	225	350	350	700	800
30	175	125	250	400	400	800	1,000
40	200	150	300	500	500	1,000	1,200
50	...	175	350	600	600		
60	...	200	400	...	700		
70	...	225	...	...	800		
100							

SOURCE: Molded Case Circuit Breakers, NEMA AB1-1964.

ture. During short-circuit conditions the magnet becomes strong enough to attract the armature, which in turn trips the latch and opens the breaker contacts. This action is "instantaneous" in that there is no deliberately built-in time delay.

Thermal-magnetic breakers provide instant tripping action on short circuits but allow momentary overloads, such as high starting currents in motors and initial surge currents in lighting circuits. Figure 14-6e and f shows the arrangement for a combination thermal-magnetic breaker.

For those applications where a thermal-acting breaker must be located in a region of unusually high, low, or fluctuating temperature, an ambient-compensating element must be used. Compensation is obtained by using an additional bimetal element to counteract the effect of ambient temperature changes on the load-sensitive bimetal. This is shown in Fig. 14-6g, h, and i. Figure 14-6h shows a condition of no overload but a high ambient temperature. Note how the compensating bimetal changes its position to prevent tripping. Figure 14-6i shows how an overload actuates the trip bar.

The de-ion arc quencher shown in Fig. 14-5 consists of parallel steel plates partially surrounding the contacts and enclosed by a fiber wrapper or ceramic supports. When the contacts open, the arc drawn induces a magnetic field in the steel plates. The interaction of this field with that surrounding the electron stream (arc) causes the arc to move upward into the plates, where it is split into a series of smaller arcs, cooled, and extinguished. The direction and polarity of the magnetic field for the horseshoe-shaped steel plates through which the arc is drawn is shown in Fig. 14-7.

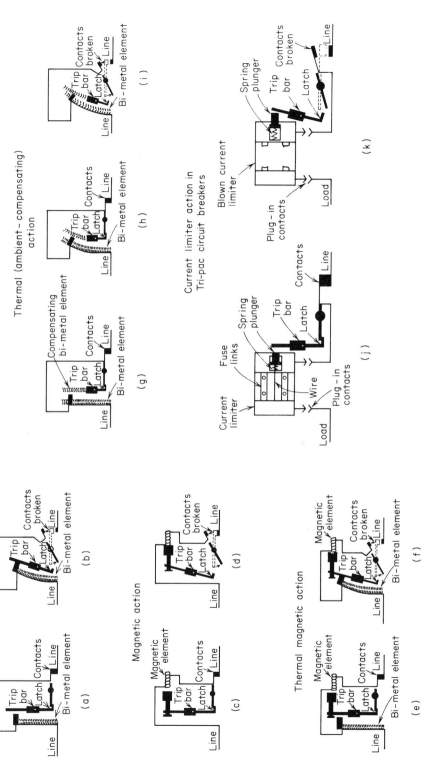

FIG. 14-6 Types of tripping actions used in molded-case circuit breakers. (*Westinghouse-Electric Corp.*)

14-5 Interrupting Capacity of a Breaker (IC Rating)

The interrupting capacity of a breaker is the maximum rms value of current at rated maximum voltage and frequency that the breaker can safely interrupt without damage to itself.

To prevent damage or complete destruction of the circuit breaker when a short circuit occurs, the breaker must be able to withstand the high mechanical forces caused by the short-circuit current. The magnitude and direction of these forces on the current carrying parts of the breaker is the same as that produced in adjacent cables of opposite polarity (see Sec. 1-16). These forces, proportional to the square of the current, cause a considerable strain on the breaker components. If the short-circuit current is greater than the interrupting capacity of the breaker, the tremendous mechanical forces that are produced on the adjacent breaker poles can literally tear it apart or, even worse, cause a violent explosion and fire.

Reclosing a breaker after a fault occurs could be dangerous if the interrupting capacity of the breaker is inadequate. The operator may be seriously injured or killed by the resulting explosion of the weakened breaker.

Figure 14-8a illustrates the damage to a switchboard in an industrial plant caused by current in excess of the interrupting capacity of a molded-

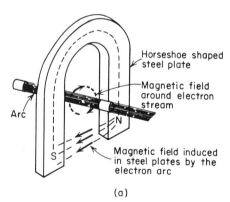

(a)

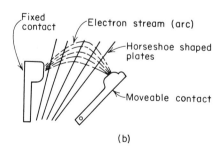

(b)

FIG. 14-7 De-ion arc quencher. (a) Development of magnetic field in steel plates; (b) upward movement of arc.

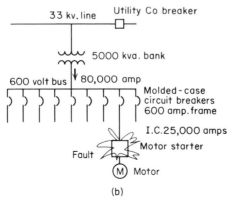

FIG. 14-8 (a) Switchboard damaged by mechanical forces caused by short-circuit current; (b) one-line diagram of the distribution system. (*Mutual Boiler and Machinery Insurance* Co.)

case circuit breaker. The switchboard consisted of sixteen 600-amp circuit breakers fed from a 5,000-kva 600-volt transformer bank. The only fault protection for the transformer bank and switchboard was a circuit breaker at the utility substation on the 33-kv line, equipped with a "three shot" automatic reclosing mechanism. No main secondary breaker was provided. A one-line diagram for the distribution system is shown in Fig. 14-8b. The available secondary-fault current was calculated to be 80,000 amp, whereas the IC rating of the distribution circuit breakers was only 25,000 amp. Evidently a heavy fault developed in a motor line starter, and when the molded-case circuit breaker opened, it exploded. Short-circuit damage and the burned-off studs on the line side of the circuit breakers are shown in Fig. 14-8a. The ¼- by 3-in. copper bus was badly distorted from the mechanical forces developed by the fault current. Ten of the sixteen circuit breakers were burned beyond repair and had to be replaced. The plant was out of service for 2 days to make temporary repairs.

In many low-voltage distribution systems, possible short-circuit currents as high as 100,000 amp are common. Because fault currents of this intensity exceed the interrupting capacity of the largest molded-case breakers, some form of backup protection is required. To avoid the use of larger and more expensive breakers capable of withstanding the large thermal and magnetic stresses, special high-interrupting-capacity current-limiting fuses are used. The CLFs restrict the short-circuit current by opening the circuit before the current rises to its peak value. This action reduces let-through current and so reduces substantially thermal and magnetic stresses on the conducting parts of the breaker and connecting cables.

A molded-case breaker with a built-in current limiter is shown in Fig. 14-9. The thermal and magnet elements are the same as previously illustrated for the standard breakers. When short-circuit currents occur, the silver links of the current limiter melt, opening the circuit; simultaneously, the magnetic trip functions to open the breaker contacts and thus aids in clearing the short circuit. A wire holding a plunger against the pressure of a spring melts when the silver links melt, causing the plunger to extend outward, holding the trip bar in the unlatched or open position. This makes it impossible to close the breaker until the blown limiter is replaced. Interlocks prevent relatching the breaker if a limiter is omitted, and will also open the circuit breaker contacts if an attempt is made to remove the limiter housing assembly with the breaker in the "on" position. Current limiter action is illustrated in Fig. 14-6j and k.

14-6 Tripping Curves

Figure 14-10 illustrates a representative average tripping curve for a 300-amp molded-case breaker of the type shown in Fig. 14-9. The curve is a plot of tripping time versus overload current. The upper left-hand segment of the curve represents the thermal tripping (average time delay) of the breaker as a result of a sustained overload condition. As indicated by the curve, the breaker will not trip at currents of 300 amp or less. For currents ranging from 325 to 3,000 amp the time delay for tripping will vary inversely with the current. Thus, with a current of 325 amp it will take over 2 hr to trip, whereas with a current of 2,500 amp the average tripping time will be approximately 12 sec.

The abrupt change in the curve to a vertical segment for currents in excess of 3,000 amp represents 'instantaneous" *magnetic tripping* of the breaker due to short-circuit currents. The broken line segment in the lower right-hand part of the characteristic curve represents operation of the current limiter. The point at which the current-limiter curve crosses the lower portion of the magnetic-trip curve is called the *crossover point,* and the magnitude of the short-circuit current at this point is called the *crossover current.* At values of current less than the crossover current

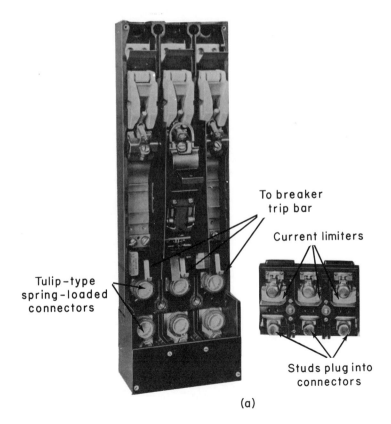

To breaker
trip bar

Current limiters

Tulip-type
spring-loaded
connectors

Studs plug into
connectors

(a)

FIG. 14-9 Molded-case circuit breaker with a built-in current limiter. (a) View of breaker showing location of plug-in limiters. (*Westinghouse Electric Corp.*)

(b)

(c)

FIG. 14-9 (b) Good limiter; (c) blown limiter, indicated by projection of plunger. (*Westinghouse Electric Corp.*)

the magnetic trip interrupts the fault without operation of the current limiter. At values of current greater than the crossover current the current limiter clears the fault.

As indicated by the characteristic curve, the protective actions of the inverse time-delay thermal trip, the "instantaneous" magnetic trip and the current limiter are coordinated so that overcurrents and low-magnitude faults are cleared by thermal action; "normal" short circuits are cleared by magnetic action, and abnormal short circuits are cleared by the current limiter.

If after tripping, the breaker cannot be reset immediately, thermal tripping due to an overload or a "high-resistance" fault is indicated; the breaker cannot be reset until cooling of the bimetal allows it to resume its original untripped position. If the breaker can be reset immediately, a "normal" fault current has been interrupted by instantaneous magnetic action. If, in the case of the breaker type shown in Fig. 14-9, it cannot be reset, then a high fault current has been interrupted by the current limiter.

Band curves (Fig. 14-11) are used to indicate the calibration limits of

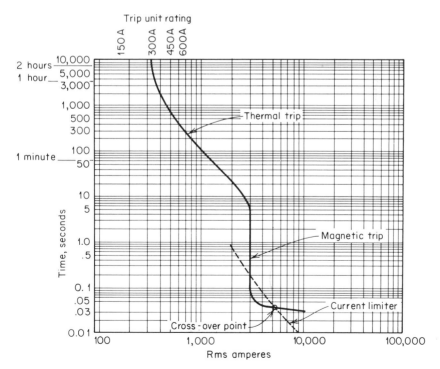

FIG. 14-10 Representative tripping curve for a 300-amp molded-case breaker with thermal, magnetic, and current-limiting action. (*Westinghouse Electric Corp.*)

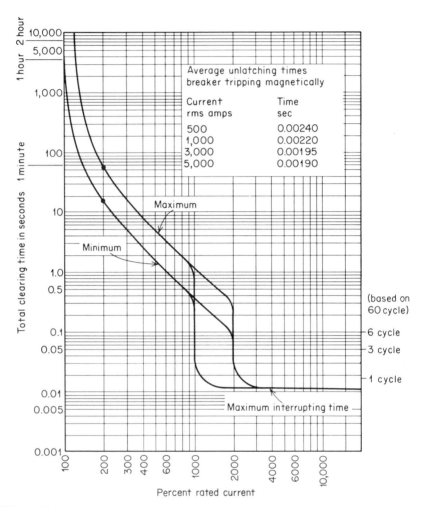

FIG. 14-11 Band curves used to indicate the calibration limits of a 15-amp ambient compensated thermal magnetic breaker. (*Westinghouse Electric Corp.*)

a thermal magnetic trip breaker. For a given overload current at 25°C, the breaker should clear the fault in some total clearing time within the band bounded by the maximum and minimum characteristics. For example, a 15-amp ambient-compensated breaker whose characteristic is shown in Fig. 14-11 should trip in not less than 16 sec and not more than 55 sec on a 30-amp current (200 percent overload). Band curves are very useful in maintenance work for checking the calibration of overload trips.

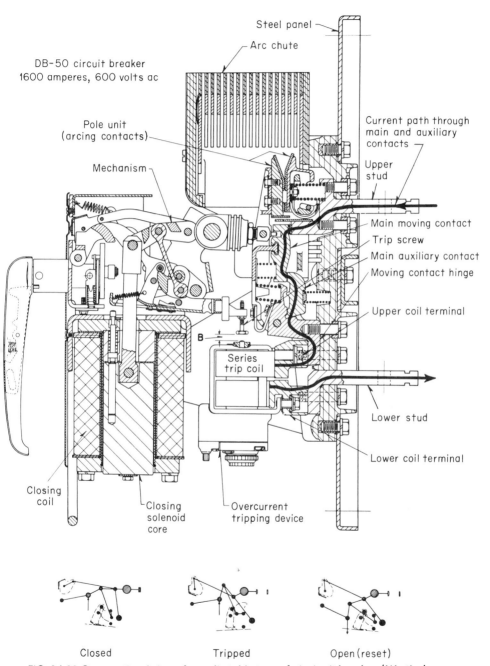

DB-50 circuit breaker
1600 amperes, 600 volts ac

Steel panel

Arc chute

Pole unit
(arcing contacts)

Mechanism

Current path through
main and auxiliary
contacts

Upper
stud

Main moving contact

Trip screw

Main auxiliary contact

Moving contact hinge

Upper coil terminal

B

Series
trip coil

Lower stud

Lower coil terminal

Closing
coil

Closing
solenoid
core

Overcurrent
tripping device

Closed

Tripped

Open (reset)

FIG. 14-12 Cross-sectional view of an adjustable type of air circuit breaker. (*Westinghouse Electric Corp.*)

14-7 Adjustable Type of Air Circuit Breaker

An adjustable type of air circuit breaker is shown in Fig. 14-12. This type of breaker has adjustable trip settings and adjustable time delays to permit coordination with other breakers, fuses, motor overload protective devices, etc. A cross-sectional view and a schematic diagram of the overcurrent tripping device with long-delay and instantaneous elements is shown in Fig. 14-13.

An overload or short-circuit current through the series-connected tripping coil D causes the moving core C to be attracted and move toward the stationary core B. At low overload currents, the moving core C carries the tripping stem F along with it, immediately closing the reset valve G, after which motion is retarded by the diaphragm E. The rate of travel of the diaphragm is determined by the rate at which air is permitted to enter the chamber T by the long-time needle valve H and by the large-orifice instantaneous valve K. At higher overload currents, when the attraction between the moving core C and the stationary core B is greater than the spring load inside the moving core, the moving core compresses the spring and travels independently of the tripping stem F. This spring ensures a constant force pattern acting on the diaphragm E. Valve H is the long-delay valve and is open to a calibrated setting. This setting, which controls the tripping time, can be changed by means of the dial S. The magnitude of current at which the long delay begins to operate is determined by the long-delay pickup adjustment knob P.

The magnitude of current at which the instantaneous trip operates is determined by the instantaneous pickup adjustment knob N. A short-circuit current causes the magnetic pull on the armature L to be greater than that provided by the restraining spring, causing valve K to open. The rapid flow of air through valve K permits the 'instantaneous" movement of tripping stem F.

Figure 14-14 is a typical time-current characteristic for a tripping unit of the type shown in Fig. 14-13. The heavy line marked "total clearing time" is a plot of fault current versus total tripping time, including arcing, that will elapse before the fault is cleared and the current ceases. The upper section of the curve represents the time-current characteristic for the long-time-delay pickup and exhibits an inverse time characteristic up to approximately 500 percent of the trip-coil rating. Within this range, the moving core C and tripping stem F move at the same speed; the magnetic pull has not increased sufficiently to overcome the loaded spring inside the moving core. Above 500 percent current the magnetic force is sufficient to overcome the spring force, slamming the movable core against the stationary core. When this occurs, the force pattern provided

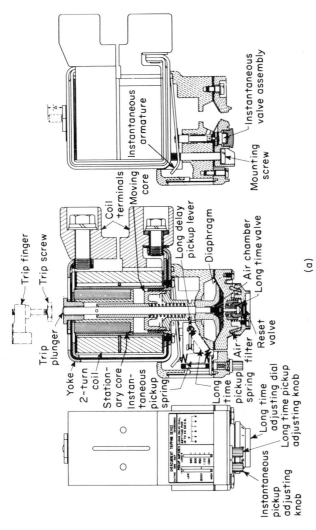

FIG. 14-13 (a) Cross-sectional view of a series overcurrent tripping device with long delay and instantaneous elements; (b) schematic diagram of overcurrent tripping device showing internal arrangement of the adjustable long delay and the adjustable instantaneous elements. (A) yoke; (B) stationary core; (C) moving core; (D) series trip coil; (E) diaphragm; (F) tripping stem; (G) reset valve; (H) valve (long delay); (K) valve (instantaneous); (L) armature; (N) instantaneous pickup adjustment; (P) long-delay pickup adjustment; (S) long-time adjusting dial; (T) air chamber. (Westinghouse Electric Corp.)

(a)

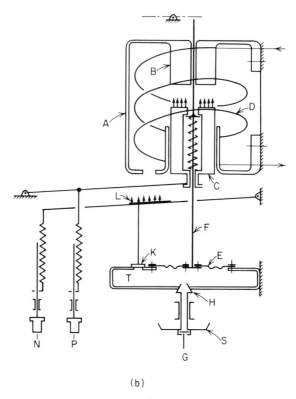

(b)

FIG. 14-13 (Continued)

by the spring causes the total clearing time to remain essentially constant for values of fault current in excess of 500 percent. This is indicated on the characteristic by a leveling off of the curve. The total clearing time remains constant for all higher fault currents up to approximately 850 percent rated current, at which value the instantaneous pickup becomes operative; this behavior is indicated by the vertical section of the characteristic. For fault currents in excess of 3,000 percent[1] rated current, the total clearing time is 2 cycles of a 60-hertz wave or 0.033 sec (2 × 1/60 = 0.033). This is the minimum total clearing time and is represented by the flat portion on the bottom of the curve.

If a fault current subsides rapidly enough, the movable core will reset itself and the breaker will not trip. The *resettable-delay curve* in Fig. 14-14 indicates the time limit for a given fault current to subside to the rated value or below if tripping is not to occur. For example, if the fault current is 200 percent of rated and is sustained, the breaker

---

[1] To prevent damage or destruction of the breaker, the interrupting capacity of the breaker must be equal to or greater than the maximum available short-circuit current.

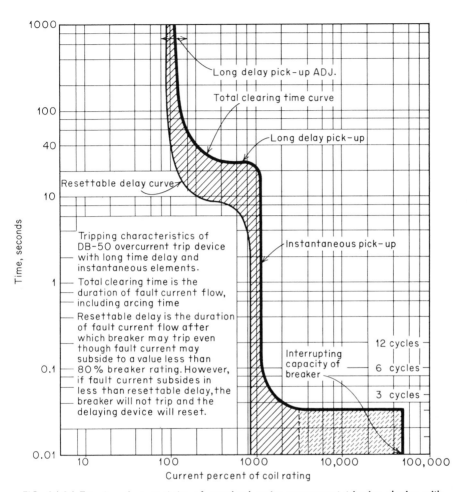

FIG. 14-14 Tripping characteristics of standard series overcurrent tripping device with long-delay and instantaneous elements. (*Westinghouse Electric Corp.*)

will clear the fault in no more than 40 sec, as determined from the total-clearing-time curve. If this magnitude of current subsides to less than the rated value within 10 sec after the start of the fault, the tripping unit will reset, and the breaker will not open (determined from the reset-table-delay curve). However, if there is an appreciable delay before the current subsides, say 20 sec, the breaker may still trip, even though the current drops below 80 percent of rated tripping current. This behavior can be explained by reference to Fig. 14-13*b*. As the movable core *C* approaches the fixed core *B*, the lessened distance between the two results in a stronger magnetic attraction. (This effect is similar to that produced when two magnets are made to approach each other, north to south.

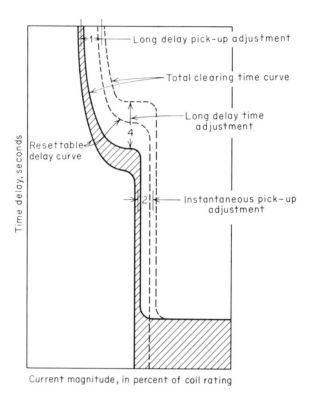

FIG. 14-15 Effect of pickup adjustments and delay-time adjustments on the characteristic curves. (*Westinghouse Electric Corp.*)

The magnetic pull becomes greater as the poles approach one another.) If the movable core travels a sufficient distance toward the fixed core, the increased magnetic attraction may cause the circuit breaker to open even if the fault current reduces to less than 80 percent of the trip-coil rating.

Figure 14-15 illustrates the effect of pickup adjustments and delay-time adjustments on the characteristic curves. The long-delay pickup adjustment knob *P* can change the position of the upper part of the curve through the range indicated by the number 1. The long-delay time adjustment dial *S* can be used to shift the knee of the curve over the range indicated by the number 4. The instantaneous pickup adjustment knob *N* can shift the vertical section of the curve to the left or right, as indicated by the number 2. As a safety measure, the breaker should be opened before making these adjustments.

Some breakers are equipped with an overcurrent tripping unit that provides adjustable long delay, adjustable short delay, and adjustable instantaneous elements. The short time delay is obtained by means similar to the long time delay except that the valve orifice is larger, allowing a greater rate of air flow.

The usual function of the long-time-delay element is to provide overload protection for conductors and connected apparatus. The function of the short-time-delay element is to provide a short delay on a fault current to give selectivity to other breakers, that is, to give other breakers closer to the fault an opportunity to trip. The instantaneous trip provides protection under high short-circuit currents. Two settings each are required to completely define the long-delay characteristic and the short-delay characteristic, namely, a pickup setting and a time-delay setting. The instantaneous characteristic is defined by the pickup setting alone. No intentional time delay is provided for the instantaneous trip. To provide adequate protection, the instantaneous pickup should be set as low as possible, without causing nuisance tripping; if the instantaneous setting is higher than the available short-circuit current, the instantaneous trip will never operate.

In addition to overcurrent tripping devices, circuit breakers are often equipped with low-voltage trips and shunt trips. The shunt trip may be actuated by auxiliary devices, such as a power-directional (reverse power) relay, a ground relay, a phase-balance relay, a remote push button, etc. For example, circuit breakers for ac generators usually have shunt trips that are actuated by a power-directional relay to trip the generator from the bus in the event that it is motorized. Similarly, breakers for dc generators in parallel use reverse-current trips to open the respective generator breaker if the machine is motorized.

14-8 Oil Circuit Breakers

An oil circuit breaker, shown in Fig. 14-16, utilizes an insulating oil to interrupt the arc. The moving contacts are resiliently mounted on heavy compression springs and attached to lift rods of specially treated wood. The movable and fixed contacts are completely immersed in a tank of insulating oil. The tank, not shown, is fabricated from heavy sheet steel and has an insulated inner liner. The breaker shown has a rating of 600 amp and 4,160 volts, a 25-Mva (million volt-amperes) interrupting rating, and an eight-cycle interrupting time. Manual or electrical tripping of the breaker causes the movable contacts to move downward. As the contacts separate, the arc produced is magnetically pulled away from the contacts (by the de-ion arc interrupter), lengthened, and forced into a wall of cool oil. The de-ionizing action of the cool oil quickly extinguishes the arc. The solenoid mechanism, shown in Fig. 14-16a, provides remote control or manual operation. Dashpots provide a time-delay feature. When making a maintenance check of circuit breakers, dashpots should be examined for dust, sludge, and other contaminants that can prevent their operation. Oil dashpots should be kept filled at the proper level.

Although insulating oil is an excellent medium for extinguishing arcs, it requires frequent testing to ensure that its quality has not been affected by the absorption of moisture or the accumulation of carbon and sludge. Moisture can enter the tank through normal breathing. During the heat of the day the air inside the tank expands, forcing some internal air through the seals to the outside. In the evening, when the tank cools, a partial vacuum is formed inside the tank, and cool, moist air is sucked in. The moisture condenses inside the tank, and after many heating and cooling cycles, a small pool of water may collect. If enough water accumulates, the resulting internal short circuit will cause a violent explosion, destroying the breaker and starting a fire.

The insulating oil should be tested every 3 to 6 months, depending

(a)

Dashpot

FIG. 14-16 Oil circuit breaker. (a) Operating mechanism; (b) internal construction. (Westinghouse Electric Corp.)

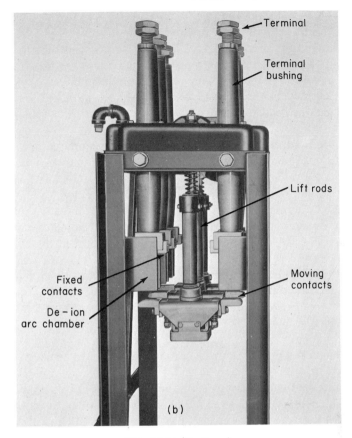

Terminal

Terminal
bushing

Lift rods

Moving
contacts

Fixed
contacts

De – ion
arc chamber

(b)

FIG. 14-16 (Continued)

on the frequency of breaker operation. It is also advisable to test the oil each time the breaker trips because of an exceptionally heavy short-circuit current. The dielectric strength of the oil should be 22,000 volts or higher, as measured by the standard test between 1-in. disks spaced 0.1 in. apart (see Sec. 2-16). If the oil shows signs of moisture, dirt, or carbonization, it should be filtered (see Sec. 2-17). Proper oil level should be maintained at all times, and the oil-level gauge should be checked periodically for accuracy.

14-9 Overcurrent Protection of Motors

The selection of an overcurrent protective device is determined by the type of overload the machine may be subject to. Motors that drive fans, pumps, and most machine tools require protection against sustained overloads that may cause excessive heating of the insulation; such motors are generally protected by time-delay thermal overload relays, whose tripping time decreases with increased overloads. In other applications, such as motor-driven elevators, winches, traction motors, and other loads that are subject to jamming, instantaneous overload relays are used to disconnect the motor from the line immediately; the jamming of a motor will cause the current to rise to its locked-rotor value, and if not immediately disconnected, the insulation may be destroyed.

Thermal overload relays. Thermal overload relays are used to protect a motor from overheating because of overcurrents caused by running overloads, a stalled rotor, low voltage, and low frequency. Overcurrents caused by the above are of relatively low magnitude compared with short circuits. One type of thermal overload relay, shown in Fig. 14-17a, consists of a heating element, bimetal, and a set of normally closed contacts. The heater is connected in series with the motor, as shown in the circuit diagrams in Fig. 13-4, so that its heating element increases in temperature with increasing values of current. In ac applications involving high motor currents a current transformer (CT) is used to provide a proportional but lower value of current to a proportionally lower rated heater. This is shown in Fig. 14-17c. The bimetal is mechanically linked but insulated from the normally closed contacts. A reset button is provided to reset the OL contacts after the relay heater and bimetal have cooled. The overload relay provides stalled-rotor and running protection for the motor, and the fuse or molded-case breaker provides short-circuit protection for the branch circuit containing the motor. It is essential that the current-time characteristic of the relay be carefully coordinated with the current-time characteristic of the breaker or fuse, so that the breaker trips (or fuse blows) on short circuits only and not on overloads or starting surges. The starting current (stalled-rotor current) drawn by a motor may be as high as 400 to 1,000 percent of its rated value. Current in excess of this is considered short-circuit current, and protection against damage due to short circuits is provided by molded-case breakers or fuses.

The approximate current-time characteristics for tripping and resetting are shown in Fig. 14-17b. Note that the tripping time decreases with increased current (inverse time characteristic), but the reset time increases with increased current. For the characteristics shown, an overload of 400 percent rated heater current causes tripping somewhere between 20 and 30 sec. However, the relay cannot be reset until at least 75 sec of cooling has elapsed, during which time the motor is also at rest and cooling. The overcurrent relay shown in Fig. 14-17a is provided with an adjustable setting to permit tripping at approximately 15 percent above or below the heater rating.

Another type of thermal overload relay, not shown, uses a low-melting-temperature alloy (eutectic alloy) to "solder" a small shaft to the inside of a fixed cylinder called a solder pot. The soldered shaft keeps a spring-loaded latch from tripping the overload contacts. A heater, attached to the cylinder, is connected in series with the motor. Excessive currents, caused by overloads, cause the heater to melt the alloy. This in turn permits the shaft to turn, tripping the relay.

Frequent tripping of the relay is reason for concern. Such tripping may be caused by misapplication or defective relays. The cause must

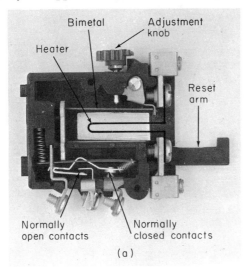

(a)

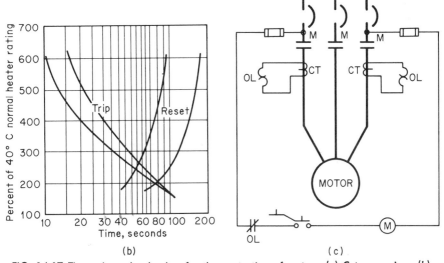

(b)   (c)

FIG. 14-17 Thermal overload relay for the protection of motors. (a) Cutaway view; (b) typical tripping and resetting characteristics; (c) typical circuit that uses current transformers in conjunction with overload heaters. (*General Electric* Co.)

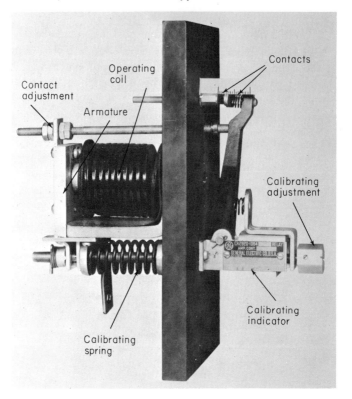

FIG. 14-18 Instantaneous overcurrent relay. (*General Electric Co.*)

be determined. Never bypass the relay; do not block it in the closed position; do not cool it with a fan, or arbitrarily readjust the setting. Such action may result in a burned-out motor or an electrical fire.

Instantaneous overcurrent relay. An instantaneous overcurrent relay, also called a jamming relay or overtorque relay, is shown in Fig. 14-18. Overcurrents in the operating coil, which is connected in series with the motor, cause the contacts to open. This in turn deenergizes the motor controller, stopping the machine. The armature (moving member) is shown in the tripped position. The tripping-current calibration spring is adjusted from the front of the panel, and the spacing of the contact tips from the back.

14-10 Coordination of Overcurrent Protective Devices

The complete protection of an electrical system against overcurrents and short circuits requires the coordination of all breakers, fuses, and overcurrent relays so that only the protective device nearest the fault will trip.

This is accomplished through a selective tripping arrangement whereby the delay time and current pickup settings of the breakers are adjusted to avoid overlapping of the time-current characteristics. In addition, each breaker and fuse in the system must be fully rated; that is, they must have interrupting ratings equal to or greater than the available short-circuit current in series with it. To assure adequate protection, the selection and coordination of protective-device settings must be done by engineers knowledgeable in the field.

Figure 14-19 illustrates a one-line diagram and coordination curves for a typical system involving air circuit breakers and fuses in a fully protected selective tripping arrangement. Breakers A and B have long-time and short-time trips, and breaker C has a long-time and instantaneous trip. If a fault of approximately 24,500 amp occurs between breaker C and the motor, only breaker C will open. As indicated in Fig. 14-19b, breaker C will clear the fault in 0.03 sec or less. Thus continuity of service will be maintained for the rest of the system. Similarly, if a fault occurs on a branch feeder of the motor control center, only the fuse in the particular motor circuit will open. This is indicated by the time-current characteristics in Fig. 14-19c.

Figure 14-20 illustrates the coordination of thermal overload relays that provide protection against running overloads and stalled rotors, and current-limiting fuses that provide branch-circuit protection for currents in excess of 1.1 times the locked-rotor current. The composite characteristic is shown with single cross-hatching and the crossover region is shown with double cross-hatching. For currents in the crossover region, the overload relay may open, the fuse may blow, or both may occur. The selection of the crossover point is determined by the locked-rotor current requirements. To account for variations in power-system voltage that may cause somewhat higher than rated locked-rotor current, a value of 1.1 times the rated locked-rotor current is generally selected as the basis for crossover determinations. Thus a current equal to 1.1 times the rated locked-rotor current should intersect the overload-relay curve at a point lower than the minimum-melting-time curve of the fuse. For the application shown, the pertinent motor data are as follows:

Motor full-load running current = 151 amp
Locked-rotor current (starting or stalled current) = 800 amp
Locked-rotor current times 1.1 = 880 amp
Normal motor starting time = 2 sec

Assuming that the motor does not start and the locked-rotor current is 880 amp, the thermal overload trip will open the motor circuit in approximately 4 to 10 sec.

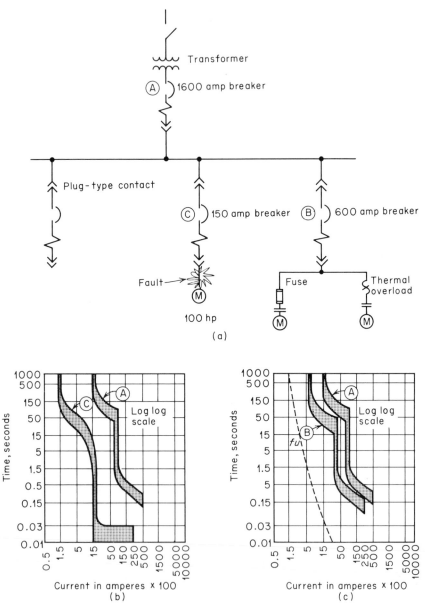

FIG. 14-19 One-line diagram and coordination curves for a typical selective tripping system. (*General Electric* Co.)

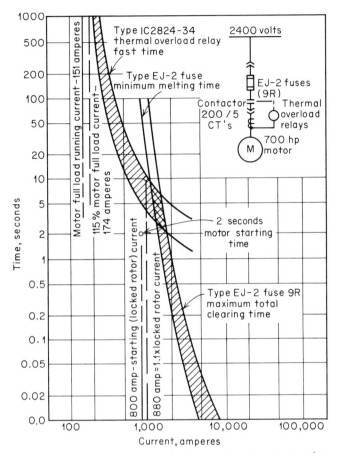

FIG. 14-20 Coordination curves for motor and branch-circuit protection. (*General Electric Co.*)

14-11 Short-circuit Current (Fault Current)

The magnitude of the short-circuit current, in the case of the simple system shown in Fig. 14-21*a,* depends on the overall impedance of the circuit as measured from the place of short circuit to the source and includes the impedance of the source itself. However, if induction motors are running at the time a short occurs (see Fig. 14-21*b*), they too will contribute to the short-circuit current. The current contributed by induction and synchronous motors to the fault at the instant the short circuit occurs is approximately equal to 3.6 times and 4.8 times, respectively, their full load current.[1] Considering the many motors and the complexity of the

[1] Molded Case Circuit Breakers, NEMA AB1-1964.

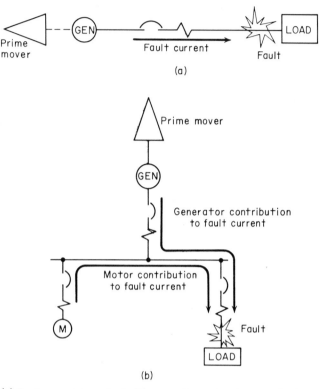

FIG. 14-21 (a) Fault current to a short; (b) generator and motor contributions to a fault.

distribution networks in many plants, the determination of the available short-circuit current in different parts of the system is not an easy problem and is beyond the scope of this book.

14-12 Maintenance and Testing of Overcurrent Relays and Breakers

Unless checked periodically, there is no assurance that a properly installed and correctly rated overload relay or circuit breaker will continue to provide the same degree of protection that was afforded at the time of installation. There are cases on record where the bimetal elements in thermal overload relays became insensitive to heat from metal fatigue and failed to operate under sustained overloads of 300 percent of rated heater current. The maintenance technician should check for corrosion, accumulation of dirt and other foreign matter that block free movement of the tripping element, loose or missing parts, and improper substitution of heater element or trip coil. All indicated troubles must be corrected.

Overcurrent relays and breakers may be tested by passing a specified overcurrent through the heater or overload coil and observing the time required for it to trip. The actual tripping time should then be compared with the manufacturer's current-time characteristic, such as that shown in Fig. 14-17b. It is recommended that breakers and overload relays be tested for proper tripping at 150, 300, and 600 percent of rated current.

To test an overcurrent relay, it should be disconnected from the power line and connected to a low-voltage high-current power source, as shown in Fig. 14-22a. A variable autotransformer, such as a Variac, may be used to provide the adjustable high current for the test, and a stopwatch should be used to determine the tripping time. A preliminary adjustment of the current should be made by connecting the two heavy test leads 1 and 2 together, closing the Variac breaker, and adjusting the Variac until the ammeter indicates slightly higher than the desired test current (the heater, when inserted, will cause the test current to

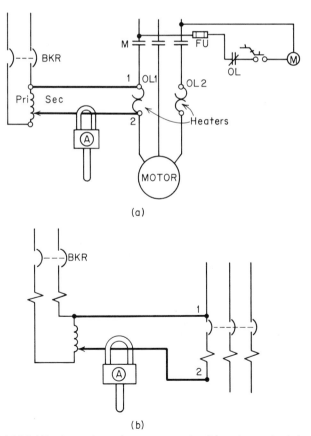

(a)

(b)

FIG. 14-22 (a) Testing a thermal overcurrent trip; (b) testing a circuit breaker.

drop slightly). Then open the Variac breaker, and connect the leads 1 and 2 to overload heater OL_1. To start the test, simultaneously close the Variac breaker and start the stopwatch, quickly readjust the current if needed, and stop the watch when the contacts open. Repeat the test for OL_2. If the tripping time is not within the manufacturer's specified limits, as indicated by the time-current curve, the heater may be defective or incorrectly sized, or the latching device may be damaged.

Magnetic overload trips and circuit breakers may be tested in a similar manner. A simple testing circuit for circuit breakers is shown in Fig. 14-22b. Each pole should be tested separately, even though the common tripping bar should cause all breaker poles to open when any one overcurrent trip is actuated.

A commercially available portable breaker testing unit complete with high-current power source, ammeter, and timer is shown in Fig. 14-23. It obtains its power from a standard 120-volt 60-hertz receptacle. Self-contained testing instruments of this type are very useful in maintenance programs. The one shown in Fig. 14-23 also has a built-in megohmmeter for insulation-resistance tests. Larger, self-contained breaker test sets are also available.

In those applications where circuit breakers are kept in either the closed or open position for extended periods of time, they should be "exercised" periodically by operating the tripping mechanisms electrically several times in succession. However, before exercising the breaker, be sure that all power lines leading to the breaker are deenergized.

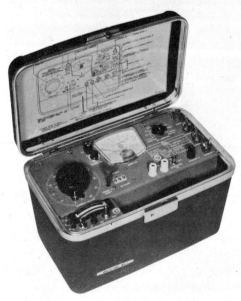

FIG. 14-23 Test set for checking the calibration of overload relays and breakers. (*Multi-Amp Corp.*)

The main contacts of circuit breakers should be inspected for wear and misalignment. Contacts roughened by service should not be filed smooth unless large projections of metal or beads of metal, formed by arcing, prevent good contact. If filing is necessary, it should be done with a smooth mill file, and should closely follow the original shape of the contact. Alignment can be checked by obtaining a print of the contacts; placing carbon paper and a thin sheet of tissue paper between the contacts and then closing and opening the breaker will leave an imprint on the tissue. If the impression indicates that less than three-fourths of the normal area of the contacts is touching, an adjustment is needed. The manufacturer's manual should be consulted for information on adjustment.

The maintenance check of a motor controller should always include a comparison of the installed heater element with the maximum ratings recommended in the National Electrical Code. The heater element shall be rated or selected to trip at no more than the following percentage of the motor full-load current rating:[1]

Motors with a marked service factor not less than 1.15	125%
Motors with a marked temperature rise not over 40°C	125%
Sealed (hermetic-type) motor compressors:	
Overload relays	140%
Other devices	125%
All other motors	115%

For a multispeed motor, each winding connection shall be considered separately.

When a separate motor running overcurrent device is so connected that it does not carry the total current designated on the motor nameplate, such as for wye-delta starting, the proper percentage of nameplate current applying to the selection or setting of the overcurrent device shall be clearly designated on the equipment or the manufacturer's selection table shall take this into account. (Consult the National Electrical Code for exceptions.)

Bibliography

Beeman, D. L.: "Industrial Power Systems Handbook," McGraw-Hill Book Company, New York, 1955.

Electric Power Distribution for Industrial Plants, IEEE No. 141, 1964.

Electrical Systems for Commercial Buildings, IEEE No. 241, 1964.

Low Voltage Cartridge Fuses, NEMA FU1-1966.

Low Voltage Power Circuit Breakers, NEMA SG3-1965.

Mason, C. R.: "The Art and Science of Protective Relaying," John Wiley & Sons, Inc., New York, 1956.

[1] National Electrical Code 1968, National Fire Protection Association.

Molded Case Circuit Breakers, NEMA AB1-1964.

National Electrical Code 1968, National Fire Protection Association.

Questions

14-1. What are some of the injurious effects that may be caused by excessive current?

14-2. How does a fuse provide protection against sustained overcurrents? How is a fuse rated?

14-3. What is meant by the interrupting capacity of a fuse? What can happen if the available short-circuit current is greater than the IC rating of the "protecting" fuse?

14-4. Describe the characteristics of the following: an NEC fuse, a time-lag fuse, and a current-limiting fuse. How are they constructed?

14-5. What are the common causes for blown fuses other than short circuits?

14-6. Describe a method that may be used to check for blown fuses without energizing the circuit.

14-7. Describe a method that utilizes the circuit voltage when testing for blown fuses in a low-voltage circuit. That is, the fuses are tested while the power is on.

14-8. What are molded-case breakers? How are they rated, and where are they used?

14-9. How does a molded-case breaker provide protection against (a) sustained overloads and (b) short circuits?

14-10. Explain the operation of the de-ion arc quencher.

14-11. Define the interrupting capacity of a breaker.

14-12. What can happen if a circuit breaker is called upon to clear a short-circuit current that is considerably in excess of its interrupting capacity?

14-13. How does the current limiter, used with some molded-case breakers, provide a limiting action?

14-14. Referring to the tripping curve in Fig. 14-10, what is the tripping time for (a) 350 amp, (b) 500 amp, and (c) 8,000 amp?

14-15. What are circuit-breaker band curves? Of what use are they in maintenance testing?

14-16. What adjustments are possible on adjustable types of air circuit breakers?

14-17. What information is provided by the resettable-delay curve?

14-18. What are the respective functions of long-time delay, short-time delay, and instantaneous elements in an air circuit breaker?

14-19. What practical considerations should determine the setting of the instantaneous pickup element?

14-20. What is the function of the oil in an oil circuit breaker?

14-21. Why should the insulating liquid in an oil circuit breaker be checked every 3 to 6 months? What type of test is required?

14-22. What types of relays are used to protect motors from damage due to (a) sustained overloads and (b) jamming? How do they work?

14-23. Referring to the curves in Fig. 14-17b, if the motor current is 300 percent of rated current, what is the longest time it will take for the overload relay to trip? What is the minimum time delay required for cooling before the relay can be reset?

14-24. Frequent tripping of a motor overload relay was "cured" by resetting the trip adjustment. Discuss the pros and cons of this action.

14-25. What are the advantages of selective tripping?

14-26. Referring to Fig. 14-19a and c, which protective device will operate if the fault current is 1,500 amp?

14-27. Referring to Fig. 14-19a and b, what would happen if the fault current were 5,000 amp, and for some reason, perhaps a mechanical defect, breaker C did not trip?

14-28. What factors determine the magnitude of the short-circuit current that can occur at some point in a system?

14-29. Describe a test procedure that may be used to check the calibration of a circuit breaker.

14-30. What should a maintenance technician check for when inspecting (a) overcurrent relays and (b) circuit breakers?

14-31. What simple test can be used to determine the area of contact between the movable and fixed contacts of a breaker?

fifteen

distribution systems

15-1 Ungrounded Distribution System

An ungrounded distribution system has no deliberate electrical connection to ground. Its most significant advantage is that accidental contact between one line and ground does not cause an outage. Although a single ground fault on an ungrounded system does not cause an interruption in service, it is important that the fault be located and cleared immediately. If not remedied and a second ground occurs on another phase or line of opposite polarity, the resultant short-circuit current may trip one or more circuit breakers. An example of a double ground-fault condition is shown in Fig. 15-1.

Ground-detecting circuits for use in ungrounded systems to indicate the presence of a ground are shown in Fig. 15-2. In the two-wire or single-phase system (Fig. 15-2a), two identical lamps are connected in series and tied to the two lines. The junction point of the two lamps is connected to ground through a normally closed spring-return push button. A ground on either side of the line will cause its respective lamp to go dark and the other to burn brightly; with no grounds both lamps will be dim. The two lamps should have identical wattage ratings and a voltage rating equal to the line voltage. A ground fault on line *B*, shown with dotted lines in Fig. 15-2a, short-circuits lamp *B*, causing it to burn dimly or go out, depending on the severity of the ground, while lamp *A* burns more brightly. The push button provides a means for comparing the normal and ground indications.

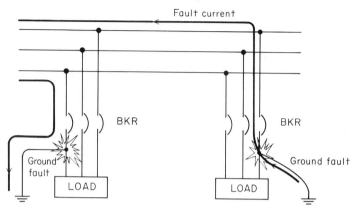

FIG. 15-1 Path of current in an ungrounded system resulting from accidental grounds in two different phases.

Figure 15-2*b* shows a ground-detecting system for a low-voltage three-phase ungrounded system. A ground fault on phase *C*, shown with dotted lines, short-circuits lamp *C*, causing it to burn dimly or go out, while lamps *A* and *B* burn more brightly.

Figure 15-2*c* illustrates a ground-detecting circuit for a high-voltage ungrounded three-phase system. A step-down isolating transformer is used to provide low voltage for the ground-indicating lamp. The primary is wye-connected with the neutral grounded, and the delta-connected secondary is closed through the lamp and resistor. With no ground fault on the system, the three secondary voltages are balanced, and no voltage appears across the lamp. However, a ground fault on phase *C*, for example, will effectively short primary coil P_C and cause the voltage of secondary coil S_C to be zero. The resultant voltage unbalance in the secondaries will cause the lamp to light, indicating the presence of a ground.

The wattage rating of the lamps used in the ground-detecting circuits determines the degree of sensitivity of the ground-detecting system. In general, 25-watt lamps provide the best overall performance. Lower-wattage lamps tend to be too sensitive, and higher-wattage lamps are not sensitive enough.

Accidental grounds should be repaired as soon as possible, for even one ground has a bad effect on the insulation. A single ground doubles the electrical stresses in the remaining insulation, thus increasing the possibility of an insulation breakdown in the line of opposite polarity. The doubled electrical stresses cause twice the electron leakage through the insulation, hastening the deterioration and shortening its life. This is illustrated in Fig. 15-3. The insulation resistance between each conductor and ground is *R* ohms. With 120 volts applied between the two conductors,

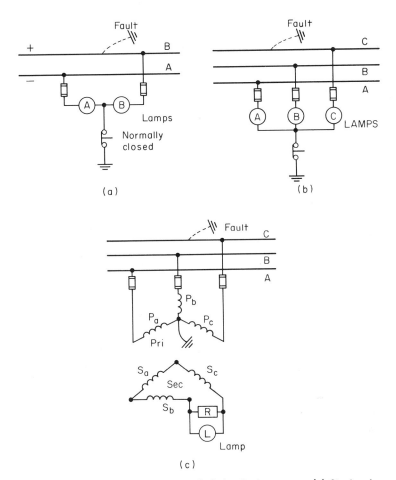

FIG. 15-2 Ground detection for an ungrounded distribution system. (a) Single phase or dc; (b) three-phase low-voltage; (c) three-phase high-voltage.

the voltage between each conductor and ground is 60 volts. However, if one conductor is grounded, as indicated by the dotted line in Fig. 15-3b, the voltage difference between the other conductor and ground will rise to 120 volts. Hence the voltage stress on the insulation of the ungrounded conductor is doubled, and if a weak spot in the ungrounded insulation causes it to blow, a short circuit will result.

Trouble shooting in ungrounded distribution systems. Troubles in distribution systems generally manifest themselves by failure of a device to operate, blown fuses, an indication on the ground-detecting apparatus, smoke, or the overheating of a cable. Shorts and opens are relatively easy to locate and are indicated by blown fuses and voltage failure respectively.

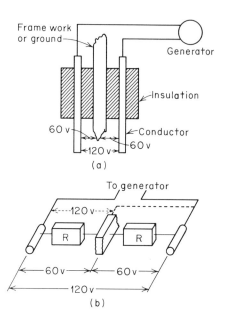

FIG. 15-3 Distribution of voltage between conductors and ground. (a) Two conductors of opposite polarity separated from each other and ground by insulation resistance; (b) elementary series circuit formed by insulation resistance R, metal framework against which insulation rests, conductors, and generator.

Grounds, on the other hand, unless accompanied by a short or open, are generally located by the process of elimination. The grounded circuit may be determined by pulling switches on the distribution panel, one at a time, until the ground-detecting device on the switchboard indicates normal. Closing each switch before opening the next keeps the interruption of service to a minimum. The opening of switches supplying vital auxiliaries should be avoided until standby equipment is placed in operation. If this procedure fails, either the ground is in the generator, or more than one ground is present.

Multiple grounds may be located by opening the switches, one at a time, and leaving them open until the ground detector indicates normal. Then, with the grounded switch left open, the others should be closed one after the other until another ground is indicated. The switches to the grounded circuits should be left open, and the procedure continued until every switch is tested. A grounded generator may be identified by transferring the load to another machine and tripping the machine in question from the line. If the generator is grounded, opening its circuit breaker and disconnect switch will cause the ground detector to indicate normal.

Tracking down the actual location of the grounded conductor is done best with a megohmmeter, magneto, or battery and buzzer set. When doing so, the switch to the grounded circuit should be blocked open, the fuses removed, and a "Do Not Close—Man Working on Line" sign

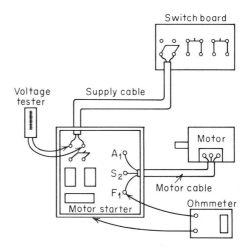

FIG. 15-4 Tracking down a ground by the process of elimination.

hung over the switch. Figure 15-4 illustrates the megohmmeter method for tracking down a ground that may be in the supply cable, starter, motor cable, or in the motor itself. Power should be removed from the grounded circuit, and the disconnect switch in the motor starter should be tested to ensure that the circuit is dead. The supply cable may then be tested by applying a megohmmeter between the metal framework of the starter and the top of the switch. A zero reading of the megohmmeter will indicate a grounded switch or a grounded cable. Assuming the cable tests clear, as indicated by the absence of a zero reading, the test should be made between the metal framework and the motor connections at the starter. If a ground is indicated, the motor leads should be disconnected from the starter, and the starter and motor tested individually. Grounds within the motor may be located by additional tests explained elsewhere.

15-2 Grounded Distribution Systems

A grounded distribution system has one conductor of the system solidly connected to a common system ground. The grounded system offers advantages of greater safety to personnel and equipment, reduced exposure to overvoltages, and easier location of ground faults.

Accidental grounds in a grounded system generally cause the opening of breakers or the blowing of fuses. However, in the case of high-resistance ground faults, the ground current may not be high enough to operate the fault-interrupting device but sufficient to cause sustained arcing and burning of electrical apparatus. The best protection against damage due to ground faults may be obtained with a ground-sensing relay that operates a tripping coil in the breaker. In the absence of a ground relay, the trip setting of the breaker should be set as low as possible; however, the trip setting must be high enough to prevent false tripping that may be caused by motor starting current.

Figure 15-5 illustrates different grounding arrangements for normally grounded distribution systems. In Fig. 15-5c the neutral is grounded through the primary of a current transformer, and the CT secondary is

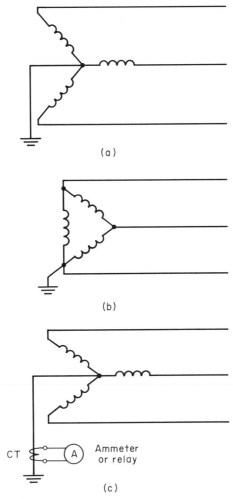

(a)

(b)

CT   (A)   Ammeter or relay

(c)

FIG. 15-5 Different grounding arrangements for normally grounded distribution systems. (a) Grounded neutral; (b) grounded delta; (c) neutral grounded through a current transformer.

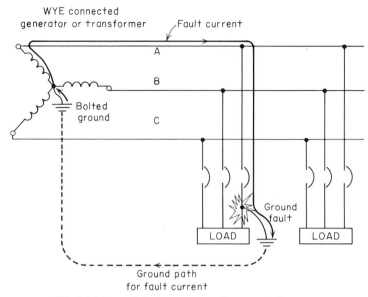

FIG. 15-6 Ground fault on a normally grounded system.

connected to a ammeter or relay. The ammeter will indicate the presence and severity of a ground, even though the ground current may not be high enough to operate the fault-interrupting equipment. The procedure for locating high-resistance ground faults in a system that is grounded through a current transformer and ammeter is similar to that described for ground-detecting lamps in an ungrounded system; the only difference is that an ammeter is used as the ground indicator.

Figure 15-6 shows the path of ground current resulting from a single low-resistance ground fault on a normally grounded system. The generator or transformer neutral is bolted to ground.

Figure 15-7 illustrates a single-phase three-wire distribution system for lighting and power circuits in domestic and general-service applications. The center tap of the transformer secondary is solidly grounded to a water pipe or to a driven ground. The ground connection is carried directly to each branch circuit and is not fused or in series with a circuit breaker; only the "hot" lines have series-connected fault-interrupting equipment. Ground faults in these circuits will cause blown fuses or tripped breakers.

15-3 System Grounding

The framework of all electrical equipment, large or small, should be connected with low-resistance metallic conductors to a common point, usually

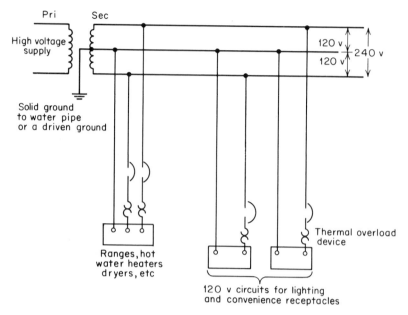

FIG. 15-7 Three-wire distribution system for lighting and power circuits.

the earth. Such connections, called *ground connections,* eliminate the buildup of static electricity, eliminate the shock hazard to operating personnel caused by normal leakage current and normal capacitive effects, and reduce the hazard to personnel if an insulation fault should occur within the apparatus; very little or no voltage can occur between properly grounded frameworks or between properly grounded frameworks and waterpipes, supporting structures, etc. Proper grounding of electrical apparatus is also essential to the effective operation of circuis used in process control systems and in communication.

The equipment ground conductors in normally grounded distribution systems described in Sec. 15-2 must be large enough in cross-sectional area to safely carry a ground-fault current. If the resistance of equipment ground conductors is too high, the ground-fault current may not be high enough to trip the breaker but still be high enough to cause severe damage or destruction by sustained arcing and burning within the equipment. Furthermore, under severe ground-fault conditions, a high-resistance ground connection may cause the voltage between the equipment frame and ground to rise high enough to present a serious shock hazard.

The grounding path may include solidly connected metallic conduit, metallic piping, one-piece copper conductor of appropriate size,[1] the steel structure of a building, the hull of a ship, etc.

[1] National Electrical Code, USAS C1-1968.

The connection to earth may be accomplished by connection to a metallic underground water-supply system or to "driven grounds" using unpainted galvanized pipes, rods, or metal plates as electrodes.[1] The pipes (¾ in. in diameter) or rods (⅝ in. in diameter) should be driven to a depth of at least 8 ft. Each plate electrode should be at least ¼ in. thick and have a cross-sectional area of at least 2 sq ft. Where over one electrode is used, spacing between them should be no less than 6 ft.

15-4 Power-factor Improvement of a Distribution System

Depending on the relative proportions of heating elements, lamps, induction motors, etc., connected to an ac system, the power factor may be somewhere between 0.35 and 0.90 lagging; the lagging power factor is caused by the kilovars drawn by the inductive loads (see Sec. 4-4). Although the lagging kilovars drawn by the inductive loads do no work,

[1] National Electrical Code, USAS C-1-1968.

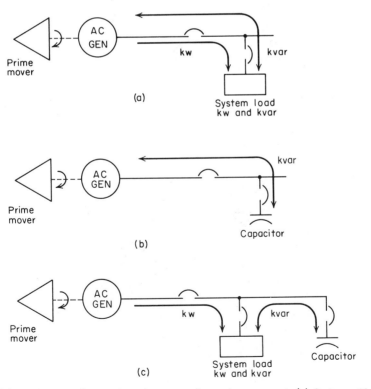

FIG. 15-8 Application of capacitors for power-factor improvement. (a) System without capacitors; (b) kvars drawn by capacitor alone; (c) capacitor in parallel with load.

they must be supplied. The kva supplied by a generator to the distribution system is equal to

$$kva = \sqrt{(kw)^2 + (kvar)^2}$$

If the generator did not have to supply the kvars, it would have more kva available for additional productive loads.

Fortunately, capacitors can be used to free the generator from its lagging kvar load. A capacitor connected to an ac generator seesaws energy between itself and the generator but, unlike the induction motor, draws very little active power. In effect it does no work. Furthermore, the capaci-

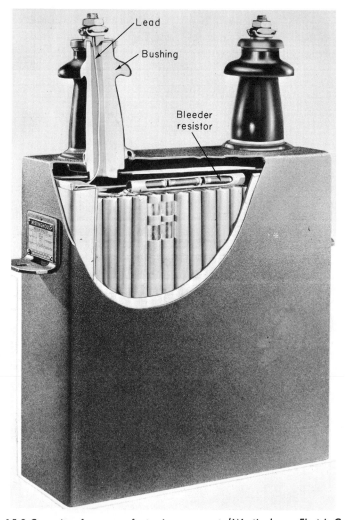

FIG. 15-9 Capacitor for power-factor improvement. (*Westinghouse Electric Corp.*)

tor draws leading kvars, and the induction motor draws lagging kvars. The significance of this is that, although they both seesaw energy with the generator, when the system inductance is drawing energy, the capacitor is releasing energy, and vice versa. Hence, if a capacitor is connected to the system, and has a kvar rating equal to the kvar rating of the inductive loads on the system, the seesaw of energy will be between the inductive loads and the capacitor and not with the generator. This is shown in Fig. 15-8. Furthermore, because of the reduced kva required, a smaller size of feeder cable may be used if the capacitor is connected as close as possible to the low-power-factor load terminals.

A capacitor for power-factor improvement is shown in Fig. 15-9. It is composed of a number of series-connected sections immersed in an insulating liquid. The internal bleeder resistor is used to discharge the capacitor to 50 volts or less in 1 to 5 min after disconnecting from the line.

A synchronous motor may also be used for power-factor improvement. The leading kvars drawn by an overexcited synchronous motor have a capacitive effect on the system. (See Sec. 5-9.)

15-5 Maintenance and Testing of Power Capacitors

A maintenance inspection of power capacitors used for power-factor improvement should always include a check of the fuses, ventilation, voltage, and ambient temperature; the life of a capacitor is shortened by overheating, overvoltage, and physical damage. Ceramic bushings and other insulated surfaces should be kept clean.

> **Warning.** Before cleaning or touching any conductors, disconnect the capacitor from the line and discharge it through a heavy-duty 50,000-ohm resistor; do not depend entirely on internal bleeder resistors. Discharging should be done between terminals and between terminals and ground. Although small capacitors may be discharged by short-circuiting their terminals with a copper strap or heavy cable, large capacitors may be damaged by this practice; the very large mechanical forces caused by the high discharge currents may destroy them. Under no circumstances should a bank of large capacitors be discharged through short-circuiting. The sudden release of a large amount of energy could vaporize or explode the shorting device, causing injury to personnel as well as damage to the capacitors.

Power capacitors that are leaking insulating liquid should be patched by soldering a patch over the hole, using a rosin-flux solder. After patching, leaky capacitors should be returned to the manufacturer for repair.

Capacitors used for power-factor improvement may be tested at rated

voltage for internal shorts or opens by using the circuit shown in Fig.
15-10a. Switch S_1 should be closed before closing the breaker; this will
prevent a high transient charging current from damaging the ammeter
at the instant the breaker is closed. After closing the breaker, open the
switch and read the ammeter. The normal current for a good capacitor,
tested at rated nameplate voltage, may be determined from

$$I_{\text{normal}} = \frac{(\text{kva})_{\text{NP}} \times 1,000}{v_{\text{NP}}} \quad \text{amp}$$

where $(\text{kva})_{\text{NP}}$ = nameplate kva of capacitor
       v_{NP} = nameplate voltage of capacitor

If rated voltage is not available, the capacitor may be tested at below
rated voltage by including a voltmeter, as shown in Fig. 15-10b. The
"normal" current for a good capacitor operating at below rated voltage

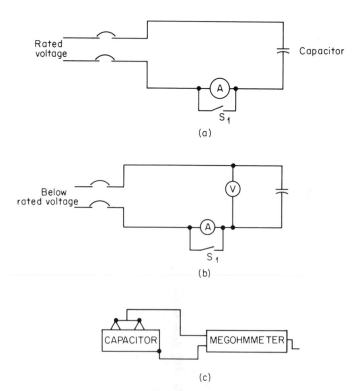

FIG. 15-10 Testing power capacitors. (a) Testing for shorts or opens at rated voltage; (b)
testing for shorts or opens at below rated voltage; (c) insulation-resistance test between
conductors and case.

may be determined from

$$I_{normal} = \left[\frac{V_{VM}}{V_{NP}}\right] \cdot \left[\frac{(kva)_{NP} \times 1,000}{V_{NP}}\right] \quad amp$$

where V_{VM} = voltmeter reading

Ammeter readings greater than the calculated normal currents indicate a short in one or more sections.[1] Ammeter readings less than the calculated normal currents indicate an open.

To determine the insulation resistance between the conductors and the metal case, connect one terminal of a megohmmeter to the line terminals of the capacitor and the other terminal of the megohmmeter to the case, as shown in Fig. 15-10c. The resistance measured should be not less than 1,000 megohms.[2]

15-6 Lightning Protection

Although lightning is an unavoidable hazard to electrical systems, its adverse effects on the system can be minimized through proper application of lightning arresters. Direct lightning strokes and even induced voltages due to lightning strokes elsewhere in the system may cause damage or complete destruction to electrical apparatus. Figure 15-11 illustrates the

[1] Above-normal currents may also be caused by a nonsinusoidal voltage.
[2] Shunt Capacitors, NEMA CPI–1968.

FIG. 15-11 Motor damaged by lightning. (*Mutual Boiler and Machinery Insurance* Co.)

damage that a nearby lightning stroke can inflict on unprotected electrical equipment. The induced voltage caused severe arcing and burning of the stator coils at the connection end of the machine.

Direct or induced voltages caused by lightning strokes can cause a high-voltage wave to travel along a power line. Unless this high surge voltage is reduced before reaching the terminals of an electrical machine, severe damage will result. The best protection for electrical machinery may be achieved with a combination of lightning arresters and protective capacitors, as shown in Fig. 15-12. The arresters should be installed as close as possible to the machine they protect. To obtain the greatest margin of protection, the lightning arrester should have the lowest voltage rating consistent with the normal line-to-ground voltages and abnormal fault voltages that may occur on the apparatus it protects; an arrester with too high a voltage rating will not provide adequate protection, and one with too low a voltage rating may be damaged. The protective capacitor should have a voltage rating equal to the line-to-line voltage rating of the machine and a capacitance of about 1.0 μf per phase for rms line-to-line voltages of 650 or less, 0.5 μf for voltages of 2,400 to 6,900, 0.25 μf for voltages of 11,500 to 13,800, and 0.125 μf for 24,000 volts. The arresters limit the magnitude of line-to-ground surge voltages, thereby protecting the line-to-ground insulation of the machine. The capacitors round out the steepness of the limited surge-voltage wave, thereby preventing excessively high voltages from being inductively generated between adjacent coil turns in the motor.

poor concept — no "induction" is involved

Even though the transmission line, substation, and power transformers have installed lightning protection, additional protection should be provided for rotating machinery.

One common type of lightning arrester uses a column of lead peroxide pellets and a series gap assembly. The pellet column forms the valve

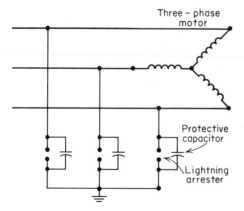

FIG. 15-12 Lightning arresters and protective capacitors used in combination to protect a motor.

element, preventing a sustained flow of system power current that would otherwise occur immediately after the lightning discharge; the resistance of the pellet column is very high at normal line voltages but drops to a very low value while under the influence of a high-voltage surge. The series gap isolates the valve element from the line until sparked over by a lightning impulse.

The pellets, which have a thin porous coating of litharge, are assembled in a porcelain-tube container with metal electrodes in contact with each end of the pellet column. The series gap assembly is contained within a gap chamber containing dry nitrogen (N_2).

Lightning arresters and protective capacitors should be inspected periodically for overheating and cracked or broken insulators and should be kept free of dirt and other foreign matter. Ground and line connections to the arrester must be tight.

> **Warning.** Before cleaning or otherwise working on a lightning arrester, it must be disconnected from the power line, and the protective capacitors must be discharged.

15-7 Maintenance of High-voltage Insulators and Bushings

A high-voltage *insulator* is a highly glazed ceramic support used to separate high-voltage lines from each other and from ground. High-voltage *bushings* are hollow ceramic tubes used to insulate conductors that pass through a metal tank or enclosure. Figure 15-13 illustrates a high-voltage bushing for a transformer (see also Fig. 15-9). The petticoat construction increases the creepage distance (the shortest distance measured on the surface of the insulator, between parts of different polarities).

Deposits of dirt on high-voltage insulators may cause a flashover, and should be removed by periodic cleaning. This is particularly important in areas where there is a prevalence of salts, conducting dusts, or other contaminants in the surrounding air. However, before attempting to test or clean a bushing, it must be deenergized and then grounded to discharge any stored energy.

Cleaning of bushings and insulators may be accomplished by rubbing with a cloth moistened with water or an ammonia solution. Special porcelain cleaning compounds, such as Lockebrite, are available for stubborn cases that do not respond to simple cleaning methods. The insulator must be thoroughly rinsed with clean fresh water before returning to service.

Chipped or abrased porcelain can be repaired. The sharp edges should be honed smooth, and the defective areas painted with an insulating varnish to provide a glossy finish, as shown in Fig. 15-14. Fine hairlike cracks in the surface of an insulator must be repaired as soon as detected;

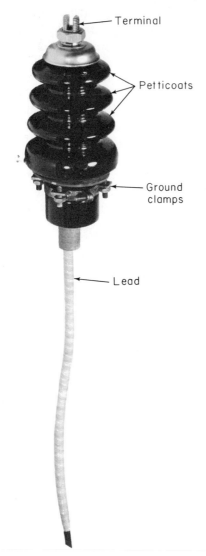

Terminal

Petticoats

Ground
clamps

Lead

FIG. 15-13 High-voltage bushing with cable
for a transformer. (*General Electric* Co.)

the accumulation of dirt and moisture in the cracks can cause leakage
current that may result in a flashover.

The insulation resistance of a bushing may be tested with a 2,500-volt
megohmmeter that has a 20,000-megohm range. Bushing resistance should
be high, and any value below 20,000 megohms indicates the need for
reconditioning. Low megohm readings indicate moisture in the bushing
or a deposit of dirt on the poreclain surface. To make the test, the bushing
or standoff insulator should be disconnected from the power circuit, the

FIG. 15-14 Varnishing a chipped bushing that is still serviceable. (*General Electric* Co.)

terminal grounded to discharge any stored energy, and then the megohm-meter leads connected between the terminals and grounds.

15-8 Cable Maintenance and Testing

Periodic inspections and insulation-resistance tests should be made on all cable installations to prevent unscheduled outages. The inspector should check for excessive cable temperatures, discolorations caused by excessive temperatures, the accumulation of water in cable ducts, the dripping of water, oil, or other liquids onto cables, indications of insulation swelling, brittleness, corrosion of lead sheaths, or other signs of deterioration. The inspector should also check for excessive mechanical stresses caused by stretching during installation, lack of support, etc.

Figure 15-15 shows sections of type R–rubber–insulated cable used as the main leads to a 100-hp motor driving a forced-draft fan. The cable was undersized for the particular load and had been in service for over 19 years, during which time it had been exposed to excessive operating temperatures. The long period of excessive temperature caused the insulation to deteriorate.

The allowable current-carrying capacities of insulated copper conductors are given in Appendixes 1 and 2; cables that are overheating should have their current-carrying capacity checked against the actual cable current, and the correct size cable installed.

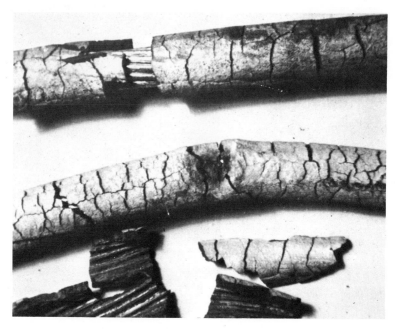

FIG. 15-15 Cable insulation badly deteriorated by excessive temperature. (*Mutual Boiler and Machinery Insurance* Co.)

Potheads, which are used to seal the ends of a cable insulation against moisture and other contaminants, should be checked for signs of deterioration. Hermetically sealed potheads are necessary for permanent protection of all paper-insulated cables.

The insulation resistance of low-voltage power cables may be tested with a megohmmeter, as shown in Fig. 15-16. Both ends of the cable should be disconnected from the apparatus or switchgear and discharged before making this test. Only one conductor should be tested at a time;

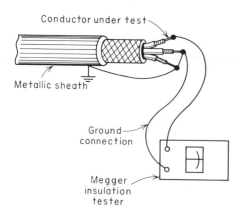

FIG. 15-16 Insulation test of a three-conductor cable. (*James G. Biddle* Co.)

the other conductors should be connected together and grounded. For comparison purposes, periodic insulation-resistance measurements should be made with the same test instrument, the same duration and value of test voltage, and at the same cable temperature.

The insulation resistance of a cable depends on its length and operating temperature. A longer cable has more leakage paths than a shorter one; hence the insulation resistance of a longer section of the same cable is less than that of a shorter section. This inverse relationship is expressed by the following formula:

$$\text{Overall insulation resistance} = \frac{\text{insulation resistance per foot of cable}}{\text{length of cable in feet}}$$

Unfortunately, there is no general standard for the minimum acceptable insulation resistance for cable in service. Each operating company establishes its own standards based on the cables and operating conditions peculiar to its system.

However, a good maintenance test of cables can be made with a dc high-potential test set, following the instructions outlined in Sec. 2-13. The maximum test voltage used should range between 1.7 times the operating line-to-line voltage of the cable to about 60 percent of the factory test voltage.

15-9 General Maintenance of Switchgear

Switchgear should be scheduled out of service on a 1- or 2-year basis for a thorough inspection, cleaning, checking, and tightening of all bus and cable connections and for the testing and calibration of critical components, such as circuit breakers, relays, instruments, etc. The frequency of inspection should be dictated by the severity of operation and the degree of atmospheric contamination. Insulation-resistance tests and high-potential tests should be made from line-to-ground and from line-to-line; voltmeters and other line-to-line and line-to-ground indicators should be in the "off" position before making such tests. During inspection and cleaning a careful check should be made of all bus and cable connections for evidence of overheating.

Switchgear operating in very hot atmospheres with high sulphur concentrations should be checked for the growth of metallic whiskers. Figure 15-17 shows a filament type of growth resembling steel wool on the silver-plated spring-loaded finger contacts of a circuit breaker. Continued growth of the whiskers would eventually result in a connection between the contacts of opposite polarity. The resultant spark could ionize the air between the contacts, causing a flashover.

FIG. 15-17 Fine filament type of growth of silver, resembling steel wool, on the silver-plated spring-loaded finger contacts of a circuit breaker. (*Mutual Boiler and Machinery Insurance Co.*)

For instructions on the testing, calibration, cleaning, and adjusting of specific apparatus, consult the Index.

15-10 Fluorescent Lights

The component parts that make up a fluorescent lighting system are the lamp, starter, and ballast. The starter is either a manually operated push button or an automatic switch. The ballast usually consists of a composite unit containing a transformer, inductance coil, or both, and most often a capacitor. The maintenance procedures detailed below apply to fluorescent lighting fixtures that have been properly installed in accordance with the manufacturer's connection diagram.

When a fluorescent lamp fails to light, the trouble may be in the lamp, starter, ballast, or supply circuit. The lamp is easily checked by substituting one that is known to be good or by trying it in another fixture that is operating properly. Sometimes removing the lamp and replacing it again serves to wipe clean any dirty contacts at the lamp sockets and thus permits restarting the lamp. Bent pins on the lamp also cause poor contact and should be straightened. If a good lamp will not light when placed in the fixture, the starter should be replaced by one of the correct size that is known to be good. If a good starter fails to light a good lamp and no fuses are blown or circuit breakers open, the power connections to the fixture should be exposed and a voltage tester used to check the supply voltage. The presence of rated voltage is a sure sign of a defective ballast that should be replaced.

Flashing on and off, generally accompanied by darkened coatings on the ends of the tube and an orange-red glow before flashing, is an indication that the lamp has reached the end of its useful life. Such lamps should be replaced immediately, or the resultant flashing will damage the starter and the ballast. If a relatively new lamp flashes on and off, a defective or wrong-sized starter may be the trouble. If both a good tube and a good starter fail to stop the flashing and a voltmeter indicates proper operating voltage, the ballast is at fault.

Flickering or swirling of the light, as contrasted with flashing, may be caused by a defective starter, a defective ballast, or even a defective lamp. When this occurs in a new lamp, it can often be remedied by turning the lamp off for a few minutes; however, such lamps often correct themselves after several days of service.

Some starters are equipped with a miniature circuit breaker that opens the circuit if the lamp continues to flash but does not start. This type of starter is called a cutout, push-button, or reset starter. The circuit breaker is reset by pressing a little plastic button on the top of the starter casing.

Great care should be taken not to break the lamp. A fluorescent lamp may explode when dropped, causing glass to fly in all directions. The mercury vapor contained in the lamp is highly toxic and should not be inhaled. Furthermore, the phosphor coating on the inside of the glass is highly poisonous, and any person cut with this glass should seek medical treatment.

Bibliography

Beeman, D. L.: "Industrial Power Systems Handbook," McGraw-Hill Book Company, New York, 1955.

Electric Power Distribution for Industrial Plants, IEEE No. 141, 1964.

Electric Systems for Commercial Buildings, IEEE No. 241, 1964.

"Electrical Transmission and Distribution Reference Book," Westinghouse Electric Corp., 1944.

Kurtz, E. B.: "The Lineman's and Cableman's Handbook," 4th ed., McGraw-Hill Book Company, New York, 1964.

Lightning Arresters, NEMA LA1-1968.

McPartland, J. F.: "Electrical Systems Design," McGraw-Hill Book Company, New York, 1960.

Questions

15-1. Why is it important that a single ground fault in an ungrounded distribution system be located and cleared immediately?

15-2. Compare the effects of a single ground fault on (*a*) an ungrounded distribution system and (*b*) a grounded distribution system.

15-3. What are the advantages and disadvantages of (*a*) a grounded distribution system and (*b*) an ungrounded distribution system?

15-4. How does the wattage rating of the ground-indicating lamps affect the sensitivity of ground detection?

15-5. State how you would determine the exact location of a ground in an ungrounded distribution system.

15-6. Why can a high-resistance ground fault in a grounded distribution system be very damaging to electrical apparatus? What can be done to protect the system against this type of fault?

15-7. Explain why it is necessary that the framework of all electrical apparatus be grounded to a common point. How may the connection to earth be accomplished?

15-8. Explain how capacitors can improve the power factor of a system.

15-9. What benefits may be derived from an improved system power factor?

15-10. What should be included in a routine maintenance check of power capacitors? What should be done about a capacitor that is leaking insulating liquid?

15-11. What precautions should be observed when discharging power capacitors?

15-12. State the functions of lightning arresters and protective capacitors for the protection of electrical machinery. Why is this equipment necessary if the local outdoor transformer has installed lightning protection?

15-13. What determines the voltage rating of a lightning arrester?

15-14. What should be included in a maintenance check of lightning arresters?

15-15. Describe the construction details of a lightning arrester.

15-16. What should be included in a routine maintenance check of lightning arresters?

15-17. Differentiate between high-voltage insulators and high-voltage bushings. What is the purpose of petticoat construction?

15-18. How should porcelain bushings and insulators be cleaned? Why should hairline cracks or abrasions in porcelain insulators be repaired as soon as detected? How should such repairs be made? What precautions should be observed before working on or testing a high-voltage insulator or bushing?

15-19. What is a pothead used for?

15-20. Describe a test procedure for checking the condition of cable insulation.

15-21. Why should a long section of cable indicate a lower insulation resistance than a shorter section of the same cable for the same operating temperature and condition of insulation?

15-22. Outline an inspection program for checking switchgear. Include a list of the apparatus to be inspected and the type of tests to be made.

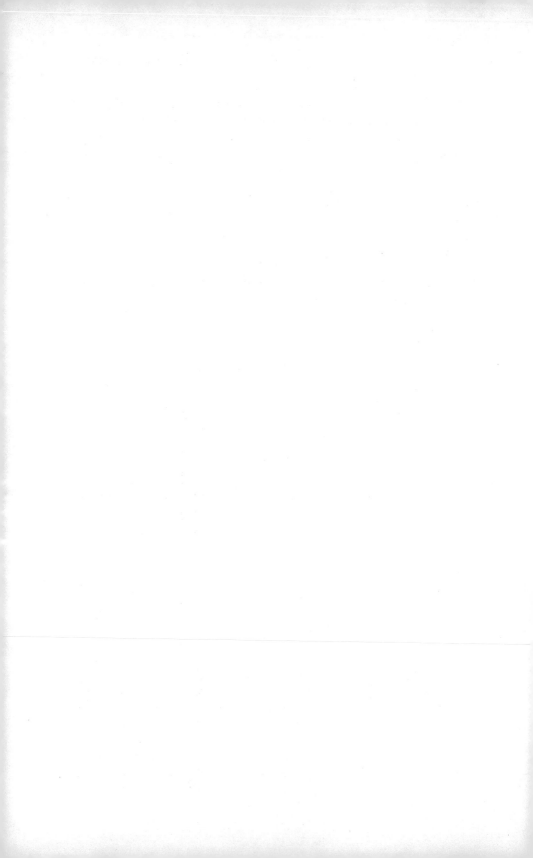

sixteen

economics, staffing, and tooling in electrical maintenance

16-1 Economic Factors

The objective of a preventive maintenance program is to reduce the number of avoidable breakdowns. Breakdowns in factory production lines add tremendously to the unit cost of production and may result in an unprofitable operation. Ships may be laid up for days or even lost because of faults resulting from little or no maintenance. Faults in locomotives can tie up railroad traffic for hours and even result in a major rail disaster. Add to this the probability of law suits that may arise from injury to personnel and failure to keep commitments, because of an avoidable breakdown, and the cost of not having a preventive-maintenance program becomes staggering. Preventive maintenance is expensive; it requires shop facilities, skilled labor, keeping records, and stocking of replacement parts. However, the cost of downtime resulting from avoidable outages may amount to ten or more times the actual cost of repair. The high cost of downtime, especially in factories using mass-production techniques, where one outage may stop the entire plant, makes it imperative to economic operation that maintenance be scheduled into the production line or operating schedule.

When preparing routine inspection schedules, it must be recognized that too frequent inspections are a waste of money and that insufficient inspections place vital equipment in jeopardy. The frequency of inspection should depend on the equipment's contribution to profitable production, its duty cycle, age, overload, and other pertinent factors. The manufacturer of electrical equipment will recommend an inspection cycle that may be modified later by experience. A good maintenance program provides for scheduling equipment out of service for routine overhauls during periods of least usage, rather than risk a breakdown at an inopportune time. Thus maintenance can be done at a time convenient to the production or operating schedule. Some companies find it convenient to close down completely during the vacation periods to provide time for planned overhauls.

When planning a preventive-maintenance program, all equipment should be classified as to its role in the manufacturing or operating process. Machinery and control equipment that could halt production if a breakdown occurred should be classified as *vital equipment* and should be provided with a very comprehensive preventive-maintenance program. Other equipment, such as room air-conditioners, electric fans, drinking-water coolers, etc., should be classified as *nonvital equipment*. A breakdown of nonvital equipment will have little or no adverse effect on the production line, railroad operation, ships' safety, etc., and a simplified maintenance program may be used. To provide the same degree of maintenance for nonvital and vital equipment is a waste of time and money. Planning a preventive-maintenance program for vital equipment must be done with as much consideration as was given to the selection of the equipment itself. The cost of such programs should be considered as part of the unit cost of operation or production. The essentials of an effective preventive-maintenance program include record taking and interpretation, acquisition of adequate replacement parts, skilled labor to carry out the program, and cooperative management.

16-2 Record Taking and Interpreting

The most important phase of a preventive-maintenance program is the recording and interpretation of data pertinent to each piece of equipment. Two or more file cards will be needed for each item. The *equipment record card* should include the motor nameplate data, location in the plant, purchase price and installation cost, application, and the manufacturer's replacement part number for each replaceable part. The *history card* should include the date of each periodic test, the test data, the type of repairs, the man-hours, the cost of each repair, and the initials of the workman in charge of the job.

16-3 Staffing

Maintenance personnel should be selected on the basis of qualifications for the job. Assignments should not be made on a hit-or-miss basis to those who happen to be idle at the moment. Much equipment has been ruined by overzealous but inexperienced personnel. The maintenance department must not be a catchall for personnel who could not adjust elsewhere. The complexity of control systems in automatic and semiautomatic plants requires well-educated and skilled personnel in the maintenance department. Very often, because of the newness and complexity of the maintenance requirements, special training must be had in company schools. A skilled maintenance force must be kept abreast of the latest techniques and developments in the field. This may be done by visiting plant maintenance shows, subscribing to appropriate periodicals, and attending specialized classes, panel discussions, and lectures. A file of equipment manufacturers' bulletins and instructions should be available to all maintenance personnel. Only modern testing equipment, good tools, and the latest methods should be used. Whenever possible, maintenance standards and testing procedures recommended by the IEEE, USAS, NEMA, and NEC should be used.[1]

A good preventive-maintenance program represents insurance against avoidable outages that can completely halt production.

16-4 Renewal Parts

A carefully selected assortment of spare parts is essential to a good maintenance program and represents insurance against prolonged shutdowns. Even though repairs may not be made on the premises, keeping an adequate stock of spare parts will avoid delay in placing the machine back in service. Overstocking and understocking should be avoided. Overstocking results in excessive carrying costs as well as losses due to obsolescence when equipment is replaced. Understocking places production in jeopardy if an outage occurs. Very often replacement parts for one machine may be used on many other units in the same plant. In such instances it is unwise to stock spares for each machine alone, but to follow the manufacturer's recommendations concerning the minimum that is permissible for the group of machines.

Rotating machinery. A recommended guide for the minimum renewal parts that should be stocked for all vital machinery is listed in Table 16-1.

---

[1] IEEE, Institute of Electrical and Electronics Engineers; USAS, United States of America Standard (formerly American Standard); NEMA, National Electrical Manufacturers Association; NEC, National Electrical Code.

Table 16-1   Recommended Minimum Spare-parts Stock for Identical Machines

Description of part	Recommended minimum stock number of units in operation			
	1 to 4	5 to 9	10 to 20*	10 to 20
Synchronous machines:				
Motor complete	0	0	0	1
Rotor complete	0	0	1	0
Stator coils with winding supplies	⅓ set	1 set	2 sets	1 set
Rotor field coil	1 set	1 set	2 sets	1 set
Brushholders	1	2	4	1
Brushholder springs	2	4	8	2
Brushholder stud insulation†	1 set	2 sets	2 sets	1 set
Brush	1 set	2 sets	4 sets	4 sets
Bearings	1 set	2 sets	2 sets	1 set
Oil rings (if used)	1 set	1 set	2 sets	1 set
Retainer rings (antifriction bearings)	1 set	2 sets	2 sets	1 set
Squirrel-cage induction motors:				
Motor complete	0	0	0	1
Stator coils with winding supplies	1 set	1 set	2 sets	1 set
Bearings	1 set	2 sets	2 sets	1 set
Oil rings (if used)	1 set	1 set	2 sets	1 set
Retainer rings (antifriction bearings)	1 set	1 set	2 sets	1 set
Wound-rotor induction motors:				
Motor complete	0	0	0	1
Wound rotor complete	0	1	1	0
Rotor coils with winding supplies	1 set	1 set	2 sets	1 set
Stator coils with winding supplies	1 set	1 set	2 sets	1 set
Brushholders	1	2	4	1
Brushholder springs	2	4	8	2
Brushholder stud insulation†	1 set	2 sets	2 sets	1 set
Brushes	1 set	2 sets	4 sets	4 sets
Bearings	1 set	2 sets	2 sets	1 set
Oil rings (if used)	1 set	1 set	2 sets	1 set
Retainer rings (antifriction bearings)	1 set	2 sets	2 sets	1 set
Dc-machines:				
Complete motor	0	0	0	1
Armature‡	0	1	1	0
Armature coils with winding supplies	1 set	1 set	1 set	1 set
Main field coil	1	1 set	1 set	1 set
Shunt field coil	1	1 set	1 set	1 set
Series field coil	1	1 set	1 set	1 set
Commutating field coil	1	1 set	1 set	1 set
Brushholder	1	1	4	2
Brushholder springs	2	4	8	2
Brushholder stud insulation†	1 set	2 sets	2 sets	1 set
Brushes	1 set	2 sets	4 sets	4 sets
Bearings	1 set	2 sets	2 sets	1 set
Oil rings (if used)	1 set	1 set	2 sets	1 set

* This column to be used when complete motor is not stocked.
† One set—sufficient insulation for one stud. In some cases this item will cover an insulating yoke.
‡ Some manufacturers may recommend that "armature" for split-frame motors include frame heads and bearings.
SOURCE: Guide for Recommended Renewal Parts, NEMA RP1-1968.

Control equipment. A recommended guide for the minimum renewal parts that should be stocked for all important control equipment is:

One fuse for each fuse installed
One spare operating coil of each type
One spare push button of each type
One spare resistor of each type
One set of spare contacts, springs, arc chutes, etc., for each type of relay
    and controller

16-5 Safety

Safe operation of machines and safety to personnel should be uppermost in all considerations when planning a preventive-maintenance program.

All electrical wiring should conform to the recommendations of the National Board of Fire Underwriters as presented in the National Electrical Code. Machinery and other equipment should be installed with a view toward easy accessibility for maintenance. Joints and valves for water, steam, waste, and other liquid-carrying lines should not be located over electrical apparatus unless such equipment is drip-proof or splash-proof as required. Only approved safety solvents should be used for cleaning insulation, bearings, and other parts.

Switches and circuit breakers should always be closed completely and with a swift firm action. This will enable the quick operation of a fuse or breaker if a fault exists. Partly closing the switch or tickling the contacts "to see if everything is alright" is dangerous. The high-resistance path of a partly closed switch will cause severe burning and arcing at the contacts.

Adequate overload and short-circuit protection is an absolute necessity to safe operation. Obsolete switchgear can cripple an industrial plant by failing to clear a short circuit. The interrupting capacity of all breakers and fuses must be sufficient to clear the fault current rapidly and without damage to itself.

The incoming electrical service and the plant distribution system should be restudied whenever sizable additions are made to the electrical load. The ampacity of the feeder cables should be checked to determine whether or not it will carry the additional load without overheating, and a new coordination study of all protective devices in the system should be made to ensure a high degree of reliability and protection.

Before attempting repair work on electrical equipment, it must be disconnected from the power supply. The incoming lines to the apparatus must be tested with a voltmeter or voltage tester of the correct range for positive proof that the circuit is dead. All capacitors connected to the apparatus must be discharged.

16-6 Electrical Fires

The standard procedure for fighting electrical fires is to open the circuit and then apply an approved extinguishing agent.

A carbon dioxide (CO_2) extinguisher offers the advantage of extinguishing the fire, cooling the apparatus, leaving no residue, and having no adverse affect on the insulation and metal parts, and it may be used on live circuits. Hence it is the preferred extinguishing agent for most electrical fires. However, when applied in confined spaces, such as in the engine room of a ship, the CO_2 should be purged with air before workmen are allowed to enter.

The dry-chemical extinguisher is satisfactory and may be used on live circuits. However, it leaves a residue and does not cool the apparatus so effectively as does the CO_2 extinguisher.

A carbon tetrachloride extinguisher will put out the fire and cool the apparatus, but excessive use of this agent will be detrimental to the insulation and metal parts. Carbon tetrachloride is also very noxious and should be used only in well-ventilated places.

Water sprinklers and steam smothering apparatus may be used only after the circuit is deenergized. Such systems have been built into large machines and are very effective. The serious drawback to their extended usage is the time required to clean and dry the apparatus before it can be placed back in service.

However, regardless of the type of extinguishing agent used, the apparatus should be thoroughly inspected and tested after the fire is out. The cause of the fire should be determined, and corrective action should be taken before the apparatus is put back in service. If no damage is apparent, and an insulation-resistance test indicates normal, the apparatus may be started at reduced load and carefully observed as load is applied.

16-7 Instruments and Tools for Maintenance and Trouble Shooting

Table 16-2 lists the instruments necessary for diagnostic testing and trouble shooting. For typical applications of these instruments see the associated applications section. The first nine instruments represent a suggested minimum for effective maintenance testing.

Table 16-3 lists the maintenance tools and materials that should be available to the maintenance technician. Such items as wrenches and screwdrivers are standard shop tools and hence are not included in the list.

Table 16-2   Instruments for Maintenance

Instrument	Typical application (book section)
1. Volt-ohmmeter ac/dc	1-6, 6-10
2. 500-volt megohmmeter	2-5
3. Circuit-breaker and relay test set	14-12
4. Vibration-amplitude meter	12-6
5. Centigrade thermometers, spirit type	5-10
6. Tachometer	6-14, 11-3
7. Hook-on ammeter	5-10, 6-7
8. Magnetic compass	6-13, 11-5
9. Mechanical voltage tester	14-3
10. Dc external shunt ammeter with 50-mv shunts	11-9
11. Growler	6-2, 11-9
12. Phase-sequence indicator	4-13
13. Motor-rotation indicator	5-2
14. Hook-on wattmeter	5-10
15. Hook-on power-factor meter	5-10
16. High-potential test set	2-12, 2-13
17. Insulating-liquid test set	2-16
18. Hydrometer	3-2
19. Industrial analyzer	5-10

Table 16-3   Maintenance Tools and Materials

Maintenance tools:
    Spring scale, 0–5 lb
    Feeler gauges (tapered) for air-gap measurements
    Smooth mill file for contacts
    Commutator stones
    Brush-seating stones
    Undercutting tools
    Canvas commutator wiper
    Soldering iron
    Insulated fuse puller
    Two-cell flashlight (insulated)
    Small mirror with extension handle and knuckle joint
Maintenance materials:
    Sandpaper, grades $1\frac{1}{2}$, 0, and 0000
    Cleaning solvent, approved safety type
    Cambric, plastic, and cotton tapes
    Insulating varnish
    Rosin-core solder

Bibliography

Guide for Recommended Renewal Parts, National Electrical Manufacturers Association PUB. NO. RP 1-1963.

Killin, A. M.: How Plant Management Evaluates Electrical Maintenance Performance, *IEEE Transactions on Industry and General Applications,* vol. IGA-3, no. 2, Mar./Apr., 1967.

Morrow, L. C.: "Maintenance Engineering Handbook," McGraw-Hill Book Company, New York, 1966.

Questions

16-1. What is the objective of a preventive-maintenance program?

16-2. What factors should be considered when planning a preventive-maintenance program?

16-3. Should the same degree of maintenance be provided for vital and nonvital equipment? Explain.

16-4. What basic records are essential to a good preventive-maintenance program?

16-5. What factors should be considered when determining the minimum number of replacement parts that should be stocked? Explain why overstocking or understocking should be avoided.

16-6. How should maintenance personnel be selected?

16-7. What factors must be considered when a considerable increase in the installed electrical load is to be made?

16-8. State the correct procedure to follow when an electric motor catches fire. What should be done before the motor is placed back in service?

16-9. Under what conditions would it be appropriate to use a sprinkler system to put out an electrical fire? What is the disadvantage of this method of fire fighting?

Appendix 1 Allowable Ampacities of Insulated Copper Conductors

Single conductor in free air (based on ambient temperature of 30°C, 86°F)

Size	Temperature rating of conductor (see Appendix 3)							
AWG MCM	**60°C (140°F)**	**75°C (167°F)**	**85°C (185°F)**	**90°C (194°F)**	**110°C (230°F)**	**125°C (257°F)**	**200°C (392°F)**	
	Types RUW (14-2), T, TW	Types RH, RHW, RUH (14-2), THW, THWN XHHW	Types V, MI	Types TA, TBS, SA, AVB, SIS, FEP, FEPB, RHH, THHN, XHHW*	Types AVA, AVL	Types AI (14-8), AIA	Types A (14-8), AA, FEP†, FEPB†	Bare and covered conductors
14	20	20	30	30‡	40	40	45	30
12	25	25	40	40‡	50	50	55	40
10	40	40	55	55‡	65	70	75	55
8	55	65	70	70	85	90	100	70
6	80	95	100	100	120	125	135	100
4	105	125	135	135	160	170	180	130
3	120	145	155	155	180	195	210	150
2	140	170	180	180	210	225	240	175
1	165	195	210	210	245	265	280	205
0	195	230	245	245	285	305	325	235
00	225	265	285	285	330	355	370	275
000	260	310	330	330	385	410	430	320
0000	300	360	385	385	445	475	510	370
250	340	405	425	425	495	530	. . .	410
300	375	445	480	480	555	590	. . .	460
350	420	505	530	530	610	655	. . .	510
400	455	545	575	575	665	710	. . .	555
500	515	620	660	660	765	815	. . .	630
600	575	690	740	740	855	910	. . .	710
700	630	755	815	815	940	1005	. . .	780
750	655	785	845	845	980	1045	. . .	810
800	680	815	880	880	1020	1085	. . .	845
900	730	870	940	940	. . . .	. . . .	. . .	905
1000	780	935	1000	1000	1165	1240	. . .	965
1250	890	1065	1130	1130				
1500	980	1175	1260	1260	1450	. . . .	. . .	1215
1750	1070	1280	1370	1370				
2000	1155	1385	1470	1470	1715	. . . .	. . .	1405

* For dry locations only (see Appendix 3). These ampacities relate only to conductors described in Appendix 3.

† Special use only (see Appendix 3).

‡ The ampacities for types FEP, FEPB, RHH, THHN, and XHHW conductors for sizes AWG 14, 12, and 10 shall be the same as designated for 75°C conductors in this table.

Correction Factors for Ambient Temperatures over 30° C, 86° F

°C	°F	60°C (140°F)	75°C (167°F)	85°C (185°F)	90°C (194°F)	110°C (230°F)	125°C (257°F)	200°C (392°F)
40	104	0.82	0.88	0.90	0.90	0.94	0.95	
45	113	0.71	0.82	0.85	0.85	0.90	0.92	
50	122	0.58	0.75	0.80	0.80	0.87	0.89	
55	131	0.41	0.67	0.74	0.74	0.83	0.86	
60	140	....	0.58	0.67	0.67	0.79	0.83	0.91
70	158	....	0.35	0.52	0.52	0.71	0.76	0.87
75	167	....	....	0.43	0.43	0.66	0.72	0.86
80	176	....	....	0.30	0.30	0.61	0.69	0.84
90	194	....	....	....	....	0.50	0.61	0.80
100	212	....	....	....	....	....	0.51	0.77
120	248	....	....	....	....	....	....	0.69
140	284	....	....	....	....	....	....	0.59

SOURCE: National Electric Code 1968, National Fire Protection Association.

Appendix 2   Allowable Ampacities of Insulated Copper Conductors

Not more than three conductors in raceway or cable or direct burial (based on ambient temperature of 30°C, 86°F)

Size	Temperature rating of conductor (see Appendix 3)						
AWG MCM	60°C (140°F)	75°C (167°F)	85°C (185°F)	90°C (194°F)	110°C (230°F)	125°C (257°F)	200°C (392°F)
	Types RUW (14-2), T, TW	Types RH, RHW, RUH, (14-2), THW, THWN, XHHW, THW-MTW	Types V, MI	Types TA, TBS,SA, AVB, SIS, FEP, FEPB, RHH, THHN, XHHW*	Types AVA, AVL	Types AI (14-8), AIA	Types A (14-8), AA, FEP†, FEPB†
14	15	15	25	25‡	30	30	30
12	20	20	30	30‡	35	40	40
10	30	30	40	40‡	45	50	55
8	40	45	50	50	60	65	70
6	55	65	70	70	80	85	95
4	70	85	90	90	105	115	120
3	80	100	105	105	120	130	145
2	95	115	120	120	135	145	165
1	110	130	140	140	160	170	190
0	125	150	155	155	190	200	225
00	145	175	185	185	215	230	250
000	165	200	210	210	245	265	285
0000	195	230	235	235	275	310	340
250	215	255	270	270	315	335	
300	240	285	300	300	345	380	
350	260	310	325	325	390	420	
400	280	335	360	360	420	450	
500	320	380	405	405	470	500	
600	355	420	455	455	525	545	
700	385	460	490	490	560	600	
750	400	475	500	500	580	620	
800	410	490	515	515	600	640	
900	435	520	555	555			
1000	455	545	585	585	680	730	
1250	495	590	645	645			
1500	520	625	700	700	785		
1750	545	650	735	735			
2000	560	665	775	775	840		

* For dry locations only (see Appendix 3). These ampacities relate only to conductors described in Appendix 3.

† Special use only (see Appendix 3).

‡ The ampacities for types FEP, FEPB, RHH, THHN, and XHHW conductors for sizes AWG 14, 12, and 10 shall be the same as designated for 75°C conductors in this table.

Correction Factors for Ambient Temperatures over 30° C, 86° F

°C	°F	60°C (140°F)	75°C (167°F)	85°C (185°F)	90°C (194°F)	110°C (230°F)	125°C (257°F)	200°C (392°F)
40	104	0.82	0.88	0.90	0.90	0.94	0.95	
45	113	0.71	0.82	0.85	0.85	0.90	0.92	
50	122	0.58	0.75	0.80	0.80	0.87	0.89	
55	131	0.41	0.67	0.74	0.74	0.83	0.86	
60	140	....	0.58	0.67	0.67	0.79	0.83	0.91
70	158	....	0.35	0.52	0.52	0.71	0.76	0.87
75	167	....	....	0.43	0.43	0.66	0.72	0.86
80	176	....	....	0.30	0.30	0.61	0.69	0.84
90	194	....	....	....	....	0.50	0.61	0.80
100	212	....	....	....	....	....	0.51	0.77
120	248	....	....	....	....	....	....	0.69
140	284	....	....	....	....	....	....	0.59

SOURCE: National Electrical Code 1968, National Fire Protection Association.

Appendix 3   Conductor Applications

Trade name	Type letter	Max. operating temp.	Application provisions
Rubber-covered fixture wire, solid or 7-strand	RF-1*	60°C 140°F	Fixture wiring; limited to 300 volts
	RF-2*	60°C 140°F	Fixture wiring, and as permitted in Sec. 310-8, NEC
Rubbered-covered fixture wire, flexible stranding	FF-1*	60°C 140°F	Fixture wiring; limited to 300 volts
	FF-2*	60°C 140°F	Fixture wiring, and as permitted in Sec. 310-8, NEC
Heat-resistant rubber-covered fixture wire, solid or 7-strand	RFH-1*	75°C 167°F	Fixture wiring; limited to 300 volts
	RFH-2*	75°C 167°F	Fixture wiring, and as permitted in Sec. 310-8, NEC
Heat-resistant rubber-covered fixture wire, flexible stranding	FFH-1*	75°C 167°F	Fixture wiring; limited to 300 volts
	FFH-2*	75°C 167°C	Fixture wiring, and as permitted in Sec. 310-8, NEC
Thermoplastic-covered fixture wire, solid or stranded	TF*	60°C 140°F	Fixture wiring, and as permitted in Sec. 310-8, and for circuits as permitted in Art. 725, NEC
Thermoplastic-covered fixture wire, flexible stranding	TFF*	60°C 140°F	Fixture wiring, and as permitted in Sec. 310-8, and for circuits as permitted in Art. 725. NEC
Heat-resistant, thermoplastic-covered fixture wire, solid or stranded	TFN*	90°C 194°F	Fixture wiring, and as permitted in Sec. 310-8, NEC
Heat-resistant thermoplastic-covered fixture wire, flexible stranding	TFFN*	90°C 194°F	Fixture wiring, and as permitted in Sec. 310-8, NEC
Cotton-covered heat-resistant fixture wire	CF*	90°C 194°F	Fixture wiring; limited to 300 volts

Appendix 3 Conductor Applications (Continued)

Trade name	Type letter	Max. operating temp.	Application provisions
Asbestos-covered heat-resistant, fixture wire	AF*	150°C 302°F	Fixture wiring; limited to 300 volts and indoor dry location
Fluorinated ethylene propylene fixture wire, solid or 7-strand	PF* PGF*	150°C 302°F	Fixture wiring, and as permitted in Sec. 310-8, NEC
Fluorinated ethylene propylene fixture wire	PFF* PGFF*	150°C 302°F	Fixture wiring, and as permitted in Sec. 310-8, NEC
Silicone rubber-insulated fixture wire, solid or 7-strand	SF-1*	200°C 392°F	Fixture wiring; limited to 300 volts
	SF-2*	200°C 392°F	Fixture wiring, and as permitted in Sec. 310-8, NEC
Silicone rubber-insulated fixture wire, flexible stranding	SFF-1*	150°C 302°F	Fixture wiring; limited to 300 volts
	SFF-2*	150°C 302°F	Fixture wiring, and as permitted in sec. 310-8, NEC
Heat-resistant rubber	RH	75°C 167°F	Dry locations
Heat-resistant rubber	RHH	90°C 194°F	Dry locations
Moisture- and heat-resistant rubber	RHW	75°C 167°F	Dry and wet locations; for over 2,000 volts, insulation shall be ozone-resistant
Heat-resistant latex rubber	RUH	75°C 167°F	Dry locations
Moisture-resistant latex rubber	RUW	60°C 140°F	Dry and wet locations
Thermoplastic	T	60°C 140°F	Dry locations
Moisture-resistant thermoplastic	TW	60°C 140°F	Dry and wet locations

Appendix 3   Conductor Applications   (Continued)

Trade name	Type letter	Max. operating temp.	Application provisions
Heat-resistant thermoplastic	THHN	90°C 194°F	Dry locations
Moisture- and heat-resistant thermoplastic	THW	75°C 167°F	Dry and wet locations
Moisture- and heat-resistant thermoplastic	THWN	75°C 167°F	Dry and wet locations
Moisture- and heat-resistant cross-linked thermosetting polyethylene	XHHW	90°C 194°F 75°C 167°F	Dry locations Wet locations
Moisture-, heat-, and oil-resistant thermoplastic	MTW	60°C 140°F 90°C 194°F	Dry and wet locations, machine tool wiring (see Art. 670 and NFPA Standard No. 79) Dry locations, machine tool wiring (see Art. 670 and NFPA Standard No. 79)
Moisture-, heat-, and oil-resistant thermoplastic	THW-MTW	75°C 167°F 90°C 194°F	Dry and wet locations Special applications within electric discharge lighting equipment, Limited to 1,000 open-circuit volts or less (size 14-8 only)
Thermoplastic and asbestos	TA	90°C 194°F	Switchboard wiring only
Thermoplastic and fibrous outer braid	TBS	90°C 194°F	Switchboard wiring only
Synthetic heat-resistant	SIS	90°C 194°F	Switchboard wiring only
Mineral insulation (metal-sheathed)	MI	85°C 185°F 250°C 482°F	Dry and wet locations with type O termination fittings; for special application

Appendix 3   Conductor Applications   (Continued)

Trade name	Type letter	Max. operating temp.	Application provisions
Silicone-asbestos	SA	90°C 194°F 125°C 257°F	Dry locations; for special application
Fluorinated ethylene propylene	FEP or FEPB	90°C 194°F 200°C 392°F	Dry locations Dry locations, for special applications
Varnished cambric	V	85°C 185°F	Dry locations only; smaller than No. 6 by special permission
Asbestos and varnished cambric	AVA	110°C 230°F	Dry locations only
Asbestos and varnished cambric	AVL	110°C 230°F	Dry and wet locations
Asbestos and varnished cambric	AVB	90°C 194°F	Dry locations only
Asbestos	A	200°C 392°F	Dry locations only; only for leads within apparatus or within raceways connected to apparatus; limited to 300 volts
Asbestos	AA	200°C 392°F	Dry locations only; only for leads within apparatus or within raceways connected to apparatus or as open wiring; limited to 300 volts
Asbestos	AI	125°C 257°F	Dry locations only; only for lead-within apparatus or within races ways connected to apparatus; limited to 300 volts
Asbestos	AIA	125°C 257°F	Dry locations only; only for leads within apparatus or within raceways connected to apparatus or as open wiring
Paper		85°C 185°F	For underground service conductors, or by special permission

* Fixture wires are not intended for installation as branch-circuit conductors except as permitted in Art. 725, NEC.

SOURCE: National Electrical Code 1968, National Fire Protection Association.

Appendix 4 Locked-rotor current of
Three-phase Single-speed Squirrel-cage
Induction Motors, 220-volt, 60-hertz

Hp	Locked-rotor current, amp*	Design letters
1/2	20	B, D
3/4	25	B, D
1	30	B, D
1½	40	B, D
2	50	B, D
3	64	B, C, D
5	92	B, C, D
7½	127	B, C, D
10	162	B, C, D
15	232	B, C, D
20	290	B, C, D
25	365	B, C, D
30	435	B, C, D
40	580	B, C, D
50	725	B, C, D
60	870	B, C, D
75	1,085	B, C, D
100	1,450	B, C, D
125	1,815	B, C, D
150	2,170	B, C, D
200	2,900	B, C
250	3,650	B
300	4,400	B
350	5,100	B
400	5,800	B
450	6,500	B
500	7,250	B

* Locked-rotor current of motors designed for voltages
other than 220 volts shall be inversely proportional to
the voltages, except for motors rated at 230 volts (see
MG1-12.34.a).
SOURCE: Motors and Generators, NEMA MG1-1967.

Appendix 5   Full-load Currents in Amperes, DC Motors

The following values of full-load currents are for motors running at base speed.

HP	120V	240V
1/4	2.9	1.5
1/3	3.6	1.8
1/2	5.2	2.6
3/4	7.4	3.7
1	9.4	4.7
1½	13.2	6.6
2	17	8.5
3	25	12.2
5	40	20
7½	58	29
10	76	38
15		55
20		72
25		89
30		106
40		140
50		173
60		206
75		255
100		341
125		425
150		506
200		675

SOURCE: National Electrical Code 1968, National Fire Protection Association.

Appendix 6 Full-load Currents in Amperes, Single-phase AC Motors

The following values of full-load currents are for motors running at usual speeds and motors with normal torque characteristics. Motors built for especially low speeds or high torques may have higher full-load current, and multispeed motors will have full-load current varying with speed, in which case the nameplate current ratings shall be used.

To obtain full-load currents of 208- and 200-volt motors, increase corresponding 230-volt motor full-load currents by 10 and 13 percent, respectively.

The voltages listed are rated motor voltages. Corresponding nominal system voltages are 110 to 120 and 220 to 240 volts.

HP	115V	230V
$\frac{1}{6}$	4.4	2.2
$\frac{1}{4}$	5.8	2.9
$\frac{1}{3}$	7.2	3.6
$\frac{1}{2}$	9.8	4.9
$\frac{3}{4}$	13.8	6.9
1	16	8
$1\frac{1}{2}$	20	10
2	24	12
3	34	17
5	56	28
$7\frac{1}{2}$	80	40
10	100	50

SOURCE: National Electrical Code 1968, National Fire Protection Association.

Appendix 7   Full-load Current, Three-phase AC Motors*

The following values of full-load currents are for motors running at speeds usual for belted motors and motors with normal torque characteristics. Motors built for especially low speeds or high torques may require more running current, and multi-speed motors will have full-load currents varying with speed, in which case the name-plate current rating shall be used.

For full-load currents of 208- and 200-volt motors, increase the corresponding 230-volt motor full-load currents by 10 and 13 percent, respectively.

The voltages listed are rated motor voltages. Corresponding nominal system voltages are 110 to 120, 220 to 240, 440 to 480, and 550 to 600 volts.

HP	Induction type, squirrel-cage and wound rotor, amp					Synchronous type, unity power factor, amp*			
	115V	230V	460V	575V	2,300V	220V	440V	550V	2,300V
½	4	2	1	.8					
¾	5.6	2.8	1.4	1.1					
1	7.2	3.6	1.8	1.4					
1½	10.4	5.2	2.6	2.1					
2	13.6	6.8	3.4	2.7					
3		9.6	4.8	3.9					
5		15.2	7.6	6.1					
7½		22	11	9					
10		28	14	11					
15		42	21	17					
20		54	27	22					
25		68	34	27		54	27	22	
30		80	40	32		65	33	26	
40		104	52	41		86	43	35	
50		130	65	52		108	54	44	
60		154	77	62	16	128	64	51	12
75		192	96	77	20	161	81	65	15
100		248	124	99	26	211	106	85	20
125		312	156	125	31	264	132	106	25
150		360	180	144	37		158	127	30
200		480	240	192	49		210	168	40

* For 90 and 80 percent power factor the above figures shall be multiplied by 1.1 and 1.25 respectively.

SOURCE: National Electrical Code 1968, National Fire Protection Association.

Appendix 8   Full-load Current, Two-phase AC Motors (Four-wire)

The following values of full-load currents are for motors running at speeds usual for belted motors and motors with normal torque characteristics. Motors built for especially low speeds or high torques may require more running current, and multi-speed motors will have full-load currents varying with speed, in which case the name-plate current rating shall be used. Current in common conductor of two-phase, three-wire system will be 1.41 times value given.

The voltages listed are rated motor voltages. Corresponding nominal system voltages are 110 to 120, 220 to 240, 440 to 480, and 550 to 600 volts.

	Induction type, squirrel-cage and wound rotor, amp					Synchronous type, unity power factor, amp*			
HP	115V	230V	460V	575V	2,300V	220V	440V	550V	2,300V
½	4	2	1	0.8					
¾	4.8	2.4	1.2	1.0					
1	6.4	3.2	1.6	1.3					
1½	9	4.5	2.3	1.8					
2	11.8	5.9	3	2.4					
3		8.3	4.2	3.3					
5		13.2	6.6	5.3					
7½		19	9	8					
10		24	12	10					
15		36	18	14					
20		47	23	19					
25		59	29	24		47	24	19	
30		69	35	28		56	29	23	
40		90	45	36		75	37	31	
50		113	56	45		94	47	38	
60		133	67	53	14	111	56	44	11
75		166	83	66	18	140	70	57	13
100		218	109	87	23	182	93	74	17
125		270	135	108	28	228	114	93	22
150		312	156	125	32		137	110	26
200		416	208	167	43		182	145	35

* For 90 and 80 percent power factor the above figures should be multiplied by 1.1 and 1.25 respectively.
SOURCE: National Electrical Code 1968, National Fire Protection Association.

Appendix 9   Standard Speeds for Synchronous Motors and Generators

$$\text{Frequency} = \frac{\text{poles} \times \text{rpm}}{120}$$

Number of poles in stator	60 hertz	50 hertz	25 hertz
2	3,600	3,000	1,500
4	1,800	1,500	750
6	1,200	1,000	500
8	900	750	375
10	720	600	300
12	600	500	250
14	514	429	214
16	450	375	188
18	400	333	167
20	360	300	150

NOTE: Induction-motor speeds will be slightly below synchronous. For example, a four-pole induction motor operating at 60 hertz may have a speed of 1,750 rpm when operating at full load.

Appendix 10a   Identifying Code Letters for Motor Starting KVA

Code letter	Kilovolt-amperes per horsepower with locked rotor
A	0– 3.14
B	3.15– 3.54
C	3.55– 3.99
D	4.0 – 4.49
E	4.5 – 4.99
F	5.0 – 5.59
G	5.6 – 6.29
H	6.3 – 7.19
J	7.1 – 7.99
K	8.0 – 8.99
L	9.0 – 9.99
M	10.0 –11.19
N	11.2 –12.49
P	12.5 –13.99
R	14.0 –15.99
S	16.0 –17.99
T	18.0 –19.99
U	20.0 –22.39
V	22.4 –and up

NOTE: The above table is an adopted standard of the National Electrical Manufacturers Association.

Appendix 10b Maximum Rating or Setting of Motor-branch-circuit Protective Devices for Motors Not Marked with a Code Letter Indicating Locked Rotor KVA

| | | Percent of full-load current | |
| | | Circuit-breaker setting | |
Type of motor	Fuse rating	Instan-taneous type	Time limit type
Single-phase, all types	300	700	250
Squirrel-cage and synchronous (full-voltage, resistor, and reactor starting)	300	700	250
Squirrel-cage and synchronous (auto-transformer starting):			
Not more than 30 amp	250	700	200
More than 30 amp	200	700	200
High-reactance squirrel-cage:			
Not more than 30 amp	250	700	250
More than 30 amp	200	700	200
Wound-rotor	150	700	150
Direct-current:			
Not more than 50 hp	150	250	150
More than 50 hp	150	175	150
Sealed (hermetic type):			
Refrigeration compressor*			
400-kva locked-rotor or less	175†	. . .	175†

For certain exceptions to the values specified see Secs. 430-52 and 430-59. The values given in the last column also cover the ratings of nonadjustable, time-limit types of circuit breakers that may also be modified as in Sec. 430-52, National Electrical Code.

Synchronous motors of the low-torque low-speed type (usually 450 rpm or lower), such as are used to drive reciprocating compressors, pumps, etc., which start up unloaded, do not require a fuse rating or circuit-breaker setting in excess of 200 percent of full-load current.

For motors marked with a code letter, see Appendix 10c.

* The locked rotor KVA is the product of the motor voltage and the motor locked rotor current (LRA) given on the motor nameplate divided by 1,000 for single-phase motors, or divided by 580 for 3-phase motors.

† This value may be increased to 225 percent if necessary to permit starting.

SOURCE: National Electrical Code 1968, National Fire Protection Association.

Appendix 10c Maximum Rating or Setting of Motor-branch-circuit Protective Devices for Motors Marked with a Code Letter Indicating Locked Rotor KVA

		Percent of full-load current	
		Circuit-breaker setting	
Type of motor	Fuse rating	Instan-taneous type	Time limit type
All ac single-phase and polyphase squirrel-cage and synchronous motors with full-voltage, resistor, or reactor starting:			
Code letter A	150	700	150
Code letter B to E	250	700	200
Code letter F to V	300	700	250
All ac squirrel-cage and synchronous motors with autotransformer starting:			
Code letter A	150	700	150
Code letter B to E	200	700	200
Code letter F to V	250	700	200

For certain exceptions to the values specified see Secs. 430-52 and 430-54. The values given in the last column also cover the ratings of nonadjustable, time-limit types of circuit breakers that may also be modified as in Sec. 430-52, National Electrical Code.

Synchronous motors of the low-torque, low-speed type (usually 450 rpm or lower), such as are used to drive reciprocating compressors, pumps, etc., which start up unloaded, do not require a fuse rating or circuit-breaker setting in excess of 200 percent of full-load current.

For motors not marked with a code letter, see Appendix 10*b*.

SOURCE: National Electrical Code 1968, National Fire Protection Association.

Appendix 11   USAS and RETMA Standard Color Code for Small Tubular Resistors

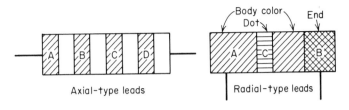

Axial-type leads

Radial-type leads

Color	Band A, 1st figure	Band B, 2d figure	Band C, remaining figures	Band D, tolerance
Black	0	0		
Brown	1	1	0	
Red	2	2	00	
Orange	3	3	000	
Yellow	4	4	0,000	
Green	5	5	00,000	
Blue	6	6	000,000	
Violet	7	7	0,000,000	
Gray	8	8	00,000,000	
White	9	9	000,000,000	
Gold	..	..	....	± 5%
Silver	..	..	....	±10%
No color	..	..	....	±20%

EXAMPLE. If the color bands of a resistor are as follows: band A yellow, band B green, band C orange, and band D gold; the value of the resistor is 45,000 ohms ±5 percent.

Appendix 12   Temperature Conversion Formulas

Degrees Fahrenheit $= (C + 40)9/5 - 40$
Degrees centigrade $= (F + 40)5/9 - 40$

Appendix 13   Multiplying Factors for Converting DC Resistance to 60-hertz AC Resistance

Size	Multiplying factor			
	For nonmetallic sheathed cables in air or nonmetallic conduit		For metallic sheathed cables or all cables in metallic raceways	
	Copper	Aluminum	Copper	Aluminum
Up to 3 AWG	1.000	1.000	1.00	1.00
2	1.000	1.000	1.01	1.00
1	1.000	1.000	1.01	1.00
0	1.001	1.000	1.02	1.00
00	1.001	1.001	1.03	1.00
000	1.002	1.001	1.04	1.01
0000	1.004	1.002	1.05	1.01
250 MCM	1.005	1.002	1.06	1.02
300 MCM	1.006	1.003	1.07	1.02
350 MCM	1.009	1.004	1.08	1.03
400 MCM	1.011	1.005	1.10	1.04
500 MCM	1.018	1.007	1.13	1.06
600 MCM	1.025	1.010	1.16	1.08
700 MCM	1.034	1.013	1.19	1.11
750 MCM	1.039	1.015	1.21	1.12
800 MCM	1.044	1.017	1.22	1.14
1000 MCM	1.067	1.026	1.30	1.19
1250 MCM	1.102	1.040	1.41	1.27
1500 MCM	1.142	1.058	1.53	1.36
1750 MCM	1.185	1.079	1.67	1.46
2000 MCM	1.233	1.100	1.82	1.56

SOURCE: National Electrical Code 1968, National Fire Protection Association.

Answers to Problems

1-1. 27.93 ohms
1-2. (a) $0.159(10)^6$ cir mils; (b) 0.013 ohm
1-3. (a) $15.56(10^{-6})$ ohm; (b) $19.84(10^{-6})$ ohm
1-4. $146.4(10^{-6})$ ohm
1-5. (a) 165.12 ohms; (b) 1.81 amp; 1.51 amp
1-6. (a) 0.492 ohm; (b) 1,771 watts; (c) 210.48 volts
1-7. 0.00555 ohm
1-8. 8,582.5 watts
1-9. 50 amp, 5 sec
1-10. (a) diagram; (b) 200 volts; (c) 4 ohms, 16 watts; 6 ohms, 24 watts; 10 ohms, 40 watts; lamp, 400 watts
1-11. 0.6 amp
1-12. (a) 2 amp; (b) 60 volts
1-13. 6, 12, and 18 volts respectively
1-14. 4.3 amp
1-15. (a) 0.1 ohm; (b) 30 amp
1-16. 4, 6, and 2 amp respectively
1-17. (a) diagram; (b) 0.5, 0.833, and 1.25 amp respectively; (c) 2.584 amp; (d) 1.334 amp
1-18. (a) diagram; (b) 7 amp; (c) 4.8 ohms; (d) 12 kw
1-19. (a) diagram; (b) 4 amp; (c) 30 ohms
1-20. (a) diagram; (b) 24 ohms; (c) 10 amp; (d) 24 ohms
1-21. (a) 10 sec; (b) 50 sec
1-22. (a) sketch; (b) 126.4 volts; (c) 1.84 amp
1-23. (a) diagram; (b) 2.5 megohms; (c) $18(10^{-6})$ joule
1-24. (a) diagram; (b) 15,000 ohms
1-25. (a) diagram; (b) 3 amp; (c) 2 sec
1-26. (a) 0.2 sec; (b) 0.2 sec; (c) 240 and 24 amp respectively
1-27. (a) 1.35 sec; (b) 12 amp; (c) 388.8 joules
1-28. (a) 37.68 ohms; (b) 30.15 ohms; (c) 6.89 amp
1-29. (a) diagram; (b) 4.24 ohms; (c) 28.3 amp

2-1. 240 megohms
2-2. graph
2-3. 1.125 megohms
2-4. 1.12 megohms
2-5. 3.3 megohms at 40°C
2-6. 30 megohms at 40°C
2-7. a, b, d, satisfactory; c, unsatisfactory
2-8. 1.083
2-9. 0.75

2-10. (*a*) 2.40; (*b*) satisfactory

4-1. (*a*) 2,025 rpm; (*b*) 67.5 hertz
4-2. 1,200 rpm
4-3. 600 rpm
4-4. 58.33 hertz
4-5. 50 hertz
4-6. 12 poles
4-7. (*a*) 800 volts; (*b*) 120 hertz
4-8. 184.97 amp
4-9. 50 amp
4-10. (*a*) 449.8 volts; (*b*) 260 volts
4-11. 500 kva
4-12. (*a*) 8 kw; (*b*) 6 kvar; (*c*) 20 kw; (*d*) 6 kvar; (*e*) 21.54 kva
4-13. (*a*) 12.4 kw; (*b*) 11.2 kvar
4-14. 3.6 kva
4-15. (*a*) 6 kva; (*b*) 0.67
4-16. 83.04 kva
4-17. (*a*) 19.9 kva; (*b*) 14.93 kw; (*c*) 12.77 kvar
4-18. (*a*) 27.21 kva; (*b*) 12.04 kvar; (*c*) 0.9
4-19. (*a*) diagram; (*b*) 25.6 kva; (*c*) 0.96; (*d*) 26.9 amp
4-20. (*a*) diagram; (*b*) 24.5 kva; (*c*) 1.0; (*d*) 25.74 amp
4-21. 61.2 hertz

5-1. (*a*) 1,800 rpm; (*b*) 1,500 rpm
5-2. (*a*) two poles; (*b*) 25 rpm
5-3. 113.42 percent
5-4. (*a*) 198.4 percent; (*b*) no

7-1. (*a*) 110 volts; (*b*) ½
7-2. 40 turns
7-3. 22.11 amp
7-4. (*a*) 102.27 amp; (*b*) 4.09:1
7-5. 184 volts, 149.5 volts
7-6. (*a*) 90 volts; (*b*) 10 amp; (*c*) 4,500 watts; (*d*) 900 watts; (*e*) 3,600 watts
7-7. (*a*) 2.95; (*b*) 58.78 amp
7-8. (*a*) 4.0; (*b*) 25 amp; (*c*) 19.9 kva
7-9. (*a*) no effect; (*b*) 11.5 kva

9-1. (*a*) diagram; (*b*) 3833.3 amp; (*c*) 2.29 amp
9-2. 224 volts
9-3. 1.59 ohms

index

index

This book was set in Times Roman by The Maple Press Company, and printed on permanent paper and bound by The Maple Press Company. The designer was Marsha Cohen; the drawings were done by J. &. R. Technical Services, Inc. The editors were Alan W. Lowe and Cynthia Newby. Morton I. Rosenberg supervised the production.